RECENT ADVANCES IN HEMATOLOGY RESEARCH

FETAL HEMOGLOBIN

THE PANACEA FOR MAJOR β-HEMOGLOBINOPATHIES

Recent Advances in Hematology Research

Additional books and ebooks in this series can be found on Nova's website under the Series tab.

RECENT ADVANCES IN HEMATOLOGY RESEARCH

FETAL HEMOGLOBIN

THE PANACEA FOR MAJOR β-HEMOGLOBINOPATHIES

NISHANT CHAKRAVORTY
EDITOR

nova
Medicine & Health
New York

Additional color graphics may be available in the e-book version of this book.

Library of Congress Cataloging-in-Publication Data

ISBN: 978-1-53617-955-2
Identifiers: LCCN 2020019471 (print) | LCCN 2020019472 (ebook) | ISBN 9781536179552 (hardcover) | ISBN 9781536179569 (adobe pdf)
Subjects: LCSH: Fetal hemoglobin. | Hemoglobinopathy--Treatment. | Hemoglobinopathy--Genetic aspects.
Classification: LCC QP96.5 .F47 2020 (print) | LCC QP96.5 (ebook) | DDC 612.1/111--dc23
LC record available at https://lccn.loc.gov/2020019471
LC ebook record available at https://lccn.loc.gov/2020019472

Published by Nova Science Publishers, Inc. † New York

Dedicated to my family

CONTENTS

PREFACE

Survival needs subsistence

Human existence depends on the exchange of resources between the body and the environment. Gaseous exchange between tissues and their surroundings is one such function that is bestowed upon the red river of blood – especially the ferroprotein component present on red blood cells, known as hemoglobin. Although the color of blood is same in every human being, the molecular details vary – ranging from minor variations to major life-threatening ones. Hemoglobin contains an assembly of four globular protein subunits (globins) associated with their heme groups. Variations in the globin sub-units form the basis of the different types of hemoglobins, the most predominant one being the adult hemoglobin form (HbA) in post-natal human life. HbA consists of two α- and two β-sub-units and is most the suited form for exchange of oxygen and carbon dioxide. While majority of the hemoglobin variants (e.g., Hb Gower I and II, Hb Portland I and II, HbF, HbA2 and HbA) are part of normal physiological development (embryonic, fetal and post-natal) process, some of the variants are pathological and disease causing (e.g., Hb D-Punjab, HbH, Hb Barts, HbS-sickle cell anemia, Hb β-thalassemia, HbC, HbE, HbE/β thalassemia etc.). Barring a few variations (e.g., α-thalassemias, HbH, Hb Barts), most of the pathological variants are associated with the β-chain. Being an intrinsic part of adult hemoglobin, any anomaly in the β-chain hampers the efficiency of gaseous transport and thereby leads to anemic manifestations.

The predominant form of hemoglobin present in fetal life is known as fetal hemoglobin (HbF) and it consists of two α- and two γ- chains. HbF is known to be an efficient carrier of gases to and from cells and tissues, especially in fetal stages of development. In fact, it has a higher binding affinity to oxygen compared HbA. While this serves as an advantage in fetal stages, where it can extract oxygen from maternal blood more efficiently, it has oft been considered as the need for fetal-to-adult switch over in post-natal life, when HbF is inactivated and HbA takes the driver's seat. The relatively lower affinity of HbA to oxygen ensures more efficient release to tissues in need than HbF. Although the higher oxygen affinity of HbF is a concern to many, it seems unlikely that high HbF level in post-natal life is detrimental. A clinical condition known as "Hereditary Persistence of Fetal Hemoglobin" (HPFH) is a living example of its capability to sustain human life in the healthy condition. Besides this, high HbF levels in patients suffering from β-hemoglobinopathies are also known to alleviate their signs and symptoms by inhibiting the precipitation of excess α-globin chains in addition to providing an alternative form of functional hemoglobin.

Modern-day hematologists and clinical researchers look up to the concept of "HbF re-activation" as a potential curative strategy for patients suffering from critical β-hemoglobinopathies like β-thalassemia and sickle cell disease. Several drugs like hydroxyurea, 5-azacytidine, cytosine arabinoside, natural products etc. have been tried in clinics to elevate HbF levels in such patients with limited success and poor understanding on the mechanisms of their action. Associated side-effects and complications of using cytotoxic agents like these restrict their use in most instances. Fortunately, with the advent of newer molecular tools and techniques, researchers are focusing their attention to re-engineer the molecular machinery and thus re-activate the gamma-globin gene. This book brings together a selection of chapters dedicated to fetal hemoglobin – its physiological role, regulation, methodologies to manipulate and future strategies. Researchers and scientists interested in the topic will have a comprehensive understanding of the current concepts on fetal hemoglobin modulation and therefore will serve as a launching pad for their research ideas.

ACKNOWLEDGMENTS

At the very outset I'd like to express my respect and gratitude to all the patients and their families surviving with such hemoglobinopathies. Their strength and determination constantly served as my motivation to conceptualize and work on this book and I am truly grateful to them. Very special thanks to my niece, Isha who is an inspiration to me. Hematological disorders like beta-thalassemia and sickle cell disease are often considered as "rare diseases" and thus have always received less attention compared to other life-threatening hematological diseases like leukemias. Thanks to all the participants who have selflessly volunteered for the clinical and research studies in this field. Our knowledge and understanding would have been extremely limited without their help and support.

This book would not have been possible without the experts agreeing to share their expertise and so I wish to convey my heartfelt gratitude to all the authors who have contributed to this book. I would like to thank all the reviewers who have critically reviewed the contents and provided their valuable comments and suggestions.

I wish to thank all the members of the Regenerative Medicine lab at the School of Medical Science and Technology, IIT Kharagpur. I would especially like to thank Dr. Suman Kumar Ray and Ms. Liza Changkakoti for their help and support in compiling the book.

I wish to thank my adoring family - my parents; elder brother, Dip; my sister-in-law, Ananya; my nephew, Neil, and niece, Meghan for their encouragement. Thanks to my wife, Rashmi and my son, Divit for their endless support at every step in life and constant belief in me. Thanks for always standing by my side through thick and thin.

List of Abbreviations

ACH	Active Chromatin Hub
ActR	Activin Receptors
Adox	Adenosine-2′, 3′ Dialdehyde
AHSCT	Allogenic Hematopoietic Stem Cell Transplantation
AHSP	Alpha Hemoglobin Stabilizing Protein
ALA	Aminolevulinic Acid
ALT	Alanine Transaminase
AMP	Adenosine Monophosphate
ATF2	Activating Transcription Factor 2
ATF4	Activating Transcription Factor 4
ATP	Adenosine Triphosphate
BASO-EB	Basophilic Erythroblast
BCL11A	B-Cell Lymphoma/ Leukemia 11A
BGLT3	Beta-Globin Locus Transcript 3
BMD	Bone Mineral Density
BMP	Bone Morphogenetic Protein
BMT	Bone Marrow Transplantation
CBC	Complete Blood Count
CHD	Chromodomain Helicase DNA Binding Protein
CLL	Chronic Lymphocytic Leukemia

CML	Chronic Myeloid Leukemia
CO	Carbon Monoxide
CO2	Carbon Dioxide
CREB	Cyclic AMP Reaction Component Restricting Protein
CReP	Constitutive Repressor of Eif2α Phosphorylation
CRISPR	Clustered Regularly Interspaced Short Palindromic Repeats
CTCF	Chromatin Structure Regulating Transcription Factor
CYGB	Cytoglobin Gene
DEXA	Dual Energy X-Ray Absorptiometry
DFO	Desferrioxamine
DFP	Deferiprone
DFX	Deferasirox
DHS	DNAse I Hypersensitive Sites
DNA	Deoxyribonucleic Acid
DNMT	DNA Methyl Transferase
DNMT3A	DNA Methyltransferase 3A
DPG	Diphosphoglycerate
DRED	Direct Repeat Erythroid Specific Definitive
EB	Erythroblast
ECG	Electrocardiogram
EKLF	Erythroid Krüppel Like Factor
EPO	Erythropoietin
ERK	Extracellular Signal Regulated Kinases
FDA	Food & Drug Administration
FKLF	Fetal Krüppel-Like Transcription Factor
FOG1	Friend of GATA 1
FoP	Friend of PRMT1
GADD34	Growth Arrest and DNA Damage-Inducible Protein 34
GATA1	GATA Binding Protein 1
GMP	Guanosine Monophosphate
GVHD	Graft Versus Host Disease
GWAS	Genome Wide Association Study
HAT	Histone Acetyltransferase
Hb	Hemoglobin

HbA	Adult Hemoglobin
HBA1	Hemoglobin subunit Alpha-1
HBA2	Hemoglobin subunit Alpha-2
HBB	Hemoglobin Beta subunit
HBD	Hemoglobin Delta subunit
HBE	Epsilon Globin Gene
HBE1	Hemoglobin Subunit Epsilon 1
HbF	Fetal Hemoglobin
HBG	Hemoglobin Subunit Gamma
HBG1	Hemoglobin Subunit Gamma 1
HBG2	Hemoglobin Subunit Gamma 2
HBM	Hemoglobin Subunit Mu
HBQ1	Hemoglobin Subunit Theta 1
HbS	Sickle Hemoglobin
HBZ	Hemoglobin Subunit Zeta
HDAC	Histone Deacetylase
HDR	Homology Directed Repair
HFE	Human Hemochromatosis Protein
HIF1	Hypoxia Inducible Factor-1
HIV	Human Immunodeficiency Virus
HLA	Human Leukocyte Antigen
HMIP	HBS1L-MYB Intergenic Polymorphism
HMT	Histone Methyltransferase
HPFH	Hereditary Persistence of Fetal Hemoglobin
HRI	Heme Regulated eIF2α Kinase
HS	Hypersensitive Sites
HSC	Hematopoietic Stem Cells
HSCT	Hematopoietic Stem Cell Transplantation
HSPC	Hematopoietic Stem Progenitor Cell
HTLV	Human T-Cell Lymphotropic Viruses
HU	Hydroxyurea
HUDEP	Human Umbilical cord blood-Derived Erythroid Progenitor cells
IC	Iron Chelation

IPP	Inositol Pentaphosphate
ISR	Integrated Stress Response
JAK2	Janus Kinase 2
JCML	Juvenile Chronic Myeloid Leukemia
JNK	C-Jun Kinase Enzyme
KLF1	Krüppel Like Factor 1
LARC	LCR Associated Chromatin Remodeling Complex
LCR	Locus Control Region
LCRB	Locus Control Region Beta
LDB1	LIM Domain Binding Protein 1
LFT	Liver Function Tests
LIC	Liver Iron Concentration
LMO2	LIM Domain Only 2
LOVD	Leiden Open Variation Database
LRF	Leukemia/Lymphoma Related Factor
LSD1	Lysine-Specific histone Demethylase 1
MAPK	Mitogen Activated Protein Kinase
MBD	Methyl Cytosine Binding Domain
MCH	Mean Corpuscular Hemoglobin
MCV	Mean Corpuscular Volume
MeCP1	Methyl CpG Binding Complex 1
MEL	Murine Erythroleukemia
MEP	Megakaryocyte Erythroid Progenitor
MRI	Magnetic Resonance Imaging
MSD	Matched Sibling Donors
MUD	Matched Unrelated Donor
NF-E2	Nuclear Factor Erythroid-2
NGB	Neuroglobin Gene
NHEJ	Non Homologous End Joining
NO	Nitric Oxide
NTDT	Non Transfusion Dependent Thalassemia
NuRD	Nucleosome Remodelling and Deacetylase
ODC	Oxygen Dissociation Curve
OR	Olfactory Receptor

ORTHO-EB	Orthochromatic Erythroblast
PAM	Protospacer Adjacent Motif
PBG	Porphobilinogen
PBS	Primer Binding Site
PO2	Partial Pressure of O2
POLY-EB	Polychromatophilic Erythroblast
PRMT	Protein Arginine Methyltransferases
PRO-EB	Proerythroblast
PTH	Parathyroid Hormone
QTL	Quantitative Trait Loci
RBC	Red Blood Cells
RNA	Ribonucleic Acid
ROS	Reactive Oxygen Species
SCA	Sickle Cell Anemia
SCD	Sickle Cell Disease
SF	Serum Ferritin
sGC	Soluble Guanylate Cyclase
SNO-Hb	S-Nitrosohemoglobin
SNP	Single Nucleotide Polymorphism
TAL1	T-Cell Acute Lymphoblastic Leukemia 1
TALEN	Transcription Activator Like Effector Nucleases
TDBT	Transfusion Dependent β-Thalassemia
TDT	Transfusion Dependent Thalassemia
TFBS	Transcription Factor Binding Sites
TR	Testicular Receptor
TRV	Tricuspid Regurgitation Velocity
TSH	Thyroid Stimulating Hormone
TSS	Transcriptional Start Site
UCBT	Umbilical Cord Blood Transplant
UTR	Untranslated Regions
WBC	White Blood Cells
YAC	Yeast Artificial Chromosome
ZBTB7A	Zinc Finger And BTB Domain Containing 7A
ZFN	Zinc Finger Nucleases

ZFPM1	Zinc Finger Protein Multitype 1

In: Fetal Hemoglobin
Editor: Nishant Chakravorty
ISBN: 978-1-53617-955-2

Chapter 1

HEMOGLOBIN: A BIOCHEMICAL AND MOLECULAR OVERVIEW

Motiur Rahaman[1], Sankha Subhra Das[1], Suman Kumar Ray[1], Ronak Reshamwala[2,3] and Nishant Chakravorty[1,*]

[1]School of Medical Science and Technology,
Indian Institute of Technology Kharagpur, West Bengal, India
[2]Menzies Health Institute Queensland,
Griffith University, Gold Coast, Australia
[3]Griffith Institute of Drug Discovery,
Griffith University, Nathan, Australia

ABSTRACT

The field of hemoglobin research has always been a major area of thrust and has revealed a plethora of knowledge about human physiology, pathology, and molecular genetics. Hemoglobin is regarded as one of the most studied molecules over the last 60 years. Ever since its discovery it has been unraveling the complex molecular regulation of globin genes.

* Corresponding Author's Email: nishant@smst.iitkgp.ac.in.

Recently, some exhilarating advancements in the field of hemoglobin biology have induced researchers to shift their focus to understand the hemoglobin's structural and functional relationship and its significance in disease and treatment. In this chapter, we have focused on the various hemoglobin proteins, their evolutionary relationships and how different ligands alter oxygen-binding efficiency of hemoglobin. We have attempted to focus upon the biochemical and molecular basis of structure-function relationships. Lastly, different genes and their regulatory elements have been described to accurately depict their developmental regulation and we believe that this will have significant insights to study different hemoglobinopathies.

Keywords: hemoglobin, fetal hemoglobin, structure, function, globin genes

INTRODUCTION

In 1949, Linus Pauling and his associates elucidated the basis of "sickle cell anemia" as differences in electrical charges in the hemoglobin of diseased and normal individuals. This landmark publication in the journal, "Science" marked sickle cell anemia as the first 'Molecular Disease', and opened the doors to peek into the various genetic disorders. Over the past few decades, research in the field of hemoglobinopathies (inherited disorders that lead to abnormal production/ structure of the hemoglobin) has been expanding by leaps and bounds, unraveling newer findings in molecular medicine and developmental biology. Hemoglobin is the respiratory pigment found in erythrocytes which carries oxygen to the tissues and facilitates the return of carbon dioxide from tissues to the lungs. In arterial circulation, hemoglobin has a higher affinity for oxygen than carbon dioxide, hydrogen, chloride, and other gases. In contrast to this, the relative affinity of binding with these gases is reversed in venous circulation (Thomas and Lumb 2012). Spellbound by the remarkable properties of hemoglobin, French Nobel laureate, Jacques Monod conferred it with the title of an "honorary enzyme." The molecular structure of hemoglobin was elucidated by Max Perutz using X-ray crystallographic images (Perutz et al. 1960), earning him the Noble Prize in Chemistry in 1962. Our understanding of hemoglobin biology is primarily based on bench research deciphering its chemistry and biology. Pioneering work of stalwarts like Linus Pauling, Vernon Ingram, Max Perutz, Titus Huisman, Ernest Beutlar and others have contributed extensively to our present day knowledge and have laid the foundation of modern day translational research in the field.

HEMOGLOBIN RESEARCH AND ERA OF MOLECULAR MEDICINE

Over the last century, advancements in the field of X-ray Crystallography, molecular cloning and sequencing techniques have revealed enormous amounts of information, pertaining to hemoglobin genetics, its structural and functional properties and how changes in its genetics and structure can influence its functions. Furthermore, these molecular details have allowed us to predict and decode the different clinical phenotypes associated with different mutations, thus enriching our understanding of hemoglobinopathies. Research conducted on rare hemoglobin types has further privileged us to uncover the regulation of different globin genes and the crosstalk among them. In the last few decades, the interaction of nitric oxide (NO) with hemoglobin has opened a new door in functional aspects of hemoglobin research (Helms and Kim-Shapiro 2013).

A Brief History of Hemoglobin Research

Hemoglobin is an iron-containing tetramer consisting of four globin chains (two α- and two non-α chains in the post-embryonic stages of life) which plays a pivotal role in the gaseous transport of the body. Abnormalities in the genetic make-up, and consequently the protein sequence and structure, of these globin chains can lead to serious disease manifestations which are collectively termed as hemoglobinopathies. The keys milestones in the field of hemoglobin research have been highlighted in Table 1.

Table 1. Key milestones and discoveries in hemoglobin research

Year	Milestones	References
1910	First Observation of sickling of RBCs from Anemic patients.	(James 1910)
1927	First description of Thalassemia disease.	(Cooley 1927)
1948	Study revealed blood smear of child's blood shows less sickling when compare with adult blood due to presence of high amount of fetal hemoglobin (HbF).	(Watson 1948)
1949	Sickle cell anemia (SCA) is a molecular disease.	(Pauling L, Itano HA, Singer SJ 1949)
1957	Study revealed the structural differences between sickle hemoglobin (HbS) and adult hemoglobin (HbA1).	(Ingram 1957)
1960	Elucidated Molecular Structure of myoglobin, which is similar to hemoglobin by X-ray Crystallography.	(Perutz et al. 1960)
1977	The sickle cell mutation identified as a GAG to GTG mutation in codon 7, causing a Glu to Val substitution.	(Marotta et al. 1977)
1984	Role of Hydroxyurea as HbF inducer has been reported.	(Platt, Orkin, and Dover 1984)
1984	Molecular Basis of non deletional hereditary persistence of fetal hemoglobin (HPFH).	(Collins et al. 1984)
1987	Role of Butyrate as an HbF inducer.	(Perrine1987)
2008	BCL11A identified as major HbF repressor by genome-wide association study (GWAS).	(Albai et al. 2008)
2016	ZBTB7A reported as second major HbF repressor.	(Masuda 2016)
2018	BCL11A and ZBTB7A independently regulate fetal hemoglobin expression through direct promoter binding.	(Martyn et al. 2018)

HEMOGLOBIN HETEROGENEITY

Adult human hemoglobin comprises of a heterogeneous group of metalloproteins which arise at different stages of ontogeny to meet the strict oxygen requirement of the body. In erythrocytes of adult individuals, hemoglobin A1 ($\alpha_2\beta_2$) reaches over 96% of the total hemoglobin. The rest of the hemoglobin consists of fetal hemoglobin ($\alpha_2\gamma_2$) (about 1%) and hemoglobinA2 ($\alpha_2\delta_2$) (around 2%) of the molecules (Schechter 2014). The quaternary structure of different hemoglobin molecules is similar and consists of two α-like and two β-like globin chains which are transcribed from different chromosomes. In the course of evolution, two globin gene loci have diverged dramatically, giving rise to multiple genes and pseudogenes that are located in close vicinity and are expressed in a sequential manner during different developmental stages (Huehns et al. 1964). In the fetal stage, ζ globin (encoded by HBZ gene) is expressed initially during the early developmental stages and gets repressed within the first trimester of pregnancy. From this stage onwards, the α globin (encoded by HBA1 and HBA2 genes) starts getting expressed and it persists throughout life (Yagi et al. 1986). Sequential expression of different globin genes that reside on the β-globin loci, is a complex phenomenon. During the developmental stages, two globin switching processes take place simultaneously in a tandem manner: first one happens during ζ to α switching and the second one occurs around the time of birth (Sankaran, Xu, and Orkin 2010). The β-globin like chain ε (encoded by HBE gene) is expressed initially, forming a tetramer with two α-like chains (ζ and α), resulting in embryonic hemoglobin Hb Gower-1 ($\varepsilon_2\zeta_2$) and Hb Gower-2 ($\varepsilon_2\alpha_2$) production respectively. During the course of development the ε chain is eventually silenced and this leads to synthesis of the γ globin chain. The γ-globin chains (encoded by γ^A and γ^Ggenes) bind with α-globin chains to form fetal hemoglobin (HbF). HbF is the form of hemoglobin that predominates in the later part of gestation period. At birth, the γ globin gene expression is slowly down-regulated and the expression β-globin like genes, β and δ are simultaneously enhanced to form adult HbA1 and HbA2, respectively (Stamatoyannopoulos 2005). Interestingly, in some individuals, the γ globin silencing mechanism gets faltered, giving rise to physiologically asymptomatic condition in which elevated fetal hemoglobin level persists in adulthood. This condition is termed as hereditary persistence of fetal hemoglobin (HPFH) (Poncz et al. 1987).

Evolutionary Relationship between the Various Hemoglobin Genes

Hemoglobin is a red-colored respiratory pigment which is abundantly present in red blood cells (RBCs) of mammals and other jawed vertebrates (gnathostomes). A detailed depiction of the evolution of hemoglobin genes in scientific literature strengthened our understanding of hemoglobin function and the various regulatory mechanisms involved in its gene expression. Comparative studies on the structural and functional aspects of hemoglobin between species enlightened us about different adaptations and led us to explore various biochemical and physiological aspects of hemoglobin which are conserved (Sanyal et al. 2013). An evolutionary tree which depicts the ontogeny of the globin gene family from bacteria to vertebrates was developed based on the fold characteristics of polypeptide chains and heme prosthetic group (Hardison 2012). Truncated globins are known to be present in bacteria, plants, and unicellular eukaryotes. These globins are involved in gaseous transport, detoxification, and storage of oxygen (Wittenberg et al. 2002). In bacteria, flavohemoglobin appears to be responsible for detoxification of Nitric oxide (NO) by acting as NO dioxygenase; whereas, in plants and protists, single-chain globins are largely involved in electron transport, oxygen storage, and scavenging (Fridovich 1998). Globin as a respiratory pigment appeared first in the invertebrate clade. In vertebrates, these globin genes have undergone duplication, deletion or translocation, leading to generation and diversification of globin gene family which is in the forefront of molecular evolution study of protein family (Hardison 1996). The human globin family predominantly consists of five major types of globin proteins which are distributed in different tissues throughout the body (shown in Table 2).

Table 2. Different human hemoglobin types and their tissue distribution

Hemoglobin types	Chromosomal location	Gene	Tissue type
α- globin	16p13.3	HBA1 and HBA2	Erythrocytes
β- globin	11p15.4	HBB	Erythrocytes
Neuroglobin	14q24.3	NGB	Brain tissue
Myoglobin	22q12.3	MYB	Skeletal and heart muscles
Cytoglobin	17q25.1	CYGB	Different human cells, such as esophageal cells, smooth muscles cells, melanocytes, etc.

The major adult hemoglobin, hemoglobin A1 is composed of four globin chains which are encoded by different globin genes. The α-globin is encoded by duplicated genes, HBA1 and HBA2, whereas β-globin is encoded by

HBB gene. In contrast to hemoglobin, myoglobin is a related monomeric heme-containing globin which is predominantly expressed in skeletal and heart muscles. It is specifically associated with oxygen storage and delivery of oxygen to mitochondria by diffusion (Wittenberg and Wittenberg 2007). However, Cytoglobin, encoded by CYGB, is present in different human tissues and found to be closely related in structure to myoglobin (Ebner et al. 2002). In addition, another noble class of globin molecules is Neuroglobin (encoded by NGB gene). NGB has high oxygen affinity and is found predominantly in brain tissue of human and mouse (Burmester et al. 2000). Sequence analysis has revealed that Neuroglobin is distantly related to other human globins and found to diverge early in metazoan evolution. Its proximity to invertebrate nerve globin advocates the presence of an ancestral gene before the divergence of vertebrates and invertebrates over 800 million years ago (Hardison 2012). The Cyclostomes (hagfish and lampreys) (an early diverging species from vertebrate clade), use heme bound globin protein for oxygen transport. Astonishingly, this globin is more closely related to CYGB than to the gnathostome (jawed vertebrate) hemoglobin, indicating convergent evolution in these two major branches of vertebrate evolution (Hoffmann, Opazo, and Storz 2010).

Hemoglobin Structure

The Structure of Adult Hemoglobin

Hemoglobin is a heterotetramer of two α-like and two β-like globin subunits; each one is associated with a heme, an iron-containing prosthetic group (compounds which are permanently associated with protein and responsible for its function).

Primary Structure

Primary structure of a protein refers to strings of amino acids which are joined together by peptide bonds via condensation reactions, releasing water molecules. This sequence-based linear structure dictates folding of the polypeptide chains into globular proteins such as hemoglobin.

Adult hemoglobin, HBA1 (accounts for over 96% of total hemoglobin) consists of two α-globin (141 residues each) and two β-globin (146 residues each) chains, which are folded into a globular three-dimensional structure to form α and β subunits respectively. These two subunits are nearly similar in structure, interacting with each other to form hemoglobin molecule.

Secondary Structure

The polypeptide chain of globin family proteins (hemoglobin, myoglobin, etc.) is mainly composed of eight α-helical regions (except the two α chains of hemoglobin which contain seven helical regions only) interconnected by short stretches of bends. Conventionally, these helical segments are named alphabetically: A through H, starting from amino (N terminus) to carboxy (C terminus) end of each polypeptide.

Tertiary Structure

The nascent polypeptides, immediately after their synthesis, bind with the heme prosthetic group, which imparts structural stability to these globin chains. Heme consists of a protoporphyrin ring which is chemically bound with an iron atom in its ferrous (Fe^{2+}) valence state. Iron atom is capable of forming six coordination bonds; four of which are coordinated with nitrogen atoms of the flat porphyrin ring and the remaining two reside perpendicular to the porphyrin plane. Electron donating properties of the nitrogen atom ensures the stability of the ferrous (Fe^{2+}) valence state, preventing oxidation to form ferric (Fe^{3+}) ions which has detrimental consequences over oxygen-carrying capacity of hemoglobin molecules. The iron atom can reversibly bind with oxygen and help in carrying out transporter function of hemoglobin, only in its ferrous state. Free heme, which is not bound to any protein, has two open coordination bonds that can react with oxygen molecules, resulting in irreversible conversion of Fe^{2+} to Fe^{3+}. In hemoglobin, the heme molecule is deeply buried in a hydrophobic core made by folded polypeptides to prevent this oxidation reaction. The molecular structure of hemoglobin has revealed that one of the remaining two coordination bonds of iron is occupied by nitrogen atom of the imidazole ring (histidine residue at position 8 of F helix). The other one is bound predominantly by molecular oxygen in the normal physiological condition of the body (Berg JM, Tymoczko JL 2002). However, some other gases such as Nitric oxide (NO), carbon monoxide (CO), etc. also can combine with heme iron of hemoglobin (Antonini and Brunori 1971).

Quaternary Structure

The quaternary structure of adult hemoglobin reveals the interaction between two heterodimeric subunits ($\alpha_1\beta_1$ and $\alpha_2\beta_2$) which get weakly associated with each other and thus form the tetramer (shown in Figure 1). Within the tetramer, each chain interacts with the other corresponding chain specifically through two interfaces $\alpha_1\beta_1$ (or $\alpha_2\beta_2$) and $\alpha_1\beta_2$ (or $\alpha_2\beta_1$).The $\alpha\beta$ dimers show greater stability than individual monomers under normal physiological condition (Manning et al. 2007). Within the Hb tetramer, $\alpha_1\beta_2$ (or $\alpha_2\beta_1$) interfaces are less stable compared to $\alpha_1\beta_1$ (or $\alpha_2\beta_2$). This explains why mild urea treatment disassembles the tetramer, but $\alpha\beta$ dimer remains intact. When oxygen binds to the heme prosthetic group of globin subunit, the $\alpha\beta$ subunits slide pass each other, rotating around this flexible $\alpha_1\beta_2$ (or $\alpha_2\beta_1$) interface. This mechanism triggers the transition from tensed T state to relaxed R state (Thom CS et al. 2013). The destabilization of $\alpha_1\beta_2$ interfaces upon oxygen binding relaxes the constrained tetramer which facilitates more oxygen-binding that can be marked by the S-shaped curve, indicating cooperativity.

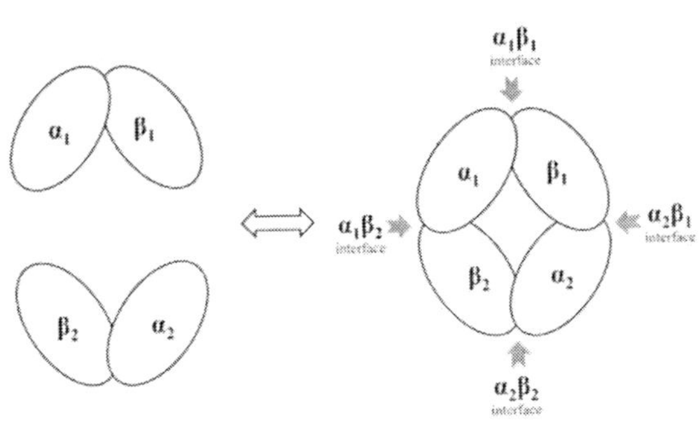

Figure 1. The quaternary structure of hemoglobin, showing two heterodimeric subunits ($\alpha_1\beta_1$ and $\alpha_2\beta_2$) assembled to form the Hb tetramer. The black arrows indicate different interfaces (such as $\alpha 1\beta 2$, $\alpha_1\beta_1$, etc.) of globin chain interactions that are crucial for conformational changes (R or T state) and ligand binding of the hemoglobin molecule.

The Structure of Fetal Hemoglobin

Fetal hemoglobin (HbF) is the major form of hemoglobin in the developing fetus, which starts to express within a few weeks after conception and continues to be expressed in significant quantities up to 6 months postnatally. HbF binds to oxygen with higher affinity compared to adult hemoglobin (HbA1), enabling transfer of oxygen to the fetus from the mother (Thein et al. 2009). HbF is a tetramer which consists of two α chains and two γ chains. HbF has the same α chains like adult hemoglobin, HbA1. The γ chains of HbF are structurally homologous to β chain of HbA1; however their N-terminal segments differ. The major structural difference between HbF and HbA1 is in their respective 2, 3- DPG binding regions. This binding region is located between γ_1-γ_2 interface and β_1-β_2 interface of HbF and HbA1 respectively. The primary structure of globin chain reveals that β1-Val and β143-His of β-globin chain are replaced by γ1-Gly and γ143-Ser of γ chain respectively. These changes are proposed to be the primary reason for higher oxygen affinity of HbF (Adachi et al. 1997). These structural differences allow 2, 3- DPG to bind less tightly to HbF in comparison to HbA1, resulting in higher affinity of HbF to oxygen (Benesch, Benesch, and Yu 1969).

The Structure of Embryonic Hemoglobin

In humans, the early embryos express three different types of embryonic hemoglobin forms during the pre-placental phase, namely: Gower-1(ζ_2 ε_2), Gower-2 ($\alpha_2\varepsilon_2$) and Portland ($\zeta_2\gamma_2$). After the placenta formation, these hemoglobin forms are replaced by fetal hemoglobin (HbF) which is later superseded by HbA1 and HbA2 at around the time of birth. The embryonic hemoglobin types are composed of two unique sets of polypeptide chains. The 141 residues ζ chain (embryonic counterpart of α chain) is present in Gower-1 and Portland hemoglobin.

Similarly, Gower-1 and Gower-2 hemoglobin molecules contain β- like ε chain (146 residues). In addition, Portland hemoglobin contains the γ

chain, which is the same as fetal hemoglobin. X-ray crystallographic study of different embryonic hemoglobin types have revealed the fact that ε subunits of Gower-1 and Gower-2 hemoglobin are structurally identical to adult β and fetal γ subunits with minor differences at N-terminal region and helix "A" of the polypeptide chain (Sutherland-Smith et al. 1998).

HEMOGLOBIN SYNTHESIS

Heme synthesis occurs in the cytosol and mitochondria of erythrocytes, and involves multiple steps (Ogun AS 2019). This process is executed by eight enzymes, four of which are known to function inside the mitochondria and the remaining ones work in the cytosol. The enzyme, Aminolevulinic acid (ALA) synthase, initiates the process inside the mitochondria by linking glycine and succinyl coenzyme A to form Aminolevulinic acid (ALA). In the next step, the action of ALA dehydratase converts ALA to porphobilinogen (PBG), which is then converted to hydroxymethylbilane in presence of porphobilinogen deaminase. In the subsequent steps, hydroxymethylbilane is converted to uroporphyrinogen III and then to coproporphyrinogen III. Coproporphyrinogen III is then converted into protoporphyrinogen IX by coproporphyrinogen oxidase (Dailey et al. 2017). The final phase of this reaction cascade leads to the addition of iron (Fe^{2+}) to protoporphyrin IX by ferrochelatase creating a molecule of heme. The globins are a superfamily of heme-containing globular proteins having a series of eight alpha helical segments which incorporate the alpha globin protein fold. The globin fold is found in its namesake, globin families as well as in phycocyanins. The exact angle of helix packing depends on the sequence of the protein, because packing is mediated by hydrophobic interactions of the amino acid side chains near the helix interfaces. Several functionally different hemoglobins can coexist in the same species. Fold of the globin superfamily is highly conserved in course of evolution. After synthesis of globin, heme joins and finally forms hemoglobin. The step by step process has been depicted in Figure 2.

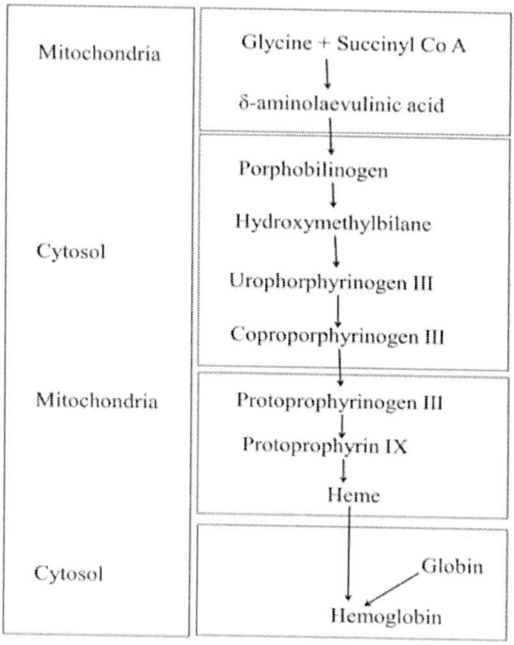

Figure 2. Flow chart representation of hemoglobin synthesis.

HEMOGLOBIN FUNCTION

Gaseous Transport

The well-elucidated function of hemoglobin is the transportation of respiratory gases (O_2 and CO_2) in vertebrates. Unlike O_2 transport, CO_2 is not directly bound with heme iron. It reacts with the amino-terminal end of polypeptide chains, forming carbamino complex that can stably transport CO_2 from body tissues to the lungs (Baier and Schroder 1972). X-Ray crystallographic study of different hemoglobin variants has shown spatial arrangement of different atoms and how they are intricately associated with oxygen binding, revealing the molecular mechanism of oxygen transport (Thom CS et al. 2013). Over the past 50 years, the understanding pertaining to hemoglobin's structural and functional relationship has attained a great deal of paradigm shift due to the advancement of existing techniques of protein chemistry and molecular biology. The allosteric model has explained the altered oxygen binding capacity of hemoglobin due to interaction with other molecules such as hydrogen ions, 2,3- Diphosphoglycerate (2,3-DPG), various anions, etc. (Kanaori et al. 2011). We will discuss this part in details in the following section. In addition, certain gases such as Carbon monoxide (CO), Nitric oxide (NO), etc. play major role in respiratory physiology of human which we will discuss in the following sections.

Carbon Monoxide as a Ligand of Hemoglobin

Carbon monoxide (CO) binds with hemoglobin with higher affinity than oxygen (O_2) itself. When CO is present in higher concentration in the body, it can competitively exclude out O_2 from binding to the hemoglobin. This is the reason why CO is regarded as highly toxic to the aerobic organisms (Lindell K. Weaver 2019). Apart from an exogenous source, CO can be produced in significant amount endogenously as a metabolic by-product of heme degradation (Lecomte, Vuletich, and Lesk 2005). Microsomal hemeoxygenases degrade heme part of respiratory proteins (such as hemoglobin, myoglobin, etc.) to produce CO, a lethal compound at high concentration (Medical and Francisco 1969). Studies have shown that CO can activate soluble guanylylcyclase and thus has a potential pharmacological role to modulate inflammation or other pathophysiological states (Ryter and Otterbein 2004).

Nitric Oxide and Hemoglobin Interactions

Interaction between NO and hemoglobin is regarded as the main focus of the researchers studying non-oxygen related functions of hemoglobin. Endothelial cells convert arginine into NO, which acts via cGMP and other cell signaling pathways to regulate vascular tone, endocrine and neuronal functions (Kosaka 1999; Denninger and Marletta 1999). Oxyhemoglobin acts as a scavenger of NO gas, regulating its bioavailability within the body

(Azarov et al. 2005). NO diffuses into the blood and subsequently reacts with oxyhemoglobin in erythrocytes to produce methemoglobin and nitrate ions. Studies have reported that most of the circulating methemoglobin is produced by this reaction (Baier and Schroder 1972). The accumulated methemoglobin is valueless as a respiratory pigment. Its ferric ion binds tightly with oxygen and results in diminished supply of oxygen to the tissues. The enzyme methemoglobin reductase, present in erythrocytes catalyzes the reversed reaction to maintain status quo (Gladwin et al. 2000).

Nitric oxide is also known to react with deoxyhemoglobin to form nitrosyl(heme)hemoglobin(NO-hemoglobin). Nitric oxide irreversibly combines with the ferrous ion of deoxyhemoglobin and is postulated to function as an endocrine-like transporter of NO throughout the body (Schechter and Gladwin 2003). Extensive studies done in this area have revealed a third reaction of NO with oxyhemoglobin (Gow et al. 1999). Nitric oxide can react with the highly conserved β93 residue of hemoglobin to form S-nitrosohemoglobin (SNO-Hb). This SNO-Hb has been hypothesized to homeostatically control blood flow to the tissues in accordance with the oxygen tension. Under high oxygen tension hemoglobin possibly reacts with NO, forming SNO-Hb; whereas, at low oxygen tension, NO is released from the complex to promote blood supply to the low oxygenated tissue by vascular dilation. Therefore, this hypothesis reveals the possibilities of RBCs as an 'oxygen sensor' which sense hypoxic conditions and adjust blood flow accordingly (Gladwin et al. 2000).

In normal condition, the kinetics of these endocrine and paracrine functions of NO is finely tuned within the body. Studies have shown that the rapid destruction of NO within erythrocytes alters the endocrine function without affecting the paracrine-like function of the gas (Schechter et al. 2012). Free heme concentration in blood shifts the equilibrium towards scavenging of NO, whereas accumulated NO can redirect it toward endocrine delivery (Schechter and Gladwin 2003).

ALLOSTERIC EFFECT

What Is Allosteric Effect?

Allosteric effect refers to a phenomenon where properties of proteins or enzymes are controlled by other molecules by binding to a region other than the active site(s) of the proteins or enzymes. Hemoglobin of red blood cells is reversibly bound with oxygen under normal physiological condition. X-ray crystallographic study reveals the structure of hemoglobin in both oxygenated and deoxygenated condition (Harrington, Adachi, and Royer 1998; Paoli et al. 1996). As opposed to lungs which expand during inhalation, hemoglobin becomes constrained (termed as T state) when in oxygenated condition and relax when oxygen is released (denoted as R state). This paradoxical phenomenon was termed as 'Molecular Breathing' by Max Perutz. Oxygen binding varies upon its partial pressure. The oxygen-hemoglobin dissociation curve is plotted as percentage of oxygen saturation against partial pressure of oxygen. Hemoglobin becomes fully saturated at about 100mm Hg partial pressure of oxygen (Collins JA, Rudenski A, Gibson J, Howard L 2015). Oxygen-hemoglobin dissociation curve is sigmoid in shape, showing cooperative oxygen binding with ferrous iron atom of each prosthetic group. This means oxygenation of one helps to proceed it further in a much easier manner. The details about Oxygen dissociation curve and factors associated with it are discussed in another chapter (Fetal Hemoglobin: Formation and Gaseous Exchange).

Ligand Binding and Altered Oxygen Affinity

Hemoglobin is mainly involved as a two-way transporter which transports O_2 from lungs to the body tissues and carries CO_2 from tissues to the lungs. Thus the kinetics of oxygen binding and release is intricately associated and adaptable with body's metabolic perturbations and developmental ontogeny.

The functional behavior of hemoglobin can be modified by binding of other ligand produced by the body. The binding of these ligands to the hemoglobin generally distort the normal oxygen-binding capacity which in turns alter body's normal respiration. Alternation of oxygen binding shifts Hb-O_2 dissociation curve to the right or left depending upon oxygen affinity towards hemoglobin which helps in release of oxygen at particular oxygen tension in body tissues. One important modifier of oxygen-hemoglobin interaction is carbon-di-oxide (CO_2). CO_2 diffuses into erythrocytes in the peripheral tissues and is metabolized to form carbonic acid by carbonic anhydrase. This process leads to generation of H^+ ions and subsequent production of oxygen (Arthurs and Sudhakar 2008). The produced H^+ interacts with Hb, which in turn diminishes the oxygen affinity of hemoglobin and thus leads to a right-shifted Hb-O_2 dissociation curve. The binding of H^+ and subsequent reduction of oxygenation of hemoglobin is called as Bohr Effect (Keller, Pedemonte, and Willmouth 1970). In contrast, low CO_2 concentration and high pH propel the oxygen binding in lung. Besides, CO_2 interacts with the amino end of α- and β-like globins, forming carbamino compound. All these impel minor effects on oxygenation of Hb when compared to the effects of CO_2 on pH.

Another important compound, 2, 3- DPG (2, 3-Diphosphoglycerate) is known to influence the oxygen dissociation curve. This compound is a metabolic byproduct which is present in large quantities in erythrocytes and is involved in allosteric regulation. It binds to Hb and imparts alteration of polypeptides in such a manner that stabilize the tense T state of Hb, facilitating the oxygen release. The binding of these allosteric regulators, H^+, 2, 3- DPG is under strict regulation, and any perturbation of metabolic process is compensated by changes in the oxygen-binding efficiency. Such changes shifts the Oxygen dissociation curve appropriately (Benesch, Benesch, and Y 1969). Further details have been discussed in another chapter.

HEMOGLOBIN GENETICS

The human hemoglobin synthesis is an intricate interplay between two developmentally regulated gene clusters which are located on chromosomes 16 and 11. The α like globin gene clusters are dispersed near the telomeric region at chromosome 16p13.3. Similarly, the β like globin genes are situated at chromosome 11p15.5 which are expressed sequentially in accord with oxygen demand of different developmental stages (embryonic, fetal and adult). The expression patterns of these genes are complex and act as repertoire of genetic and epigenetic regulation studies. The α globin gene is strongly expressed immediately after a brief period of early embryonic development. Studies have shown that 5′ α-2 is expressed more strongly than 3′ α-1 gene (Ribeiro and Sonati 2008). In case of β-like genes, the ε gene initially expresses and is succeeded by two γ genes (γ^G and γ^A), and ultimately in adult stages, δ globin and β globin is expressed predominantly. Expression patterns of these genes shift from 5′ to 3′ as development progresses into embryonic, neonatal and the adult stages. The coordinated gene expression of these two clusters is developmentally regulated and dependent upon some cis-acting regulatory elements situated upstream to these gene clusters (Wienert et al. 2018). The α like globin gene cluster is regulated by erythroid-specific regulatory element termed as hypersensitivity site-40 (HS-40), which is situated upstream to α cluster within an intron of the adjacent housekeeping gene, C16orf35(Higgs et al. 2005). Similarly, the β cluster is under the regulation of five DNase I hypersensitivity region (5′ HSs 1to 5) called Locus Control Region (LCR). The LCRs are strong enhancer of transcription of downstream β loci genes and deletion of these regions severely reduce β globin gene expression (Li et al. 2002). The expression of α and β globin genes are very delicately balanced during erythropoiesis to produce normal healthy hemoglobin level in the erythrocytes.

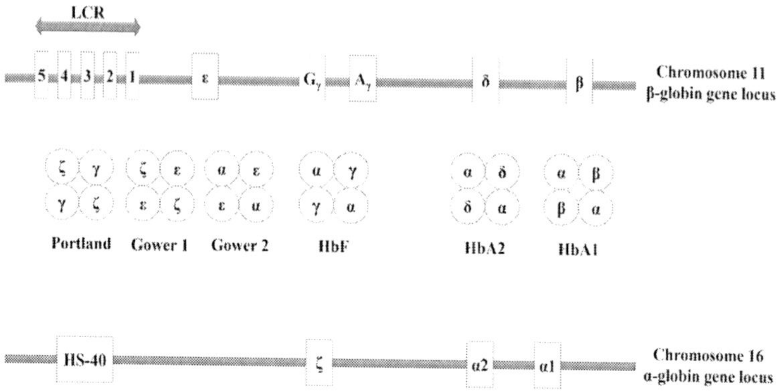

Figure 3. Schematic diagram of human globin loci and various hemoglobin types such as embryonic (Portland, Gower1 and Gower2), fetal (HbF) and adult (HbA1 and HbA2) hemoglobin. In the top panel, β globin gene locus is schematically represented having different developmental group of genes (ε, γ^G, γ^A, δ and β) and the locus control region (LCR). In the bottom, α globin locus is illustrated which contains two α genes (α_1 and α_2), ζ gene and a single upstream erythroid-specific DNaseI hypersensitive site termed as HS-40.

Each DNA segment in these two globin gene clusters (α and β) is correspondent to the specific globin chain (shown in Figure 3). The transcripts of these genes are similar in structure and consist of 3 coding exons and two intervening sequences or introns. Mutations that affect the splicing efficiency of these transcripts may give rise to different hemoglobin related disorder or hemoglobinopathies. In addition, mutations in the5′ promoter region and different trans acting region are also contributing to alternation in splicing efficiency which may lead to different hemoglobinopathies (Thein 2013).

Conclusion

Hemoglobin is perhaps the most extensively studied protein, yet it holds an enormous amount of unresolved puzzles in both clinical levels as well as structural and molecular level. Recently, despite the revelation of various unexpected new functions of hemoglobin, the main focus of biomedical hemoglobin research has shifted to the developmental regulation of α and β-like goblin gene clusters. Over the past few decades, numerous studies have been devoted to the various globin interactions and subsequent diseases manifestations caused by the alteration of these interactions. This particular field of research has attained a wider spectrum of studies from basic molecular genetics, to different "translational" models, to various clinical endeavor devoted to the treatment of the patients. It is hoped that delving deeper into the obscurity of globin gene regulation would lead us to explore new possibilities to treat various hemoglobinopathies. Despite the extensive studies done so far on hemoglobin at genetic and protein levels, existing lacunae need to be addressed in a profound and consolidated manner in the future.

ACKNOWLEDGMENTS

The authors would like to acknowledge the Indian Institute of Technology, Kharagpur for providing the infrastructural and financial support.

REFERENCES

Adachi, Kazuhiko, Patrick Konitzer, Jian Pang, Konda S. Reddy, and Saul Surrey. 1997. "Amino Acids Responsible for Decreased 2, 3-Biphosphosphoglycerate Binding to Fetal Hemoglobin." *Blood* 90 (8): 2916–20.

Albai, Giuseppe, Maria Grazia Piras, Natascia Sestu, Sandra Lai, Mariano Dei, Antonella Mulas, Laura Crisponi, et al. 2008. "Genome-Wide Association Study Shows BCL11A Associated with Persistent Fetal Hemoglobin and Amelioration of the Phenotype of Beta-Thalassemia." *Proceedings of the National Academy of Sciences of the United States of America* 105 (5): 1620-1625 %U http://www.ncbi.nlm.nih.gov/pubmed/18.

Antonini, Eraldo, and Maurizio Brunori. 1971. *Hemoglobin and Myoglobin in Their Reactions with Ligands*. Frontiers. Amsterdam: North-Holland. https://lib.ugent.be/catalog/rug01:001650991.

Arthurs, G. J., and M. Sudhakar. 2008. "Carbon Dioxide Transport." *Update in Anaesthesia* 24 (2): 26–29. https://doi.org/10.1093/bjaceaccp/mki050.

Azarov, Ivan, Kris T. Huang, Swati Basu, Mark T. Gladwin, Neil Hogg, and Daniel B. Kim-Shapiro. 2005. "Nitric Oxide Scavenging by Red Blood Cells as a Function of Hematocrit and Oxygenation." *Journal of Biological Chemistry* 280 (47): 39024–32. https://doi.org/10.1074/jbc.M509045200.

Baier, B. Y. C., and E Schroder. 1972. "*HocIschule, 3,*" 457–71.

Benesch, Ruth E., Reinhold Benesch, and Chi Ing Yu. 1969. "The Oxygenation of Hemoglobin in the Presence of 2,3-Diphosphoglycerate. Effect of Temperature, PH, Ionic Strength, and Hemoglobin Concentration." *Biochemistry* 8 (6): 2567–71. https://doi.org/10.1021/bi00834a046.

Benesch, Ruth E., Reinhold Benesch, and Chi Ing Y. 1969. "*The Oxygenation*" 8 (6): 2567–71.

Berg J. M., Tymoczko J. L., Stryer L. 2002. *Section 10.2Hemoglobin Transports Oxygen Efficiently by Binding Oxygen Cooperatively*. Biochemist. New York: W. H. Freeman and Company. https://www.ncbi.nlm.nih.gov/books/NBK22596/.

Burmester, Thorsten, Bettina Weich, Sigrid Reinhardt, and Thomas Hankeln. 2000. "A Vertebrate Globin Expressed in the Brain : Article : Nature." *Nature* 407 (6803): 520–23. https://doi.org/10.1038/35035093.

Ch, E. A. R. 2016. "*And Wild-Type Larvae. These Experiments Showed A*" 351 (6270): 285–90.

Collins, F. S., C. J. Stoeckert, G. R. Serjeant, B. G. Forget, and S. M. Weissman. 1984. "(G)Γβ+ Hereditary Persistence of Fetal Hemoglobin: Cosmid Cloning and Identification of a Specific Mutation 5' to the (G)γ Gene." *Proceedings of the National Academy of Sciences of the United*

States of America 81 (15 I): 4894–98. https://doi.org/10.1073/pnas.81.15.4894.

Collins J. A., Rudenski A., Gibson J., Howard L. 2015. "*Key Points on O2 Dissociation Curve*" 11 (3): 194–201. https://doi.org/10.1183/20734735.001415.

Cooley, T. B. 1927. "Anemia in Children. Long Bones Also Present Striking Changes . Observation of the Peculiarities." *American Journal of Disease of Children* 34 (3): 347.

Dailey, Harry A., Tamara A. Dailey, Svetlana Gerdes, Dieter Jahn, Martina Jahn, Mark R. O'Brian, and Martin J. Warren. 2017. "Prokaryotic Heme Biosynthesis: Multiple Pathways to a Common Essential Product." *Microbiology and Molecular Biology Reviews* 81 (1): 1–62. https://doi.org/10.1128/mmbr.00048-16.

Denninger, John W., and Michael A. Marletta. 1999. "Guanylate Cyclase and the .NO/CGMP Signaling Pathway." *Biochimica et Biophysica Acta - Bioenergetics* 1411 (2–3): 334–50. https://doi.org/10.1016/S0005-2728(99)00024-9.

Ebner, Bettina, Bettina Weich, Thomas Hankeln, and Thorsten Burmester. 2002. "Cytoglobin: A Novel Globin Type Ubiquitously Expressed in Vertebrate Tissues." *Molecular Biology and Evolution* 19 (4): 416–21. http://www.ncbi.nlm.nih.gov/pubmed/11919282%5Cnhttp://www.ncbi.nlm.nih.gov/sites/entrez.

Fridovich, Irwin. 1998. "*<Gardner NO Dioxygenase_Erwinia Chrysanthemi.Pdf>*" 95 (September): 10378–83.

Gladwin, M. T., F. P. Ognibene, L. K. Pannell, J. S. Nichols, M. E. Pease-Fye, J. H. Shelhamer, and A. N. Schechter. 2000. "Relative Role of Heme Nitrosylation and Beta -Cysteine 93 Nitrosation in the Transport and Metabolism of Nitric Oxide by Hemoglobin in the Human Circulation." *Proceedings of the National Academy of Sciences* 97 (18): 9943–48. https://doi.org/10.1073/pnas.180155397.

Gow, Andrew J., Benjamin P. Luchsinger, John R. Pawloski, David J. Singel, and Jonathan S. Stamler. 1999. "The Oxyhemoglobin Reaction of Nitric Oxide." *Proceedings of the National Academy of Sciences of the United States of America* 96 (16): 9027–32. https://doi.org/10.1073/pnas.96.16.9027.

Hardison, Ross C. 2012. "Evolution of Hemoglobin and Its Genes." *Cold Spring Harbor Perspectives in Medicine* 2 (12): 1–18. https://doi.org/10.1101/cshperspect.a011627.

Hardison, Ross C. 1996. "Commentary History Hemoglobins: Plant, Animal, Protist, and Bacteria." *Proc Natl Acad Sci USA* 93 (June): 5675–79.

Harrington, Daniel J., Kazuhiko Adachi, and William E. Royer. 1998. "Crystal Structure of Deoxy-Human Hemoglobin B6 Glu → Trp." *Journal of Biological Chemistry* 273 (49): 32690–96. https://doi.org/10.1074/jbc.273.49.32690.

Helms, Christine, and Daniel B. Kim-Shapiro. 2013. "Hemoglobin-Mediated Nitric Oxide Signaling." *Free Radical Biology and Medicine* 61: 464–72. https://doi.org/10.1016/j.freeradbiomed.2013.04.028.

Higgs, D. R., D. Garrick, E. Anguita, M. De Gobbi, J. Hughes, M. Muers, D. Vernimmen, et al. 2005. "Understanding α-Globin Gene Regulation: Aiming to Improve the Management of Thalassemia." *Annals of the New York Academy of Sciences* 1054: 92–102. https://doi.org/10.1196/annals.1345.012.

Hoffmann, Federico G., Juan C. Opazo, and Jay F. Storz. 2010. "Gene Cooption and Convergent Evolution of Oxygen Transport Hemoglobins in Jawed and Jawless Vertebrates." *Proceedings of the National Academy of Sciences of the United States of America* 107 (32): 14274–79. https://doi.org/10.1073/pnas.1006756107.

Huehns, E. R., N. Dance, G. H. Beaven, F. Hecht, and A. G. Motulsky. 1964. "Human Embryonic Hemoglobins." *Cold Spring Harbor Symposia on Quantitative Biology* 29: 327–31. https://doi.org/10.1101/SQB.1964.029.01.035.

James, B. 1910. "*Peculiar Physical*" VI (5): 517. https://doi.org/10.1001/archinte.1910.00050330050003.

Kanaori, K., Y. Tajiri, A. Tsuneshige, and T. Yonetani. 2011. "Allosteric Mechanism of Oxygen-Binding in Hemoglobin." *Biophysical Journal* 100 (3): 380a. https://doi.org/10.1016/j.bpj.2010.12.2262.

Keller, A., E. Pedemonte, and F.M. Willmouth. 1970. "© 1970 Nature Publishing Group." *Nature* 225: 538–39.

Kosaka, Hiroaki. 1999. "Nitric Oxide and Hemoglobin Interactions in the Vasculature." *Biochimica et Biophysica Acta - Bioenergetics* 1411 (2–3): 370–77. https://doi.org/10.1016/S0005-2728(99)00026-2.

Lecomte, Juliette T. J., David A. Vuletich, and Arthur M. Lesk. 2005. "Structural Divergence and Distant Relationships in Proteins: Evolution of the Globins." *Current Opinion in Structural Biology* 15 (3 SPEC. ISS.): 290–301. https://doi.org/10.1016/j.sbi.2005.05.008.

Li, Qiliang, Kenneth R Peterson, Xiangdong Fang, and George Stamatoyannopoulos. 2002. "Review Article Locus Control Regions." *Blood* 100 (9): 3077–86. https://doi.org/10.1182/blood-2002-04-1104. BLOOD.

Lindell K. Weaver, M. D. 2019. "Carbon Monoxide Poisoning - Carbon Monoxide Kills." *The New England Journal of Medicine* 94: 270–72. https://www.ncbi.nlm.nih.gov/pmc/articles/PMC1281520/pdf/0940270 .pdf%0Ahttps://www.nhs.uk/conditions/carbon-monoxide-poisoning/%0Ahttp://www.carbonmonoxidekills.com/are-you-at-risk/carbon-monoxide-poisoning/.

Manning, Lois R., J. Eric Russell, Julio C. Padovan, Brian T. Chait, Anthony Popowicz, Robert S. Manning, and James M. Manning. 2007. "Human Embryonic, Fetal, and Adult Hemoglobins Have Different Subunit Interface Strengths. Correlation with Lifespan in the Red Cell." *Protein Science* 16 (8): 1641–58. https://doi.org/10.1110/ps.072891007.

Marotta, A., John T. Wilson, New Haven, M. Weissman, and New Haven. 1977. "*FROM*" 252 (14).

Martyn, Gabriella E., Beeke Wienert, Lu Yang, Manan Shah, Laura J. Norton, Jon Burdach, Ryo Kurita, et al. 2018. "Natural Regulatory Mutations Elevate the Fetal Globin Gene via Disruption of BCL11A or ZBTB7A Binding." *Nature Genetics* 50 (April). https://doi.org/10.1038/s41588-018-0085-0.

Medical, San Francisco, and Xan Francisco. 1969. "*Heme Oxygenase*" 244 (23).

Ogun A. S, Valentine M. 2019. *Biochemistry, Heme Synthesis*. Brooklyn: StatPearls Publishing. https://www.ncbi.nlm.nih.gov/books/NBK 537329/.

Paoli, Massimo, Robert Liddington, Jeremy Tame, Anthony Wilkinson, and Guy Dodson. 1996. "Crystal Structure of T State Haemoglobin with Oxygen Bound at All Four Haems." *Journal of Molecular Biology* 256 (4): 775–92. https://doi.org/10.1006/jmbi.1996.0124.

Pauling L., Itano H. A., Singer S. J., Wells I. C. 1949. "Sickle Cell Anaemia, a Molecular Disease." *Science* 110 (3): 543–48. http://journals.lww.com/jpho-online/Abstract/2012/05000/Prenatal_Screening_for_Sickle_Cell_Anemia_.3.aspx.

Perutz, By M. F., M. G. Rossmann, A. N. N. F. Cullis, Hilary Muirhead, Georg Will, and A. C. T. North. 1960. "*Structure of Hjemoglobin*" 1: 416–22.

Platt, O. S., S. H. Orkin, and G. Dover. 1984. "Hydroxyurea Enhanced Fetal Hemoglobin Production in Sickle Cell Anemia." *Journal of Clinical Investigation* 74 (2): 652–56. https://doi.org/10.1172/JCI111464.

Poncz, Mortimer, Miilicent Sutton, Kathleen Delgrosso, Elias Schwartz, and Saul Surrey. 1987. "*Nucleic Acids Research*" 15 (13): 5169–79.

Ribeiro, D. M., and M. F. Sonati. 2008. "Regulation of Human Alpha-Globin Gene Expression and Alpha-Thalassemia." *Genetics and Molecular Research* 7 (4): 1045–53. https://doi.org/10.4238/vol7-4gmr472.

Ryter, Stefan W., and Leo E. Otterbein. 2004. "Carbon Monoxide in Biology and Medicine." *BioEssays* 26 (3): 270–80. https://doi.org/10.1002/bies.20005.

Sankaran, Vijay G., Jian Xu, and Stuart H. Orkin. 2010. "Advances in the Understanding of Haemoglobin Switching: Review." *British Journal of Haematology* 149 (2): 181–94. https://doi.org/10.1111/j.1365-2141.2010.08105.x.

Sanyal, Mahima, Paurnima Patil, Ekta Jaiswal, and Aparna Deshmukh. 2013. "Structural and Functional Diversity of Haemoglobin Molecule Properties amongst Different Classes and Species of Animals." *Acta Biologica Indica* 2 (2): 381–87.

Schechter, Alan N., and Mark T. Gladwin. 2003. "Hemoglobin and the Paracrine and Endocrine Functions of Nitric Oxide." *New England Journal of Medicine* 348 (15): 1483–85. https://doi.org/10.1056/nejmcibr023045.

Schechter, Alan N., Melissa A. Qazi, Fabiola Rizzatti, Barbora Piknova, Nathawut Sibmooh, and David F. Stroncek. 2012. "Effect of Storage Levels of Nitric Oxide Derivatives in Blood Components." *F1000Research* 1: 1–16. https://doi.org/10.12688/f1000research.1-35.v1.

Schechter, Alan N. 2014. "*Hemoglobin Research and the Origins of Molecular Medicine ASH 50th Anniversary Review Hemoglobin Research and the Origins of Molecular Medicine*," 3927–38. https://doi.org/10.1182/blood-BLOOD.

Schoental. 1957. "© 1957 Nature Publishing Group." *Group Nature Publishing*. https://doi.org/doi:10.1038/1791302a0.

Stamatoyannopoulos, George. 2005. "Control of Globin Gene Expression during Development and Erythroid Differentiation." *Experimental Hematology* 33 (3): 259–71. https://doi.org/10.1016/j.exphem.2004.11.007.

Sutherland-Smith, Andrew J., Heather M. Baker, Oliver M. Hofmann, Thomas Brittain, and Edward N. Baker. 1998. "Crystal Structure of a Human Embryonic Haemoglobin: The Carbonmonoxy Form of Gower II (A2ε2) Haemoglobin at 2.9 Å Resolution." *Journal of Molecular Biology* 280 (3): 475–84. https://doi.org/10.1006/jmbi.1998.1868.

Thein, Swee Lay. 2013. "The Molecular Basis of β-Thalassemia." *Cold Spring Harbor Perspectives in Medicine* 3 (5): 1–24. https://doi.org/10.1101/cshperspect.a011700.

Thein, Swee Lay, Stephan Menzel, Mark Lathrop, and Chad Garner. 2009. "Control of Fetal Hemoglobin: New Insights Emerging from Genomics and Clinical Implications." *Human Molecular Genetics* 18 (R2): 216–23. https://doi.org/10.1093/hmg/ddp401.

Thom C. S., Dickson C. F., Gell D. A., and Weiss M. J. 2013. "Hemoglobin Variants: Biochemical Properties and Clinical Correlates." *Cold Spring Harbor Perspectives in Medicine* 3 (3): 1–22.

Thomas, Caroline, and Andrew B. Lumb. 2012. "Physiology of Haemoglobin." *Continuing Education in Anaesthesia, Critical Care and Pain* 12 (5): 251–56. https://doi.org/10.1093/bjaceaccp/mks025.

Watson, Janet. 1948. "Sickling in Negro Newborns: Its Possible Relationship to Fetal Hemoglobin." *The American Journal of Medicine* 5 (1): 159–60. https://doi.org/10.1016/0002-9343(48)90029-1.

Wienert, Beeke, Gabriella E. Martyn, Alister P. W. Funnell, Kate G. R. Quinlan, and Merlin Crossley. 2018. "Wake-up Sleepy Gene: Reactivating Fetal Globin for β-Hemoglobinopathies." *Trends in Genetics* 34 (12): 927–40. https://doi.org/10.1016/j.tig.2018.09.004.

Wittenberg, Jonathan B., Martino Bolognesi, Beatrice A. Wittenberg, and Michel Guertin. 2002. "Truncated Hemoglobins: A New Family of Hemoglobins Widely Distributed in Bacteria, Unicellular Eukaryotes, and Plants." *Journal of Biological Chemistry* 277 (2): 871–74. https://doi.org/10.1074/jbc.R100058200.

Wittenberg, Jonathan B., and Beatrice A. Wittenberg. 2007. "Myoglobin-Enhanced Oxygen Delivery to Isolated Cardiac Mitochondria." *Journal of Experimental Biology* 210 (12): 2082–90. https://doi.org/10.1242/jeb.003947.

Yagi, M., R. Gelinas, J. T. Elder, M. Peretz, T. Papayannopoulou, G. Stamatoyannopoulos, and M. Groudine. 1986. "Chromatin Structure and Developmental Expression of the Human Alpha-Globin Cluster." *Molecular and Cellular Biology* 6 (4): 1108–16. https://doi.org/10.1128/mcb.6.4.1108.

In: Fetal Hemoglobin
Editor: Nishant Chakravorty

ISBN: 978-1-53617-955-2

Chapter 2

PRENATAL AND POSTNATAL HEMOGLOBINS: FORMATION AND GASEOUS EXCHANGE

Megha Shah[1,*], Ronak Reshamwala[1,2,*,†], Indra Neil Choudhury[1] and Motiur Rahaman[3]

[1]Menzies Health Institute Queensland, Griffith University, Gold Coast, Australia
[2]Griffith Institute of Drug Discovery, Griffith University, Nathan, Australia
[3]School of Medical Science and Technology, Indian Institute of Technology, Kharagpur, India

ABSTRACT

Hemoglobin (Hb) is a quaternary conformation of a globin type protein that consists of a heme part and a globin part. The globin part is made of four polypeptide chains whereas the non-peptide heme part is the ferrous (Fe^{++}) containing moiety. Each molecule of hemoglobin can carry up to 4 molecules of oxygen. Throughout life, based on the conformation of the globin tetramer, researchers have identified multiple variants of hemoglobin, although there are three most important variants – HbA,

† Corresponding Author's Email: ronak.reshamwala@griffithuni.edu.au.

HbA2, and HbF. In this nomenclature, A stands for adult and F stands for fetal, which also indicates when these types are predominantly found in the body. Synthesis of fetal hemoglobin starts in utero (10 to 12 weeks of gestation). It has unique gas exchange properties, which are critical for survival and development of the fetus. A higher affinity towards oxygen than the HbA allows it to extract oxygen from the maternal blood stream. This concept is best explained with an oxygen dissociation curve. Here we discuss oxygen dissociation and factors affecting it, such as pH, temperature, levels of diphosphoglycerate (2,3-DPG) and carbon monoxide. In this chapter, we explain synthesis of hemoglobin at different stages of life and different sites of hemoglobin production. We also describe the fetal hemoglobin gaseous exchange. Furthermore, we explain some pathological conditions in which other variants of hemoglobin are likely to be seen as a physiological response or a compensation mechanism. We specifically focus on the potential role and pitfalls of HbF as a treatment in different hemoglobinopathies, which are discussed in the following chapters in depth.

Keywords: HbF, oxygen carriage, Hb synthesis, developmental hematopoiesis, hemoglobinopathies

Hemoglobin Structure: An Overview

Hemoglobin (Hb) is an oxygen binding metalloprotein present in red blood cells. Hb constitutes to approximately 35% of the red blood cells' total contents, and about 96% of the dry contents (by weight) (Weed, Reed, and Berg 1963). It has a molecular weight of 64,458 g/mol (Mairbaurl and Weber 2012; Beekvelt et al. 2001). Hb is a tetramer, composed of 'Heme' moiety and four polypeptide globin chains. Each of the four subunits consists of a heme moiety, with the globin polypeptide chain folded around it in a specific way. The folding of each globin chain and complex of four such subunits in turn forms a compact, spherical shape in the three-dimensional space (See Figure 1).

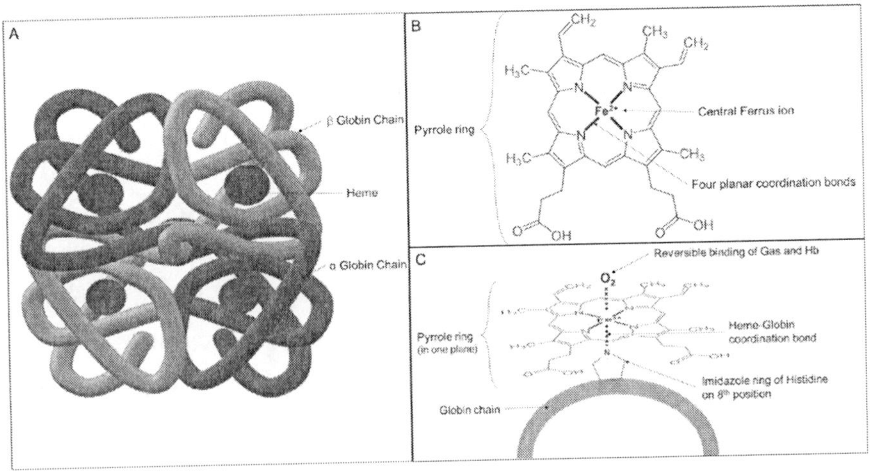

Figure 1. Hemoglobin structure. A: Structure of Hb molecule with four globin chains and four heme moieties. B: Schematic of Heme structure and four of the six coordination bonds of Ferrous ion. C: Three-dimensional integration of heme with globin and binding with gas molecules.

Heme contains a protoporphyrin ring along with central iron in the ferrous state (Fe^{2+}) (Thomas and Lumb 2012). This central ferrous ion is coordinated to six atoms in total. Four of the six atoms are nitrogen atoms belonging to the pyrrole ring and lie in a single spatial plane. One more nitrogen atom belongs to an imidazole ring of histidine amino acid which is located at the eighth position of the "F"-helix and lies on one side of the plane created by pyrrole ring. On the opposite side of this plane, the ferrous ion can bind reversibly with a gas atom from either oxygen (O_2) or carbon monoxide (CO) or nitric oxide (NO) (Schechter 2008). These non-covalent forces are responsible for conserving the physiological ferrous valence state of iron, and gaseous transport. Carbon dioxide (CO_2) is transported by the amino acids of Hb by the formation of weak carbamino complexes and by dissolving in the aqueous component of blood, but not by binding with the central iron (Schechter 2008). In addition to the gasses mentioned above, the ferrous iron can also bind to alkyl isocyanides and nitroso compounds. Physiological state of the central iron is ferrous, because if the ferrous state iron (Fe^{2+}) is oxidized into ferric state (Fe^{3+}), it cannot bind to the above-mentioned ligands. The ferric state iron, however, binds to water (H_2O), or hydroxyl (OH^-) ions, depending on the pH. It can also reversibly bind to fluoride, azide, cyanide, thiocyanate or cyanate complexes. The details of ferric state iron are discussed in depth in another chapter. This oxidized state of Hb is known as "Methemoglobin" (metHb) or "Hemiglobin" (Mairbaurl and Weber 2012). MetHb can be found in trace amounts (less than 3%) under normal conditions in healthy adults.

Binding with different ligands and the oxidation state of the central iron also has conformational effects on the Hb shape and structure. The highly specific function has led the Hb molecule to evolve into a dynamic structure, where the structure constantly changes at very minute levels depending on a number of external as well as internal factors. Some examples of the external factors are pH, ionic concentrations, and availability of certain molecules like Adenosine triphosphate (ATP) and 2, 3 – Diphosphoglycerate (2,3-DPG); whereas the internal factors would include the oxidation state of iron, oxygen saturation of heme moieties and any mutations leading to change in the normal structure. These factors are discussed in depth with the functions of Hb later in this chapter.

Types of Hemoglobin

The different types of hemoglobins are defined structurally based on the four polypeptide chains that constitute the molecule. These different types of hemoglobin occur in the human body throughout life at different developmental stages. The most common form is adult hemoglobin (HbA). HbA has two alpha and two beta globin chains ($\alpha2\beta2$), which is almost 95% of an adult's hemoglobin content. HbA2 has two alpha and two delta chains ($\alpha2\delta2$), which is 2.2-3.5% of an adult's hemoglobin. Whereas, two alpha and two gamma chains ($\alpha2\gamma2$) together form the fetal hemoglobin (Thomas and Lumb 2012). Alpha chain contains 141 amino acid residues and beta chain contains 146 amino acid residues. The delta and gamma chains also contain 146 amino acids but they have 10 and 39 amino acid residues different than beta chain, respectively. Additionally, there are three embryonic forms of hemoglobins called Gower 1 ($\zeta2\epsilon2$), Gower 2 ($\alpha2\epsilon2$) and Portland ($\zeta2\gamma2$) (K.E. Barrett et al. 2015; Schechter 2008). The details are given in Table 1.

Table 1. Different types of adult, fetal and embryonic hemoglobins, and their globin compositions

Hemoglobins	Globin chains
HbA	α2ß2
HbA2	α2δ2
HbF	α2γ2
Gower 1	ζ2ϵ2
Gower 2	α2ϵ2
Portland	ζ2γ2

Genetics Behind Globin Composition of Hb

The globin part of Hb molecule is synthesized with two different homodimers. In all of the described Hb types, one of the dimers is formed by globin chains encoded in the α cluster (α-like chains) on chromosome 16, and the other homo-dimer is always formed by globin chains encoded by the β cluster (β-like chains) on the chromosome 11 (Deisseroth et al. 1978; Deisseroth et al. 1977; H.B. Zhang, Liu, and Liang 2002). The α cluster houses ζ, ψζ, ψα, along with α2 and α1 genes in the telomeric region, and spans over 45 kilo-base pairs (Flint et al. 1997; Vyas et al. 1992; Smith and Higgs 1999; Brown et al. 2001; Hardison et al. 1997). The α-like chains include α and ζ chains (Brittain 2002). Similarly, the β cluster on the chromosome 11 houses ε, γ^G, γ^A, ψβ, δ and β genes encoding for the β-like chains and spans over 70 kilo-base pairs region (Hardison 1998; Engel and Tanimoto 2000). The β-like chains include β, γ, δ and ε chains (Brittain 2002). The genes of α cluster are expressed in a wider variety of tissues whereas the DNA of β cluster shows properties suggestive of tissue-specific expression (Brittain 2002).

Hemoglobin Functions

The functions of Hb can be described as gaseous transport related and others. Although gaseous transport constitutes for its major and primary function, Hb also plays critical roles in pH balance, maintenance of the shape of RBCs, regulation of vascular dilation and more. These functions are discussed in depth here.

Gaseous Transport

Transport of Oxygen

The primary function of hemoglobin is transport of oxygen from lungs to the tissues and transport of carbon dioxide from tissues to lungs. Although these respiratory gases can be transported via the dissolution into the aqueous part of the blood, Hb provides a much more efficient way for oxygen transport. Hb increases the oxygen carrying capacity of blood by 30 to 100 folds as compared to the oxygen carried in the dissolved form in blood (Hall and Guyton 2011). In other words, of the total oxygen carried by the blood from lungs to the periphery, about 97% is carried by the oxygenation of hemoglobin in the red blood cells. The water component of plasma and other blood cells only carry the remaining 3% of the oxygen in blood (Hall and Guyton 2011). The oxygen carrying capacity of the Hb is one of the most crucial factors affecting oxygen transport in the body, along with some other factors such as atmospheric oxygen pressure, pulmonary gas exchange and cardiac output (Nagalakshmi et al. 2016; K.E. Barrett et al. 2015).

As mentioned earlier, the heme moiety in hemoglobin is responsible for binding with the oxygen in a reversible manner. The oxygen that diffuses through the pulmonary membrane and into the blood, thus binds with the heme, forming oxyhemoglobin. Each heme moiety can bind with one oxygen molecule, and hence, one hemoglobin molecule can carry up to four oxygen molecules simultaneously. The environmental factors in the peripheral vascular bed are different from the pulmonary vascular bed, which in turn causes the bound oxygen to get released from the hemoglobin and diffuse into the tissue. This is known as oxygen dissociation from hemoglobin (Hsia 1998; Thomas and Lumb 2012; Dunn, Mythen, and Grocott 2016; Hall and Guyton 2011; West 2005).

Thus, the two major processes involved in oxygen transportation are oxygen binding and oxygen dissociation. The graphical representation of this entire process is called an oxygen dissociation curve (ODC). The ODC, in essence, depicts the relationship between oxygen saturation and partial pressure (PO_2) in the blood. The overall proportion of oxygen binding sites in Hb that are occupied by oxygen, is referred to as the oxygen saturation of blood (Collins et al. 2015).

Oxygen Dissociation Curve

Oxygenation is a term used to describe oxygen binding to the hemoglobin and deoxygenation is a term used to describe oxygen unloading from the hemoglobin molecule. During the oxygenation process the globin part of Hb undergoes a conformational change and the quaternary shape of Hb shifts from 'tense' state to 'relaxed' state. These two states of Hb represent distinct change in the structure of Hb in terms of spatial arrangements of the subunits and distance between iron atoms. In the 'relaxed' state, affinity of Hb for oxygen is higher than that in 'tensed' state. This means that for every oxygenated subunit of Hb, the affinity for oxygen of the remaining subunits increases (Mairbaurl and Weber 2012). Similarly, oxygen affinity to Hb reduces if there is increased carbon dioxide and hydrogen ion concentration accelerating the release of oxygen from hemoglobin (Thomas and Lumb 2012). This is the reason of sigmoid shape of oxygen dissociation curve (See Figure 2). This phenomenon, along with a number of other factors that differ from the pulmonary to peripheral capillary beds affect the hemoglobin's affinity towards oxygen and essentially govern the hemoglobin mediated oxygen transport, where the Hb binds with oxygen in the lungs and releases the oxygen in the tissues. Such dynamic nature of the affinity between oxygen and Hb molecule is also known as "allosteric effect" or "molecular breathing".

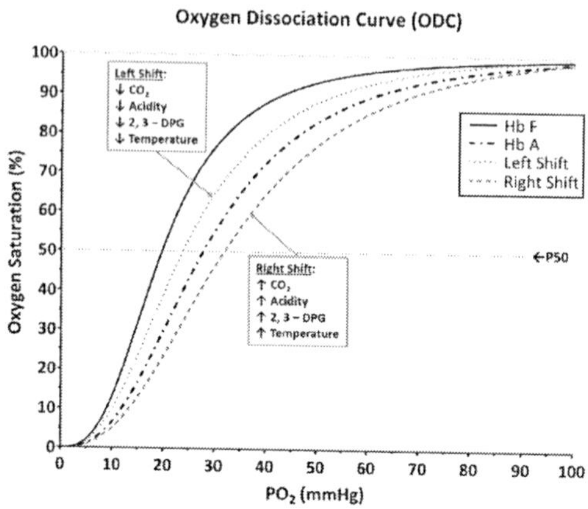

Figure 2. Oxygen dissociation curve. Depicting HbF, HbA, and different factors affecting oxygen dissociation. The plotted values are only approximations, derived from literature (Thomas and Lumb 2012; Kasinger, Huch, and Huch 1981; Greer et al. 2014).

If, for any reason, hemoglobin's affinity towards oxygen is reduced, the sigmoid curve of oxygen dissociation appears to have shifted towards the right side (See Figure 2). This means that the Hb will release more oxygen under the given conditions and the oxygen saturation will decrease. This is also known as 'right shift' of the ODC. Conversely, the left shift of ODC implies that the Hb will release less oxygen (or bind with more oxygen) under the given conditions and become more saturated with oxygen.

Factors Affecting Oxygen Dissociation

The structure and functions of a protein are greatly dependent on each other. As mentioned earlier, the highly dynamic structure of Hb is affected by various factors including the oxygenation state of Hb itself. These factors are discussed here in relation with their effects on the oxygen dissociation.

- *pH and CO_2:* Acidic pH reduces the oxygen affinity and causes the ODC to shift to right; and alkaline pH causes a left shift. An increase in carbon dioxide (CO_2) causes the pH to decrease and in turn, ODC shifts to right. The molecular explanation of this phenomenon is that the deoxygenated Hb becomes more stable in presence of high concentration of H^+ ions (acidic pH), which is why a decrease in pH reduces the oxygen affinity. Physiologically, this means that higher partial pressure of CO_2 and higher acidity in the peripheral tissue, arising from accumulation of metabolic waste, reduces the oxygen affinity of Hb and causes the Hb-bound oxygen to release. This oxygen then diffuses to the tissue where it is required for metabolism (Malte and Lykkeboe 2018). This inverse relationship between Hb's oxygen affinity and CO_2 and acidity is also known as Bohr effect (Patel S 2019).

- *2,3 Diphosphoglycerate (DPG):* This is a molecule produced as a part of glycolysis and is found within the red blood cells. Higher concentrations of 2, 3 – DPG induces a right shift in ODC by reducing the oxygen affinity, and thus promoting oxygen release from Hb (Dunn, Mythen, and Grocott 2016; Mairbaurl and Weber 2012). Another physiological implication of this phenomenon is chronic hypoxia, which causes higher concentration of 2, 3 – DPG. A shift to right in ODC would be beneficial in this situation as this would improve oxygen dissociation in the oxygen deprived tissues (Collins et al. 2015).
- *Temperature:* Increase in the temperature induces oxygen dissociation by reducing oxygen affinity and causing a right shift of the ODC. Whereas a lower temperature allows for a left shift in the ODC by improving the oxygen binding in the Hb. The physiological significance of this lies in the fact that increased temperature indicates higher metabolism (e.g., exercise), which in turn indicates an increased oxygen demand. Thus, increased temperature facilitating oxygen dissociation is physiologically important for homeostasis. (Patel S 2019; Dunn, Mythen, and Grocott 2016)
- *Fetal Hb and ODC:* The structural differences between adult Hb (HbA) and fetal Hb (HbF) are responsible for HbF to have a relatively higher affinity for oxygen. The two gamma chains of the HbF award this higher oxygen affinity as compared to the two beta chains of the HbA (Yagami et al. 2002). This means that the HbF can bind with more oxygen even at lower partial pressure of oxygen and as a result, the ODC for HbF would appear to have left-shifted

as compared to the ODC of HbA (Dunn, Mythen, and Grocott 2016). Physiologically, this phenomenon allows the fetus to survive in utero, by enabling the HbF to extract oxygen from HbA during the placental gas exchange. This is discussed in detail in the section "Unique properties of HbF". Another important difference is that 2, 3 – DPG has a more potent effect on the HbA as compared to HbF (Adachi et al. 1997), which is why the HbA in the maternal circulation releases oxygen under the influence of 2, 3 – DPG in the placenta and HbF can easily bind with the oxygen (Kaufman DP 2019 Jan- [Updated 2019 Apr 16]; Patel S 2019).

- *Ligands binding to Hb:* In addition to the above mentioned factors that do not bind with Hb (2, 3 – DPG, H^+ ions, CO_2), and the molecules that do bind with Hb (O_2, CO_2 in carbamino hemoglobin form), a number of other factors also affect O_2 binding. The non-heme ligands are called heterotropic effectors. These include chloride ions (Cl^-), organic phosphates other than DPG in mammals, inositol pentaphosphate (IPP) in species of birds, and adenosine triphosphate (ATP) in ectothermic vertebrate species. Other ligands that bind directly with the heme moiety and affect the oxygen binding and release potential of Hb are called allosteric effectors. These factors include carbon monoxide (CO), nitric oxide (NO), water (H_2O) and lactate ions (Mairbaurl and Weber 2012).

Transport of Carbon Dioxide

Carbon dioxide is carried from the tissues to the lungs by blood in any of the three ways: by directly dissolving in the plasma; as bicarbonate; or in the form of carbamino hemoglobin in the RBCs. The increased ability of deoxygenated blood to carry CO_2 is known as the Haldane effect (Hsia 1998; Thomas and Lumb 2012; Dunn, Mythen, and Grocott 2016).

Non-Gaseous Transport Functions of Hb

In addition to the above-mentioned gaseous transport functions, Hb also serves a number of other purposes within the blood. Hemoglobin is also known to induce changes in the internal cell membrane of ageing RBCs that ultimately result in destruction and removal these cells near the end of their lifespan within the blood circulation (Rifkind and Nagababu 2013). Apart from that, hemoglobin also acts as a buffer by maintaining the blood pH at 7.4. When carbon dioxide accumulates in the body, the pH reduces. This can be changed by ventilation and role of hemoglobin in respiration (Panawala 2017; Thomas and Lumb 2012).

Hemoglobin can bind with nitric oxide through high affinity ferrous binding sites on heme and acts as a carrier of nitric oxide. These nitrosothiols on the hemoglobin molecule act as vasodilators on capillaries and survive and has longer survival than free nitric oxide. This action reduces regional vascular resistance and enhances oxygen transport (Hsia 1998; Thomas and Lumb 2012).

Some recent works have discovered that hemoglobin is expressed in non-erythroid cells like epithelial cells from human tissues like lung, neurons, retina, and endometrium as well as cervicovaginal epithelium. Here, the hemoglobin and its peptides seem to be responsible for reduction in oxidative stress as well as it shows anti-microbial properties (Saha et al. 2014).

METABOLISM OF HEMOGLOBIN

Synthesis of Hb under Normal Conditions

Hemoglobin synthesis is a major part of a process called erythropoiesis, where red blood cells are synthesized within the bone marrow of flat bones in adults. The site of erythropoiesis changes during different stages of development, which is discussed later in this chapter.

The mature adult form of RBCs is called definitive non-nucleated erythrocyte, which differentiates from hematopoietic stem cells (HSCs) or hemocytoblasts after passing through a series of progenitor or precursor stages. Chronologically, this series includes HSCs, common myeloid progenitor cell, proerythroblast (PRO-EBs), basophilic erythroblast (BASO-

EBs), polychromatophilic erythroblast (POLY-EBs), orthochromatic erythroblast (ORTHO-EBs) and reticulocytes, which ultimately mature into erythrocytes in the circulation by removing the internal organelles and assuming the characteristic bi-concave shape (Ney 2011; Ponka, Koury, and Sheftel 2014; Liu, McConnell, and Ryan 2013).

Hb synthesis starts in the BASO-EBs and continues through the POLY-EB stage. The proliferation by cell division stops by the POLY-EB stage as they cannot divide anymore (Ross and Sullivan 1985). However, the process of enucleation starts in the ORTHO-EB stage. This leads to the reticulocyte formation where, despite the lack of nucleus, active synthesis of globin chains and heme moieties is still ongoing in the ribosomes and mitochondria, respectively. These cells then enter the circulation and within 24-36 hours, mature in to the RBCs (Ponka, Koury, and Sheftel 2014). Hb constitutes to over 95% of total cytosolic protein content of RBCs (Liu, McConnell, and Ryan 2013). Normal ranges of Hb content according to ages are described in Table 2.

Table 2. Normal hemoglobin values specific to age and proportion of HbF.

The values are derived from various sources in the literature (Janus and Moerschel 2010; Greer et al. 2014)

Age	Normal Hb Range (g/dl)	% Fetal Hb proportion
At birth	13.5-19.5	55–90
1 months	10.7-17.1	50–80
2 months	9.4-13	30–55
3–6 months	11.1-14.1	5–25
6 months–1 year	10.5-13.5	<5
2 years onwards	11.5-13.5	<2
Adult female	13.5-17.5	<2
Adult male	12-16	<2

Catabolism of Hemoglobin

When red blood cells are old and aged, they are destroyed from the circulation by tissue macrophages in the reticuloendothelial system comprised of spleen or liver. Globin part of the molecule is then removed, and heme molecule goes through a series of autocatalytic oxidation reaction caused by mainly the enzyme heme oxygenase. The products of this process are biliverdin, carbon monoxide and iron. The iron molecule gets recycled for the synthesis of hemoglobin. Biliverdin is converted into bilirubin and gets excreted as one of the major component of bile into intestines (Kim E. Barrett, Boitano, and Barman 2010; Kikuchi, Yoshida, and Noguchi 2005).

Hemoglobin Synthesis in Utero

Developmental Hematopoiesis/Site and Type of Hematopoiesis

Hemoglobin synthesis and erythropoiesis progress and mature in tandem in the developmental life. During the embryonic development, the first Hb appears at the third week of gestation. Until this point, the size of embryo is small enough to meet its oxygen demand by simple diffusion from the maternal interstitial fluid. However day 15 onwards, the size and metabolic oxygen demands of the growing embryo exceeds the amount extracted by diffusion, and the first Hb genes are activated in the yolk sac for the synthesis of oxygen carrier protein – Gower-1 (Fantoni, Farace, and Gambari 1981). The start of synthesis of this first embryonic Hb also marks the start of early hematopoiesis, where large (approximately 200 μ^3 in volume) nucleated spherical RBCs are produced with macrophages in the yolk sac (Stamatoyannopoulos and Nienhuis 1987; Babovic and Eaves 2014). The cardiovascular system is also in its primitive stages of development as the heart is a still a single pulsatile tubular entity, and distinct arterial and venous circulations are not yet established. Distinct arterial and venous circulations are established in the middle of week four, when the left and right cardiac chambers are demarcated by the middle septation. This is when the other two embryonic hemoglobins are synthesized - Gower-2 ($\alpha2\varepsilon2$) and Portland ($\zeta2\gamma2$) and the placenta begins to develop. Up to day 37 of development, Gower-1 remains the predominant Hb type and accounts for over 40% of the circulating Hb at this stage (Hecht et al. 1966), after this Gower-2 increases due to the ζ to α switch (genetic control of Hb synthesis is explained later). HbF first appears at about 6th week of gestation, in the liver. Soon after this (by the 8th week), erythropoiesis in the yolk sac declines and embryonic Hb starts to decrease. Liver remains the primary site of hematopoiesis from 6 to 30 weeks of gestation, along with spleen (8-28 weeks) (Brittain 2002). During this phase, there is a continuing shift of the circulatory systems where the vitelline circulation is gradually being replaced by the placental circulation, and the yolk sac disappears. As the liver, and to a lesser extent the spleen, take over the hematopoiesis, they start producing definitive discoid, non-nucleated red blood cell (approximately 125 μ^3 in volume) containing the first fetal hemoglobin (HbF, $\alpha2\gamma2$) and traces of adult hemoglobin (Bloom and Bartelmez 1940; Pataryas and Stamatoyannopoulos 1972; Greer et al. 2014). Eventually, bone marrow takes over the task of hematopoiesis from the 28th week through to birth and then through to the adult life. These RBCs are definitive biconcave in shape and non-nucleated (approximately 80 μ^3 in volume) containing adult hemoglobin (HbA, $\alpha2\beta2$)(Lewis 1974). As the fetus develops, the Hb type changes along with the changes in RBCs shape and size and this indicates the maturation process of both Hb and RBCs. This eventual maturation facilitates rapid oxygenation during passage through the placenta and lungs (Brittain and Simpson 1989). Thus, during the development, fetus produces a series of hemoglobins optimized for oxygen transport under the constantly maturing physiological conditions and intrauterine environment according to the developmental stages. At birth, the circulating blood contains approximately 70% HbF and 30% HbA. After birth, HbA starts replacing HbF and within five months, the HbF proportions reduce to 3% and by one year of age, only trace amounts (1-2%) of HbF remains in the system (Kaufman DP 2019 Jan- [Updated 2019 Apr 16]) as it is almost completely replaced by the bone marrow derived HBA, and a trace amount (<3%) of HbA2 (α2δ2)(Huehns 1974).

Genetic Controls of Hemoglobin Synthesis

As mentioned earlier, Gower – 1, Gower – 2 and Portland Hb are the embryonic Hb, and they are synthesized between 3-8 weeks of gestation by the yolk sac. Hence, ζ and ε genes (coding for ζ and ε globin chains) are expressed primarily by the yolk sac, adjacent para-aortic regions and eventually by liver (Forget 1998; Schechter 2008). There are two genetic control switches involved during development: embryonic to fetal Hb switch in utero, and HbF to HbA switch after birth.

Embryonic to Fetal Hb Switch

This switch happens when the primitive erythroid cells start to mature into the first enucleated definitive erythrocytes, and at the same time, the primary site of hematopoiesis shifts to liver (McGrath and Palis 2008) from the yolk sac. Studies have determined the timeline for this key event between five to seven weeks of gestation (Peschle et al. 1985). This involves two distinctive switches - ζ- to α-globin and ϵ- to γ-globin, both of which happen during this same time period (Peschle et al. 1985). There are two γ-genes involved in the ϵ- to γ- switch: $^{G}\gamma$ and $^{A}\gamma$, both of which code for functionally identical globin chains that are structurally different in only one amino acid. There is a glycine at position 136 in the $^{G}\gamma$ globin chain, whereas an alanine in the $^{A}\gamma$ globin chain (Forget 1998; Schechter 2008). This switch ultimately results in the first mature globin chain production - γ-globin in the liver (Sankaran, Xu, and Orkin 2010; Sankaran and Nathan 2010b). Here they form dimers and combine with dimers of mature α-globin chains to form the stable tetramer structures of HbF, which is the major Hb type throughout the life in utero (Sankaran and Orkin 2013). This is when the definitive erythrocyte production begins in the fetal liver (McGrath and Palis 2008).

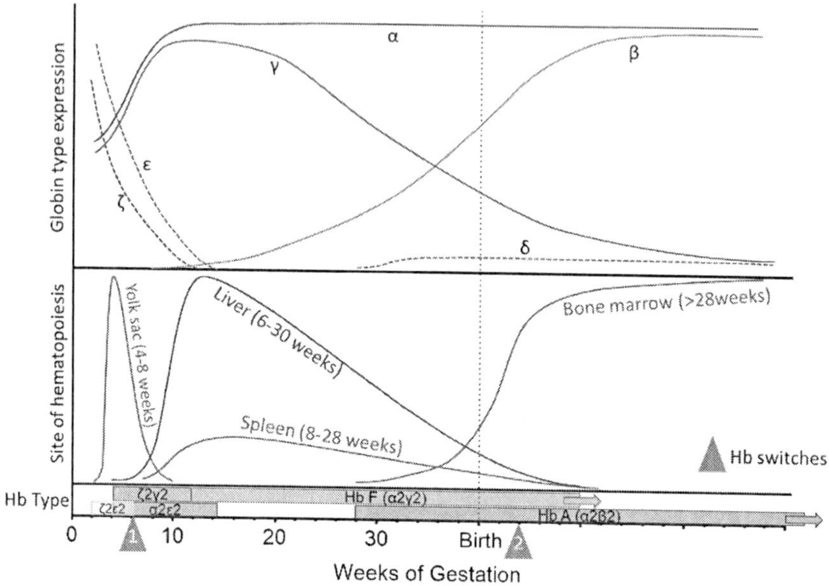

Figure 3. Hematopoiesis in utero: sites, types and switches. Switch 1 = embryonic to fetal Hb switch, includes ζ to α, and ε to γ switch. Switch 2 = fetal to adult Hb switch, includes γ to β switch. The top segment depicts the timeline of rise and fall in different globin type expressions. The middle segment depicts different sites of hematopoiesis throughout life. Bottom segment represents the types of hemoglobin that appear and disappear throughout developmental stages [Compiled from information gathered from various sources (Kaufman DP 2019 Jan- [Updated 2019 Apr 16]; Brittain 2002; Kim E. Barrett, Boitano, and Barman 2010)].

Fetal to Adult Hemoglobin Switch

This switch happens shortly after birth which leads to HbF, the primary Hb type synthesized thus far, to be replaced by the adult Hb (HbA). This is brought about by a γ- to β-globin transcriptional switch in the definitive erythroid progenitors (Sankaran and Orkin 2013). Number of different factors have been proposed to participate in this transcriptional regulation of globin gene and the switching (Sankaran, Xu, and Orkin 2010). One of the most prominent regulator in silencing γ globin gene expression is a transcriptional factor B-cell lymphoma/leukemia 11A (BCL11A) (Bauer, Kamran, and Orkin 2012). Studies have shown that the transcription factor BCL11A forms protein complexes that bind with the globin loci to suppress the γ -globin gene expression (Sankaran and Orkin 2013; Sankaran et al.

2008). The several proteins participating in this complex formation include GATA binding protein 1 (GATA1), zinc finger protein multitype 1 (ZFPM1, or FOG1), and the nucleosome remodeling and deacetylase (NuRD) complex (Sankaran and Nathan 2010b). Krüppel-like factor 1 (KLF1) is an essential transcription factor needed for β globin expression (Perkins, Sharpe, and Orkin 1995; Nuez et al. 1995; Guy et al. 1998; Bauer, Kamran, and Orkin 2012). Researchers have shown increased formation of fetal hemoglobin in knockdown of KLF1 in erythroid progenitor cells (Bauer, Kamran, and Orkin 2012; Borg et al. 2010; Zhou et al. 2010; Perkins, Gaensler, and Orkin 1996). KLF1 directly binds to the adult β-globin gene and promotes its transcription. It also positively regulates BCL11A expression, leading to silencing of the γ-gene (Sankaran and Nathan 2010b). One of the SRY-related HMG box transcription factor family genes - SOX6 has also been proposed to improve erythropoiesis by stimulating cell survival and proliferation, but the exact role of SOX6 in hemoglobin switching is yet unclear. P22NF-E4 seems to be a critical transcription factor in regulating hemoglobin switch but further research is needed to explain exact mechanism of action. Other transcription factors important for the hemoglobin switch are: COUP-TF, DRED/TR2/TR4, MBD2 and Ikaros-PYR complex (Sankaran, Xu, and Orkin 2010). These molecules could serve as important therapeutic targets to induce fetal hemoglobin as a treatment option for some hemoglobinopathies like sickle cell anemia or beta thalassemia (Sankaran 2011). Studies have shown that interfering with fetal hemoglobin silencing may be enough to treat hemoglobinopathies like sickle cell disease (Xu et al. 2011).

Unique Properties of HbF

As discussed before, the replacement of embryonic Hb by HbF starts by week 5 of gestation and this embryonic Hb is completely replaced by HbF by 8-10 weeks of gestation. The start of β-globin expression may be as early as week 8, but the rise is quite slow; the β-globin chain proportions reach a mere 10% by week 30-35 of fetal development. The sudden peak in the β-globin synthesis coincides with the start of decline in γ-globin just before birth, with the switch from fetal to adult Hb, and HbF falls to trace amounts (~1%) by 2 years of age. Thus, the fetal to adult switch is not complete, since these trace amounts of HbF persists throughout life. The HbF in adults are reported to be between 0.2-7% of circulating hemoglobin, and is found in a small sub population of RBCs known as F-cells (S. L. Thein 2011).

In addition to the obvious difference in their Hb contents, there a couple of other differences between adult and fetal RBCs. Fetal RBCs exhibit a different isoenzyme profile, and relatively lower levels of enzymes involved in glycolysis as compared to adult RBCs. Fetal RBCs express “i” type surface antigen as opposed to the “I” type antigen expressed by adult cells (Beutler and Porter 1988).

As it has been previously discussed, HbF possesses a significantly higher oxygen affinity than the HbA. The P50 for HbF is 19 mmHg whereas it is 27 mmHg for HbA. This means that the HbF can extract and retain oxygen from HbA of the maternal circulation. HbA can only derive oxygen from the inhaled atmospheric air. Additionally, the physiological fetal hematocrit is ~15 g/dl, as compared to the ~12 g/dl of the maternal blood. This 25% higher fetal hematocrit allows for a higher oxygen carrying capacity per unit volume of the blood (Kaufman DP 2019 Jan- [Updated 2019 Apr 16]; Piccin et al. 2019; Wienert et al. 2018; Zivot et al. 2018). Moreover, the relatively low sensitivity of the HbF to 2, 3 – DPG also allows it to readily take up the oxygen dissociated from HbA in presence of this glycolysis metabolite.

All these factors are responsible for a lower PO_2 in the fetal circulation, which is essential for relieving oxygen from placental circulation and unloading sufficient oxygen from the fetal Hb, given the high oxygen affinity of HbF (Kaufman DP 2019 Jan- [Updated 2019 Apr 16]; Piccin et al. 2019; Wienert et al. 2018; Zivot et al. 2018). The fetal blood has a PO_2 of merely 30 mmHg post oxygenation (Hall and Guyton 2011). The fact that fetal blood PO_2 is lower than adults is also of significance for development. The mild hypoxia triggers the expression of a transcription factor ‘hypoxia-inducible factor-1’ (HIF-1), which is largely responsible for angiogenesis in the rapidly growing fetus. Studies have also proposed that mild hypoxia/hypercapnia may be critical for the fetal stem cells to maintain their 'stemness' (Wang Y 2010).

Diffusion of Oxygen Through the Placental Membrane

Several factors enable and facilitate placental gaseous exchange, in addition to the obvious difference in the oxygen affinities of HbA and HbF. Fetal blood can carry up to 20-50% more oxygen due to higher oxygen affinity, higher hematocrit and lower PO_2 of fetal blood. According to the Bohr effect, oxygen carrying ability of Hb reduces in presence of CO_2. Fetal blood releases large amounts of CO_2 into maternal blood by diffusion at the placenta. This leads to reduced oxygen affinity of maternal blood, and improved oxygen affinity of the fetal blood. In other words, the Bohr effect causes right shift in the maternal ODC, while causing a left shift of the fetal ODC. Therefore this is referred to as the 'double Bohr effect' as compared to the simple unidirectional Bohr effect driving the pulmonary gas exchange (Hall and Guyton 2011; Thomas and Lumb 2012).

Other Properties

Due to the unique thermodynamic properties involved with the oxygenation and deoxygenation of the fetal Hb, it has been suggested to play an important role in thermoregulation of the internal temperature (Giardina et al. 1995).

Owing to the difference between β and γ globin chains, HbF shows higher tolerance to acidic or alkaline pH and does not denature easily (Duck-Chong 1983).

HbF also shows less reactivity to free radicle production and DNA cleavage compared to HbA (Chakane et al. 2017).

Hemoglobinopathies Causing Excess Hb F Formation

As discussed earlier, the fetal Hb only persists in trace amounts after 1st year of life, limited to only 0.2-7% of total circulating Hb as the production of γ globin chains declines sharply. However, there are a few pathologies where the synthesis of γ globin chain does not decline as it should, and HbF is found in higher proportions.

Hereditary Persistence of Fetal Hemoglobin (HPFH)

This pathology is characterized by persistent high levels of HbF in adults and has a multifactorial etiology (genetic as well as environmental) (Borg et al. 2010). HPFH includes a genetically and physically heterogeneous group of conditions (Swee Lay Thein and Menzel 2009).

In HPFH, the fetal Hb levels can be as high as 10-40%. If the fetal Hb is distributed homogenously across all RBCs, this distribution pattern is known as pancellular (or homocellular) HPFH. Conversely, if the HbF distribution is uneven amongst all the RBCs, it is known as heterocellular HPFH (Shaukat et al. 2018). Homocellular HPFH arises from pleomorphic deletions in β cluster of genes on chromosome 11, or from point mutations in the γ globin gene promoters (Forget 1998; J. Zhang et al. 2019). It follows a Mendelian inheritance pattern; however, inheritance pattern of heterocellular HPHF is not known (Shaukat et al. 2018).

Sickle Cell Anemia

This autosomal recessive disorder arises from a point mutation in the β-globin gene where Adenine base is replaced by Thymine base. This changes the usual GAG codon to GTG codon, and as a result, the usual glutamic acid gets replaced by valine at the 6th position of the β-globin chain (Clancy

2008). The Hb molecules containing the aberrant β chains are designated as HbS. HbS molecules, upon deoxygenation, tend to polymerize together and thereby changing the usual bi-concave shape of RBCs to elongated and rigid form. This shape of RBCs is called sickled shape, which are prone to get stuck inside narrow capillaries and small blood vessels, causing vaso-occlusion, endothelial activation and hemolysis, leading to anemia, hypoxia and several acute and chronic complications (Zivot et al. 2018).

HbF being devoid of the defective β globin chains, is unaffected by the sickle cell disease. The higher levels of HbF in sickle cell patients are associated with somewhat better prognosis, and reduced complication rates (Akinsheye et al. 2011). The HbF to HbA switch is delayed in the course of this disease to allow physiological compensation, thereby leading to higher than normal levels of HbF in most patients (Akinsheye et al. 2011; Somervaille 2001; Higgs, Thein, and Wood 2008). HbF levels have been found to be normal in carriers, however sickle cell patients have shown HbF levels between 1-30% (Swee Lay Thein and Menzel 2009). Hence, induction of HbF in sickle cell disease patients may present with an acceptable and effective management option.

Beta Thalassemia

Thalassemia is a group of syndromes arising from defective Hb production, where a defect in α chain synthesis leads to α-thalassemia; and defective β chain synthesis gives rise to the β-thalassemia group of syndromes. Since HbF and HbA both require healthy production of α chains, elevated HbF is seen in β-thalassemia, where the β chains are defective. In absence of healthy β globin chains, the α chains remain unpaired within the erythroid precursor cells. Here they precipitate and elicit unfolded protein response, thereby causing cell membrane damage and eventually cell death (Sankaran and Nathan 2010a), causing hemolysis.

Depending on the underlying genetic defects, thalassemia can present with varying levels of clinical severity, where the mildest form being β thalassemia minor (heterozygotic mutations), most severe form being thalassemia major (homozygotic mutations), and a spectrum of varying symptoms in between the two known as β thalassemia intermedia. All of these forms of β thalassemias present with varying degrees of elevated HbF.

As it can be seen from both sickle cell diseases as well as β thalassemia, increased levels of HbF do not represent a direct downstream event of the mutations, but rather a compensatory response resulting from hemolysis and stressed erythropoiesis enabling prolonged survival of RBC precursors containing HbF (Swee Lay Thein and Menzel 2009).

Other Conditions Causing Elevated HbF

A number of conditions where increased HbF levels are pathogenic include drugs, chromosomal disorders, hemoglobinopathies and malignancies, and are characterized by uneven distribution of HbF and an inconsistent rise in the levels. Hydroxyurea and Pomalidomide (Thalidomide analog) are known to induce higher HbF levels (Dulmovits et al. 2016). Chromosomal disorders such as trisomy 13 cause a delayed HbF to HbA switch and lead to high HbF levels (Pinkerton and Cohen 1967; Sankaran and Sapp 2012). Various malignancies are associated with elevated HbF levels, such as leukemias following chemotherapy, juvenile chronic myeloid leukemia (JCML) (Weatherall et al. 1975), erythroleukemia in infancy (Sheridan et al. 1976), acute myeloblastic leukemia, lymphoblastic leukemia, chronic myeloid leukemia (Weatherall et al. 1975) and following bone marrow transplants, due to proliferation of donor marrow (Alter et al. 1976; Meletis et al. 2001). In these and other conditions, high HbF levels are considered as indicators of stressed erythropoiesis. One such example is after treatment of severe iron deficiency anemia following acute blood loss (Dover, Boyer, and Zinkham 1979). Inherited bone marrow failure syndromes including Diamond-Blackfan, dyskeratosis congenita, Fanconi's anemia, Schwachman-Diamond syndrome etc. may also present with high HbF levels reflecting the same stressed erythropoiesis (Alter 1979). Solid tumors such as choriocarcinoma, adenocarcinoma of the lung and hepatoma may also present with elevated HbF, although very rarely

(Chudwin et al. 1977; Krauss, Rodriguez, and Milner 1981; Mori 1979; Wolk, Kieselstein, and Brufman 1988).

REFERENCES

Adachi, Kazuhiko, Patrick Konitzer, Jian Pang, Konda Reddy, and Saul Surrey. 1997. "Amino acids responsible for decreased 2,3-biphosphoglycerate binding to fetal hemoglobin." *Blood* 90: 2916-20. https://doi.org/10.1182/blood.V90.8.2916.

Akinsheye, Idowu, Abdulrahman Alsultan, Nadia Solovieff, Duyen Ngo, Clinton T. Baldwin, Paola Sebastiani, David H. K. Chui, and Martin H. Steinberg. 2011. "Fetal hemoglobin in sickle cell anemia." *Blood* 118 (1): 19-27. https://doi.org/10.1182/blood-2011-03-325258.

Alter, B. P. 1979. "Fetal erythropoiesis in stress hematopoiesis." *Exp Hematol* 7 Suppl 5: 200-9.

Alter, B. P., J. M. Rappeport, T. H. Huisman, W. A. Schroeder, and D. G. Nathan. 1976. "Fetal erythropoiesis following bone marrow transplantation." *Blood* 48 (6): 843-53.

Babovic, Sonja, and Connie J. Eaves. 2014. "Hierarchical organization of fetal and adult hematopoietic stem cells." *Experimental Cell Research* 329 (2): 185-191. https://doi.org/https://doi.org/10.1016/j.yexcr.2014.08.005.

Barrett, K. E., S. M. Barman, S. Boitano, and H. Brooks. 2015. *Ganong's Review of Medical Physiology* 25th Edition. McGraw-Hill Education.

Barrett, Kim E., Scott Boitano, and Susan M. Barman. 2010. "*Ganong's Review of Medical Physiology* (23rd Edition)."

Bauer, Daniel E., Sophia C. Kamran, and Stuart H. Orkin. 2012. "Reawakening fetal hemoglobin: prospects for new therapies for the β-globin disorders." *Blood* 120 (15): 2945-2953. https://doi.org/10.1182/blood-2012-06-292078.

Beekvelt, Mireille C. P. Van, Willy N. J. M. Colier, Ron A. Wevers, and Baziel G. M. Van Engelen. 2001. "Performance of near-infrared spectroscopy in measuring local O2 consumption and blood flow in skeletal muscle." *Journal of Applied Physiology* 90 (2): 511-519. https://doi.org/10.1152/jappl.2001.90.2.511.

Beutler, Ernest, and Ian H. Porter. 1988. "The molecular basis of blood diseases. G. Stamatoyannopoulos, A. W. Nienhuis, P. Leder, P. W. Majerus. Philadelphia: W. B. Saunders Company, 1987, 747 pp $95.00." *American Journal of Medical Genetics* 29 (3): 701-702. https://doi.org/10.1002/ajmg.1320290339.

Bloom, W., and G. W. Bartelmez. 1940. "Hematopoiesis in young human embryos." *American Journal of Anatomy* 67: 21-53.

Borg, J., P. Papadopoulos, M. Georgitsi, L. Gutierrez, G. Grech, P. Fanis, M. Phylactides, A. J. Verkerk, P. J. van der Spek, C. A. Scerri, W. Cassar, R. Galdies, W. van Ijcken, Z. Ozgur, N. Gillemans, J. Hou, M. Bugeja, F. G. Grosveld, M. von Lindern, A. E. Felice, G. P. Patrinos, and S. Philipsen. 2010. "Haploinsufficiency for the erythroid transcription factor KLF1 causes hereditary persistence of fetal hemoglobin." *Nat Genet* 42 (9): 801-5. https://doi.org/10.1038/ng.630.

Brittain, T. 2002. "Molecular aspects of embryonic hemoglobin function." *Mol Aspects Med* 23 (4): 293-342. https://doi.org/10.1016/s0098-2997(02)00004-3.

Brittain, T., and R. Simpson. 1989. "Stopped-flow kinetics of oxygen uptake and release by embryonic red blood cells." *Biochemical Journal* 261 (3): 1009-1013. https://doi.org/10.1042/bj2611009.

Brown, K. E., S. Amoils, J. M. Horn, V. J. Buckle, D. R. Higgs, M. Merkenschlager, and A. G. Fisher. 2001. "Expression of α- and β-globin genes occurs within different nuclear domains in haemopoietic cells." *Nature Cell Biology* 3 (6): 602-606. https://doi.org/10.1038/35078577.

Chakane, Sandeep, Tiago Matos, Karin Kettisen, and Leif Bulow. 2017. "Fetal hemoglobin is much less prone to DNA cleavage compared to the adult protein." *Redox biology* 12: 114-120. https://doi.org/10.1016/j.redox.2017.02.008.

Chudwin, D. S., D. L. Rucknagel, A. P. Scholnik, T. A. Waldmann, and K. R. McIntire. 1977. "Fetal hemoglobin and alpha-fetoprotein in various malignancies." *Acta Haematol* 58 (5): 288-93. https://doi.org/10.1159/000207839.

Clancy, S. 2008. "Genetic mutation." *Nature Education* 1 (1): 187.

Collins, Julie-Ann, Aram Rudenski, John Gibson, Luke Howard, and Ronan O'Driscoll. 2015. "Relating oxygen partial pressure, saturation and content: the haemoglobin-oxygen dissociation curve." *Breathe* (Sheffield, England) 11 (3): 194-201. https://doi.org/10.1183/20734735.001415.

Deisseroth, A., A. Nienhuis, J. Lawrence, R. Giles, P. Turner, and F. H. Ruddle. 1978. "Chromosomal localization of human beta globin gene on human chromosome 11 in somatic cell hybrids." *Proc Natl Acad Sci USA* 75 (3): 1456-60. https://doi.org/10.1073/pnas.75.3.1456.

Deisseroth, A., A. Nienhuis, P. Turner, R. Velez, W. F. Anderson, F. Ruddle, J. Lawrence, R. Creagan, and R. Kucherlapati. 1977. "Localization of the human alpha-globin structural gene to chromosome 16 in somatic cell hybrids by molecular hybridization assay." *Cell* 12 (1): 205-18. https://doi.org/10.1016/0092-8674(77)90198-2.

Dover, G. J., S. H. Boyer, and W. H. Zinkham. 1979. "Production of erythrocytes that contain fetal hemoglobin in anemia. Transient in vivo changes." *J Clin Invest* 63 (2): 173-6. https://doi.org/10.1172/jci109286.

Duck-Chong, C. G. 1983. "Differential effect of detergents on the alkaline denaturation of haemoglobin in maternal and fetal blood, with particular reference to Triton X-100." *Journal of clinical pathology* 36 (8): 910-914. https://doi.org/10.1136/jcp.36.8.910.

Dulmovits, Brian M., Abena O. Appiah-Kubi, Julien Papoin, John Hale, Mingzhu He, Yousef Al-Abed, Sebastien Didier, Michael Gould, Sehba Husain-Krautter, Sharon A. Singh, Kyle W. H. Chan, Adrianna Vlachos, Steven L. Allen, Naomi Taylor, Philippe Marambaud, Xiuli An, Patrick G. Gallagher, Narla Mohandas, Jeffrey M. Lipton, Johnson M. Liu, and Lionel Blanc. 2016. "Pomalidomide reverses γ-globin silencing through the transcriptional reprogramming of adult hematopoietic progenitors." *Blood* 127 (11): 1481-1492. https://doi.org/10.1182/blood-2015-09-667923.

Dunn, J-OC, MG Mythen, and MP Grocott. 2016. "Physiology of oxygen transport." *BJA Education* 16 (10): 341-348. https://doi.org/10.1093/bjaed/mkw012.

Engel, J. D., and K. Tanimoto. 2000. "Looping, linking, and chromatin activity: New insights into β-globin locus regulation." *Cell* 100 (5): 499-502. https://doi.org/10.1016/S0092-8674(00)80686-8.

Fantoni, A., M. G. Farace, and R. Gambari. 1981. "Embryonic hemoglobins in man and other mammals." *Blood* 57 (4): 623-633.

Flint, J., K. Thomas, G. Micklem, H. Raynham, K. Clark, N. A. Doggett, A. King, and D. R. Higgs. 1997. "The relationship between chromo-some structure and function at a human telomeric region." *Nature Genetics* 15 (3): 252-257. https://doi.org/10.1038/ng0397-252.

Forget, B. G. 1998. "Molecular basis of hereditary persistence of fetal hemoglobin." *Ann N Y Acad Sci* 850: 38-44. https://doi.org/10.1111/j.1749-6632.1998.tb10460.x.

Giardina, B., I. Messana, R. Scatena, and M. Castagnola. 1995. "The multiple functions of hemoglobin." *Crit Rev Biochem Mol Biol* 30 (3): 165-96. https://doi.org/10.3109/10409239509085142.

Greer, John P., Daniel A. Arber, Bertil Glader, Alan F. List, Robert T. Means Jr., Frixos Paraskevas, George M. Rodgers; editor emeritus, and John Foerster. 2014. *Wintrobe's clinical hematology.* 13 ed., edited by John P Greer. United States: Philadelphia: Wolters Kluwer Lippincott Williams & Wilkins Health, [2014].

Guy, L. G., Q. Mei, A. C. Perkins, S. H. Orkin, and L. Wall. 1998. "Erythroid Kruppel-like factor is essential for beta-globin gene expression even in absence of gene competition, but is not sufficient to induce the switch from gamma-globin to beta-globin gene expression." *Blood* 91 (7): 2259-63.

Hall, J. E., and A. C. Guyton. 2011. *Guyton and Hall Textbook of Medical Physiology.* Saunders/Elsevier.

Hardison, R. 1998. "Hemoglobins from bacteria to man: Evolution of different patterns of gene expression." *Journal of Experimental Biology* 201 (8): 1099-1117.

Hardison, R., J. L. Slightom, D. L. Gumucio, M. Goodman, N. Stojanovic, and W. Miller. 1997. "Locus control regions of mammalian β-globin gene clusters: Combining phylogenetic analyses and experimental results to gain functional insights." *Gene* 205 (1-2): 73-94. https://doi.org/10.1016/S0378-1119(97)00474-5.

Hecht, F., A. G. Motulsky, R. J. Lemire, and T. E. Shepard. 1966. "Predominance of hemoglobin Gower 1 in early human embryonic development." *Science* 152 (3718): 91-2. https://doi.org/10.1126/science.152.3718.91.

Higgs, D. R., Swee Thein, and W. G. Wood. 2008. "Thalassaemia: Classification, Genetics and Relationship to Other Inherited Disorders of Haemoglobin." *The Thalassemia Syndromes:* 121-132. https://doi.org/10.1002/9780470696705.ch3.

Hsia, C. C. 1998. "Respiratory function of hemoglobin." *N Engl J Med* 338 (4): 239-47. https://doi.org/10.1056/nejm199801223380407.

Huehns, E. R. 1974. The structure and function of haemoglobin: clinical disorders due to abnormal haemoglobin structure. Edited by R. M. Hardiston, Weatherall, D. J. *Blood and its Disorders.* Oxford, UK: Blackwell Scientific.

Janus, J., and S. K. Moerschel. 2010. "Evaluation of anemia in children." *Am Fam Physician* 81 (12): 1462-71.

Kasinger, W., R. Huch, and A. Huch. 1981. "In vivo oxygen dissociation curve for whole fetal blood: fitting the Adair equation and blood gas nomogram." *Scand J Clin Lab Invest* 41 (7): 701-7. https://doi.org/10.3109/00365518109090518.

Kaufman DP, Lappin SL. 2019 Jan- [Updated 2019 Apr 16]. Physiology, Fetal Hemoglobin. In: *StatPearls* [Internet]: Treasure Island (FL): StatPearls Publishing.

Kikuchi, Goro, Tadashi Yoshida, and Masato Noguchi. 2005. "Heme oxygenase and heme degradation." *Biochemical and Biophysical Research Communications* 338 (1): 558-567. https://doi.org/https://doi.org/10.1016/j.bbrc.2005.08.020.

Krauss, J. S., A. R. Rodriguez, and P. F. Milner. 1981. "Erythroleukemia with high fetal hemoglobin after therapy for ovarian carcinoma." *Am J Clin Pathol* 76 (5): 721-2. https://doi.org/10.1093/ajcp/76.5.721.

Lewis, S. M. 1974. The constituents of normal blood. Edited by R. M. Hardiston, Weatherall, D. J. *Blood and its Disorders*. UK: Oxford.

Liu, Shanrun, Sean C. McConnell, and Thomas M. Ryan. 2013. "Erythropoiesis in the absence of adult hemoglobin." *Molecular and cellular biology* 33 (11): 2241-2251. https://doi.org/10.1128/MCB.01734-12.

Mairbaurl, H., and R. E. Weber. 2012. "Oxygen transport by hemoglobin." *Compr Physiol* 2 (2): 1463-89. https://doi.org/10.1002/cphy.c080113.

Malte, H., and G. Lykkeboe. 2018. "The Bohr/Haldane effect: a model-based uncovering of the full extent of its impact on O2 delivery to and CO2 removal from tissues." *J Appl Physiol* (1985) 125 (3): 916-922. https://doi.org/10.1152/japplphysiol.00140.2018.

McGrath, K., and J. Palis. 2008. "Ontogeny of erythropoiesis in the mammalian embryo." *Curr Top Dev Biol* 82: 1-22. https://doi.org/10.1016/s0070-2153(07)00001-4.

Meletis, J., M. Dalekou, M. Samarkos, E. Paravasiliou, C. Meletis, K. Konstantopoulos, E. Apostolidou, V. Komninaka, E. Terpos, O. Benopoulou, K. Korovesis, E. Variami, and D. Loukopoulos. 2001. "Fetal Erythropoiesis after Allogeneic Bone Marrow Transplantation Estimated by the Peripheral Blood Erythrocytes Containing Hemoglobin F (F-cells)." *Hematology* 5 (6): 447-53. https://doi.org/10.1080/10245332.2001.11746542.

Mori, W. 1979. "Hepatocellular carcinoma and fetal hemoglobin." *Kurume Med J* 26 (3): 205-13. https://doi.org/10.2739/kurumemedj.26.205.

Nagalakshmi, Nagan, Ravi Madhusudhana, Nikhila Rajendra, and Abhishek Manjunath. 2016. "Hemoglobin and oxygen transport." *Karnataka Anaesthesia Journal* 2 (1): 1-6. https://doi.org/10.4103/2394-6954.190767.

Ney, Paul A. 2011. "Normal and disordered reticulocyte maturation." *Current opinion in hematology* 18 (3): 152-157. https://doi.org/10.1097/MOH.0b013e328345213e.

Nuez, B., D. Michalovich, A. Bygrave, R. Ploemacher, and F. Grosveld. 1995. "Defective haematopoiesis in fetal liver resulting from inactivation of the EKLF gene." *Nature* 375 (6529): 316-8. https://doi.org/10.1038/375316a0.

Panawala, Lakna. 2017. "*What is the Function of Hemoglobin in the Human Body.*"

Pataryas, H. A., and G. Stamatoyannopoulos. 1972. "Hemoglobins in human fetuses: evidence for adult hemoglobin production after the 11th gestational week." *Blood* 39 (5): 688-696.

Patel S, Mohiuddin SS. 2019. Physiology, Oxygen Transport and Carbon Dioxide Dissociation Curve. [Updated 2019 Apr 25]. In: *StatPearls* [Internet], Treasure Island (FL): StatPearls Publishing.

Perkins, A. C., K. M. Gaensler, and S. H. Orkin. 1996. "Silencing of human fetal globin expression is impaired in the absence of the adult beta-globin gene activator protein EKLF." *Proceedings of the National Academy of Sciences of the United States of America* 93 (22): 12267-12271. https://doi.org/10.1073/pnas.93.22.12267.

Perkins, A. C., A. H. Sharpe, and S. H. Orkin. 1995. "Lethal beta-thalassaemia in mice lacking the erythroid CACCC-transcription factor EKLF." *Nature* 375 (6529): 318-22. https://doi.org/10.1038/375318a0.

Peschle, C., F. Mavilio, A. Care, G. Migliaccio, A. R. Migliaccio, G. Salvo, P. Samoggia, S. Petti, R. Guerriero, M. Marinucci, and et al. 1985. "Haemoglobin switching in human embryos: asynchrony of zeta–alpha and epsilon–gamma-globin switches in primitive and definite erythropoietic lineage." *Nature* 313 (5999): 235-8. https://doi.org/10.1038/313235a0.

Piccin, A., C. Murphy, E. Eakins, M. B. Rondinelli, M. Daves, C. Vecchiato, D. Wolf, C. Mc Mahon, and O. P. Smith. 2019. "Insight into the complex pathophysiology of sickle cell anaemia and possible treatment." *Eur J Haematol* 102 (4): 319-330. https://doi.org/10.1111/ejh.13212.

Pinkerton, P. H., and M. M. Cohen. 1967. "Persistence of hemoglobin F in D/D translocation with trisomy 13-15 (D1)." *Jama* 200 (7): 647-9.

Ponka, Prem, Mark Koury, and Alex Sheftel. 2014. "*Erythropoiesis, Hemoglobin Synthesis, and Erythroid Mitochondrial Iron Homeostasis.*" 41-84.

Rifkind, Joseph M., and Enika Nagababu. 2013. "Hemoglobin redox reactions and red blood cell aging." *Antioxidants & redox signaling* 18 (17): 2274-2283. https://doi.org/10.1089/ars.2012.4867.

Ross, J., and T. D. Sullivan. 1985. "Half-lives of beta and gamma globin messenger RNAs and of protein synthetic capacity in cultured human reticulocytes." *Blood* 66 (5): 1149-1154. https://doi.org/10.1182/blood.V66.5.1149.1149.

Saha, Debarchana, Mandar Patgaonkar, Ankit Shroff, Kanchana Ayyar, Tahir Bashir, and K. V. R. Reddy. 2014. "Hemoglobin expression in nonerythroid cells: novel or ubiquitous?" *International journal of inflammation* 2014: 803237-803237. https://doi.org/10.1155/2014/803237.

Sankaran, V. G. 2011. "Targeted therapeutic strategies for fetal hemoglobin induction." *Hematology Am Soc Hematol Educ Program* 2011: 459-65. https://doi.org/10.1182/asheducation-2011.1.459.

Sankaran, V. G., T. F. Menne, J. Xu, T. E. Akie, G. Lettre, B. Van Handel, H. K. Mikkola, J. N. Hirschhorn, A. B. Cantor, and S. H. Orkin. 2008. "Human fetal hemoglobin expression is regulated by the developmental stage-specific repressor BCL11A." *Science* 322 (5909): 1839-42. https://doi.org/10.1126/science.1165409.

Sankaran, V. G., and D. G. Nathan. 2010a. "Thalassemia: an overview of 50 years of clinical research." *Hematol Oncol Clin North Am* 24 (6): 1005-20. https://doi.org/10.1016/j.hoc.2010.08.009.

Sankaran, V. G., and S. H. Orkin. 2013. "The switch from fetal to adult hemoglobin." *Cold Spring Harb Perspect Med* 3 (1): a011643. https://doi.org/10.1101/cshperspect.a011643.

Sankaran, V. G., and M. V. Sapp. 2012. "Persistence of fetal hemoglobin expression in an older child with trisomy 13." *J Pediatr* 160 (2): 352. https://doi.org/10.1016/j.jpeds.2011.07.036.

Sankaran, Vijay G., and David G. Nathan. 2010b. "Reversing the Hemoglobin Switch." *New England Journal of Medicine* 363 (23): 2258-2260. https://doi.org/10.1056/NEJMcibr1010767.

Sankaran, Vijay G., Jian Xu, and Stuart H. Orkin. 2010. "Advances in the understanding of haemoglobin switching." *British Journal of Haematology* 149 (2): 181-194. https://doi.org/10.1111/j.1365-2141.2010.08105.x.

Schechter, Alan N. 2008. "Hemoglobin research and the origins of molecular medicine." *Blood* 112 (10): 3927-3938. https://doi.org/10.1182/blood-2008-04-078188.

Shaukat, Irfan, Amrit Pudal, Sayf Yassin, Naseruddin Höti, and Sadaf Mustafa. 2018. "Blessing in disguise; a case of Hereditary Persistence of Fetal Hemoglobin." *Journal of community hospital internal medicine perspectives* 8 (6): 380-381. https://doi.org/10.1080/20009666.2018.1536241.

Sheridan, B. L., D. J. Weatherall, J. B. Clegg, J. Pritchard, W. G. Wood, S. T. Callender, I. J. Durrant, W. R. McWhirter, M. Ali, J. W. Partridge, and E. N. Thompson. 1976. "The patterns of fetal haemoglobin production in leukaemia." *Br J Haematol* 32 (4): 487-506. https://doi.org/10.1111/j.1365-2141.1976.tb00952.x.

Smith, Z. E., and D. R. Higgs. 1999. "The pattern of replication at a human telomeric region (16p13.3): Its relationship to chromosome structure and gene expression." *Human Molecular Genetics* 8 (8): 1373-1386. https://doi.org/10.1093/hmg/8.8.1373.

Somervaille, Tim. 2001. "Disorders of Hemoglobin: Genetics, Pathophysiology, and Clinical Management." *Journal of the Royal Society of Medicine* 94 (11): 602-603.

Stamatoyannopoulos, G., and A. W. Nienhuis. 1987. In *Stamatoyannopoulos*: 66-105.

Thein, S. L. 2011. "CHAPTER 9 - Abnormalities of the structure and synthesis of hemoglobin." In *Blood and Bone Marrow Pathology (Second Edition)*, edited by Anna Porwit, Jeffrey McCullough and Wendy N. Erber, 131-155. Edinburgh: Churchill Livingstone.

Thein, Swee Lay, and Stephan Menzel. 2009. "Discovering the genetics underlying foetal haemoglobin production in adults." *British Journal of Haematology* 145 (4): 455-467. https://doi.org/10.1111/j.1365-2141.2009.07650.x.

Thomas, Caroline, and Andrew B Lumb. 2012. "Physiology of haemoglobin." *BJA Education* 12 (5): 251-256. https://doi.org/10.1093/bjaceaccp/mks025.

Vyas, P., M. A. Vickers, D. L. Simmons, H. Ayyub, C. F. Craddock, and D. R. Higgs. 1992. "Cis-acting sequences regulating expression of the human α-globin cluster lie within constitutively open chromatin." *Cell* 69 (5): 781-793. https://doi.org/10.1016/0092-8674(92)90290-S.

Wang Y, Zhao S. 2010. Chapter 5, Oxygen Tension and Placental Vascular Development. *Vascular Biology of the Placenta.* San Rafael (CA): Morgan & Claypool Life Sciences.

Weatherall, D. J., J. B. Clegg, W. G. Wood, S. T. Callender, B. L. Sheridan, and J. Pritchard. 1975. "Foetal erythropoiesis in human leukaemia." *Nature* 257 (5528): 710-2. https://doi.org/10.1038/ 257710a0.

Weed, R. I., C. F. Reed, and G. Berg. 1963. "Is hemoglobin an essential structural component of human erythrocyte membranes?" *The Journal of clinical investigation* 42 (4): 581-588. https://doi.org/10.1172/JCI104747.

West, John B. 2005. *Respiratory physiology the essentials.* 7 ed. Philadelphia: Lippincott Williams & Wilkins.

Wienert, B., G. E. Martyn, A. P. W. Funnell, K. G. R. Quinlan, and M. Crossley. 2018. "Wake-up Sleepy Gene: Reactivating Fetal Globin for beta-Hemoglobinopathies." *Trends Genet* 34 (12): 927-940. https://doi.org/10.1016/j.tig.2018.09.004.

Wolk, M., M. Kieselstein, and G. Brufman. 1988. "Evaluation of fetal hemoglobin in various malignancies with reference to the patients' age." *Tumour Biol* 9 (2-3): 95-100. https://doi.org/10.1159/000217549.

Xu, J., C. Peng, V. G. Sankaran, Z. Shao, E. B. Esrick, B. G. Chong, G. C. Ippolito, Y. Fujiwara, B. L. Ebert, P. W. Tucker, and S. H. Orkin. 2011. "Correction of sickle cell disease in adult mice by interference with fetal hemoglobin silencing." *Science* 334 (6058): 993-6. https://doi.org/10.1126/science.1211053.

Yagami, T., B. T. Ballard, J. C. Padovan, B. T. Chait, A. M. Popowicz, and J. M. Manning. 2002. "N-terminal contributions of the gamma-subunit of fetal hemoglobin to its tetramer strength: remote effects at subunit contacts." *Protein Sci* 11 (1): 27-35. https://doi.org/10.1110/ps.30602.

Zhang, H. B., D. P. Liu, and C. C. Liang. 2002. "The control of expression of the alpha-globin gene cluster." *Int J Hematol* 76 (5): 420-6. https://doi.org/10.1007/bf02982807.

Zhang, J., Y. Yang, P. Li, Y. Yan, T. Lv, T. Zhao, X. Zeng, D. Li, X. Zhou, H. Chen, J. Su, T. Yang, J. He, and B. Zhu. 2019. "Analysis of deletional hereditary persistence of fetal hemoglobin/deltabeta-thalassemia and delta-globin gene mutations in Southerwestern China." *Mol Genet Genomic Med* 7 (6): e706. https://doi.org/10.1002/mgg3.706.

Zhou, D., K. Liu, C. W. Sun, K. M. Pawlik, and T. M. Townes. 2010. "KLF1 regulates BCL11A expression and gamma- to beta-globin gene switching." *Nat Genet* 42 (9): 742-4. https://doi.org/10.1038/ng.637.

Zivot, Andrea, Jeffrey M. Lipton, Anupama Narla, and Lionel Blanc. 2018. "Erythropoiesis: insights into pathophysiology and treatments in 2017." *Molecular medicine* (Cambridge, Mass.) 24 (1): 11-11. https://doi.org/10.1186/s10020-018-0011-z.

In: Fetal Hemoglobin
Editor: Nishant Chakravorty

ISBN: 978-1-53617-955-2

Chapter 3

AN INTRODUCTION TO FETAL HEMOGLOBIN AND ITS CHARACTERISTICS

***Suman Kumar Ray*[1], *Shalini Sanyal*[2], *Rashmi Sinha*[3,4] *and Nishant Chakravorty*[1,*]**

[1]School of Medical Science and Technology, Indian Institute of Technology Kharagpur, Kharagpur, West Bengal, India
[2]Institute for Stem Cell Science and Regenerative Medicine, Rajiv Gandhi Nagar, Kodigehalli, Bengaluru, Karnataka, India
[3]Dr. B. C. Roy Technology Hospital, IIT Kharagpur, Kharagpur, West Bengal, India
[4]Air Force Hospital, Kalaikunda, Kalaikunda Air field, Kharagpur, Paschim Medinipur, West Bengal, India

ABSTRACT

Hemoglobin is the major carrier of respiratory gases to the different tissues of the body. Although indistinguishable to the human eye, hemoglobin is actually a mix of several forms of ferroproteins that differ in their globin sequences. While the major form of hemoglobin found in

* Corresponding Author's Email: nishant@smst.iitkgp.ac.in.

adult life is HbA (95-98%) (and to some extent HbA2; 2-3%), we still have traces of its fetal form in blood (known as HbF; 0.8-2%). HbF is a functional form of hemoglobin which is capable to meet the demands of the adult human body. However, the fetal-to-adult switch mechanism restricts its expression in post-natal life. And this phenomenon is observed even in individuals with defects in HbA (e.g., beta-thalassemia and sickle cell anemia). Therefore, this form of hemoglobin and the mechanisms involved in its silencing have intrigued researchers for decades now. This chapter aims to provide a detailed introduction to this form of hemoglobin and its characteristic features. The chapter discusses the different forms of globin genes, the structural and functional features of HbF and the detection methods for HbF. Further, the chapter also touches upon the concept of "F-cells" which are believed to be the type of red blood cells (RBCs) that contain HbF. Besides these, several conditions associated with raised HbF levels in adult life have also been discussed. It is expected that this chapter will serve as the foundation stone for scientists in the field of fetal hemoglobin research.

Keywords: fetal hemoglobin, HbF, globin genes, F-cells, elevated HbF levels

INTRODUCTION

Hemoglobin is an iron containing protein found in erythrocytes that serves as a carrier of our respiratory gases (oxygen and carbon dioxide). It is known to consist of four chains of globin polypeptides (Perutz, 1978). Each globin subunit is associated with an organic protoporphyrin ring created by a heme moiety and a ferrous core iron ion (Fe^{2+}). The iron molecule can bind and unbind to oxygen in each heme moiety, enabling oxygen to be transported in the body. Some inherent properties of carbon dioxide (CO_2) are known to influence the transport of oxygen. The non-aqueous nature of heme encourages reversible oxygen binding and protects oxidation of Fe^{2+} ion to Fe^{3+} ion. The ferric form of iron (Fe^{3+}) is present in methemoglobin and is unable to bind to oxygen reversibly (David et al. 2018). Globin production and heme synthesis are the two major events in hemoglobin synthesis. The synthesis of the globin chains happens in the erythrocytic cytosol through a cascade of genetic, transcriptional and translational events.

The globin polypeptides of hemoglobin are of multiple forms (based on the sequence of amino acids found in them) (α, β, γ, δ, ε, ζ). Depending on the types of heterodimers formed, the hemoglobin forms found in humans include HbA (α2β2), HbA2 (α2δ2), HbF (α2γ2), Gower 1 (ζ2ε2), Gower 2 (α2ε2), Hemoglobin Portland I (ζ2γ2) and Hemoglobin Portland II (ζ2β2) (Manning et al. 2007).

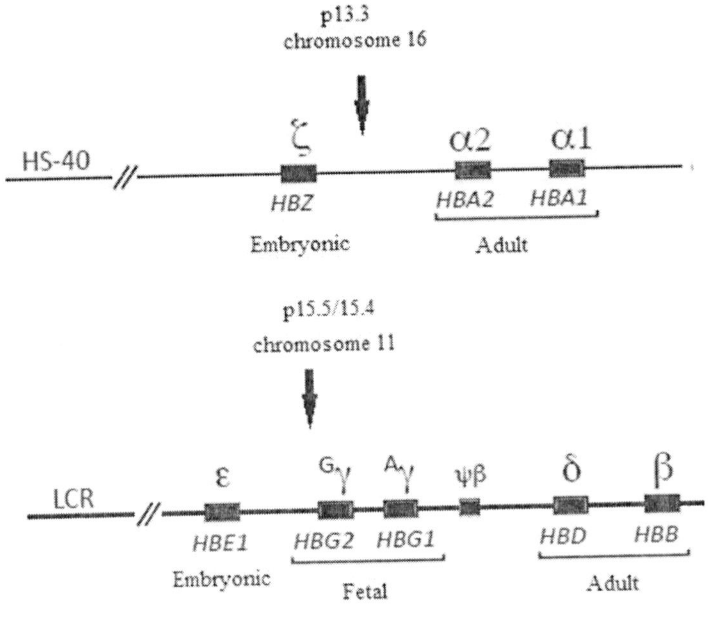

Figure 1. Schematic representation of human globin genes. The locations of the α-globin gene complex are in chromosome 16 (short arm) and β-globin gene on chromosome 11 (short arm). Genes are shown as boxes and named according to the globin polypeptide that is encoded.

As mentioned above, the adult hemoglobin form (HbA) consists of two subunits of α (alpha) and two β (beta) globin chains (Sankaran et al. 2013), whereas the fetal hemoglobin form (HbF) consists of two subunits of α and two subunits of γ (gamma) and is frequently referred to as α2γ2 (Rowe, 2006). While the alpha chain genes (and ζ) are present on chromosome 16 in humans, the non-α like chain genes including the γ genes are present on chromosome 11 (Figure 1). Interestingly, there are two comparable copies of the γ- subunit gene: the γG containing glycine (at location 136) and the γA containing alanine. This observation is similar to the α-subunit coding genes which are also found in duplicate.

The fetal form of hemoglobin (HbF) is known to be produced by erythroid precursor cells from 10 to 12 weeks of pregnancy (Cao et al. 2010) with a steady decline in the levels during the first 6 months of postnatal life (Figure 2).

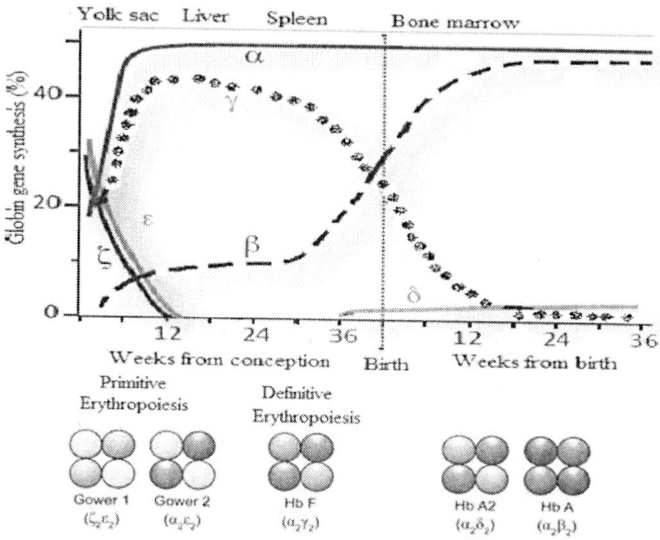

Figure 2. Diagrammatic representation of different globin gene synthesis sites and rates and presence of different hemoglobin (in percentage) during embryonic and fetal and infancy period.

Fetal Hemoglobin: A Brief Look of Early Research

In 1866 Korber found in that blood from the human placenta was more susceptible to acid denaturation than ordinary blood. Two immunologically distinguishable molecules were found by Darrow and his colleagues in 1940.

HbF varies from HbA in terms of several physical characteristics and they can be exploited to develop an easy method for its detection (Singer et al. 1951). While HbA possesses only valine at its N terminus, HbF has both glycine and valine. The molecular structures of the amino-acid clusters in horses and human hemoglobins have been determined and explained by x-ray crystallography in excellent details (Perutz 1978). A gradual shift in the form of hemoglobin from embryonic to fetal and then to adult has been noted in majority of animals. These various hemoglobin types have been suggested to depict the adaptations to the distinct ways of acquiring oxygen. This was one of the early indicators suggesting that adult and fetal/newborn hemoglobins differed (Bertles et al. 1969). The quantity of HbF present at birth is nearly 70-90% of the total hemoglobin and it gradually decreases during the first one year after birth; at 6 months it is less than 8% and at 12 months it found to be less than 2% (Huehns et al. 1971). The genetic determinants and inducers of HbF production are shown in the following tables (Table 1 and 2).

Table 1. Genetic determinants of hemoglobin F production

Chromosome Number	Locus
11	Xmn1-HBG2
11	OR5-OR6 (Olfactory receptor gene cluster)
6	HMIP (HBS1L-MYB intergenic polymorphism)
2	BCL11A

Table 2. Inducers of hemoglobin F production

Inducers of HbF production	Example
Nucleoside analogs	Azacytidine, Decitabine
Hydroxycarbamide	Hydroxyurea
Histone deacetylase inhibitors	Butyrate, Trichostatin-A, Apicidin, Scriptaid, Hydroxyamides
Others	Valproate, Thalidomide derivatives, Kit ligand

Fetal Hemoglobins in Other Animals

The existence of fetal hemoglobin is not exclusive to the humankind. In fact similar hemoglobin types have been reported in several species of mammals and other phyla. Some of the examples are listed in Table 3.

Table 3. Fetal hemoglobins in different animals, a historical perspective

Name of the animal	Research work	Reference
Rhesus monkey (*Macaca mulatta*)	Detected HbF by gel electrophoresis.	Beaven et al. (1959)
Rhesus monkey (*Macaca mulatta*)	Measured the alkali resistant Hb.	Sen et al. (1960)
Prosimian primates Lemuriformes and Lorisiformes *Cebus paella, Macaca mulatta*	Alkali resistant hemoglobins	Buettner-Janusch et al. (1961)
Bovine	The amino acid sequence of the γ chain of fetal hemoglobin. It differs from human γ chain in 40 positions.	Babin et al. (1963)
Sheep	Complete sequence for the γ chain of Hb F	Wilson et al. (1966)
Bullfrog	N terminal acetyl group distinguished two chains of embryonic Hb, similarly like human HbF	deWitt et al. (1967).
Cattle, sheep, and pig	Production of two embryonic hemoglobins analogous to human Hbs Gower 1 and 2. These hemoglobins are replaced by a fetal form and then by the adult form	Kleihauer et al. (1968)
Macaca speciosa (stump-tailed macaque)	Hemoglobin peptide mapping.	Kitchen et al. (1968)
Goats	Genotyping of the fetal hemoglobin gene.	Huisman et al. (1969)
Macaca nemestrina (pig-tailed macaque).	Hemolysates assay	Nute et al. (1971)
Chimpanzee	Discovered two types of γ chain	deJong (1971)

F-Cells

Early researchers in the field of hemoglobin and HbF have been of the opinion that fetal hemoglobin (HbF) is not found in all erythrocytes (except in certain conditions such as HPFH) (Boyer et al. 1975). The erythrocytes which contain fetal hemoglobin in adult individuals are termed as "F-cells." Synthesis of HbF is reported to decrease below 0.6% of total hemoglobin, at around the time of birth. However, in certain acquired or hereditary conditions such as HPFH, HbF synthesis remains significantly high even in adults, owing to the high prevalence of HbF-per cell and presence of higher numbers of F-cells in blood (Rochette et al. 1994). A comparative study between HPFH homozygous and heterozygous mothers showed enhanced production of F-cells in the blood during pregnancy (Boyer et al. 1975).

The amount of F-cells present in normal adults ranges from 0.5 to 7%. However, in patients with β-thalassemia or sickle cell disease, higher prevalence of F-cells has been reported (Dover et al. 1978). The clinical importance of raised HbF level is that it is reported to ameliorate the clinical severity in conditions like β-thalassemia, SCD etc. However, patients with β-hemoglobinopathies show a heterogeneous spectrum of F-cell concentration in the blood; the reason of which is not elucidated properly.

A significant correlation between percentage of F-cells and HbF concentration in blood has been confirmed and this can be utilized while designing therapeutic interventions for SCD patients (Marcus et al. 1997). Several pharmacological agents have already undergone clinical trials as HbF inducing agents. Various pharmacological agents such as hydroxyurea, 5-azacytidine, thalidomide etc. have been identified to affect the concentration of F-cells in the blood, leading to upregulated expression of HbF (Musallam et al. 2013). Detailed discussions on these agents are available in other chapters of this book.

The selection of F-cells in circulation, their regulation and maintenance of F-cell progenitors have been extensively studied in both in vivo and in vitro conditions to understand the HbF and F-cells production in human body (Gibney et al. 2008).

FUNCTION OF HBF

The developing fetus gains its nourishment and supply of oxygenated blood from the maternal blood in circulation (Kessler et al. 2008). The placenta supplies oxygenated blood to the fetus through the umbilical vein attached to the uterine wall. The chorionic membrane that surrounds the developing fetus functions as a barrier between fetal and maternal circulation and thus prevents admixture of blood (Mohan et al. 2017). Maternal blood is delivered to the chorionic plate intervillous space via open-ended arterioles and thus washes chorionic villi, enabling the exchange of gas between fetal and maternal circulation (Nye et al. 2018). The relatively low oxygen tension in the maternal blood supply reaching chorion necessitates the presence of a physiological form of hemoglobin that has higher binding affinity than adult hemoglobin. Naturally fetal hemoglobin (HbF) fits the role during this stage of human life. The affinity of HbF to oxygen is significantly higher than that of adult hemoglobin which is indicated by its P_{50} (around 19 mm Hg), in contrast to that of adult hemoglobin (26.8 mm Hg) (Kaufman et al. 2018). Therefore the presence of γ-globin sub-unit bestows a survival advantage compared to the β-subunit. This is also reflected in demanding situations like pregnancy, aplastic anemia, hemoglobinopathies, thyrotoxicosis etc. when the levels of HbF are seen to rise even in adult life.

MEASUREMENT OF FETAL HEMOGLOBIN

Assessment of the presence or absence of fetal hemoglobin is required in different situations encountered in day-to-day clinical practice (Akinsheye et al. 2011). Fortunately, we have several tests in our diagnostic toolkit. One of the simple tests that allows us to differentiate between fetal or neonatal blood from maternal blood (found occasionally newborn's stool or vomitus) is based on the differences in chemical properties between maternal and fetal hemoglobin (alkali denaturation test, A test or Apt test)

(Kaufman et al. 2018). This alkaline denaturation test employs HbF's property of resistance to alkaline denaturation, which HbA does not possess. In this test, the samples are lysed, centrifuged and subjected to sodium hydroxide treatment, leading to the formation of "hematin" which converts the sample to brown color, when HbA is present; otherwise, the sample remains pink (in presence of HbF). The Kleihauer-Betke is a test (Krywko et al. 2019) in practice for Rh-negative mothers and this is used to evaluate the magnitude of maternal-fetal hemorrhage and the necessary amount of RhoD Ig to prevent Rh antibodies. Acid based experiments of hemoglobin typically use acid baths that keep HbF cells blue while maternal cells appear black. The dose calculation for RhoD immunoglobulin is dependent upon the proportion of blue to white cells and also takes into account the projected fetal weight and blood volume. A more definitive test to check the various kinds of hemoglobins in blood is the Hb electrophoresis test (Mehdi et al. 2013). The test utilizes electrical current to separate normal and other hemoglobin types present in a blood sample.

Structure of Fetal Hemoglobin

Hemoglobin F is a tetrameric molecule consisting of a couple of α-chains as in HbA and A2 and two γ chains. The γ chain comprises of 146 amino acids and varies from the β chain of HbA in 39 locations (Thom 2013). There are nine modifications in inner DNA, including substitution of tyrosine 130 for tryptophan, which is detectable by spectroscopic technique. These replacements are primarily restrictive in nature, where an amino acid is substituted by another of a similar type.

Hemoglobin Variants, HbF and Hemoglobin Barts

Over the last forty years, extensive studies have been conducted to study the HbF type of hemoglobin and thousands of cord blood specimens with several variations of HbF have been discovered. As it is expected, any mutation in the adult hemoglobin type leads to production of abnormal Hb throughout an individual's post-natal life, however when such a mutation occurs in the γ chain, abnormal hemoglobin form significant to affect physiology, only occurs during fetal life and perhaps a few months after birth (Angastiniotis et al. 2019).

Hemoglobin Barts (Hb Barts-γ4) (Ager et al. 1968) is a type of hemoglobin encountered in α-thalassemia and is known to migrate faster than HbA upon electrophoresis. Thus, it was formerly known as 'quick fetal hemoglobin' (Fessas et al. 1959). Hb Barts can be separated by chromatography techniques (Synodinos et al. 2015). Nearly 5-10% of thalassemia cases detected have Hb Barts.

HbF in Pregnancy and Development

Tabrizi and his group (2015) noted that HbF levels maybe around 20-25% higher during pregnancy, especially in the second trimester, and six months post-delivery. Another study published recently also recorded mildly increased levels of HbF concentrations in the second trimester of pregnancy in up to 10% of the cases (Jwa et al. 2015). This study also observed high quantities of chorionic gonadotropins being excreted by pregnant women with high HbF levels. Cases which lacked a developing fetus but showing rise in HbF levels clearly indicated to the increased expression of maternal HbF. All such findings suggest the existence of some factor or factors that can induce HbF synthesis in premature pregnancy. Earlier Davies and Bull (1971) had noted that the administration of prolactin to hamsters caused the development of heme protein and also has impacts on HbF levels. There is a heterogeneous distribution of HbF among the erythrocyte populations and this observation has been documented in various hemoglobinopathies (Galanello et al. 2013).

Conditions Associated with Elevated Fetal Hemoglobin Levels

Several acquired conditions are associated with modest elevations of HbF (Figure 3). They include pregnancy, recovery from marrow hypoplasia, aplastic anemia, leukemia, thyrotoxicosis, hepatoma, and juvenile chronic myeloid leukemia. The latter condition is exceptional in that it seems to reflect a genuine reversion to fetal erythropoiesis.

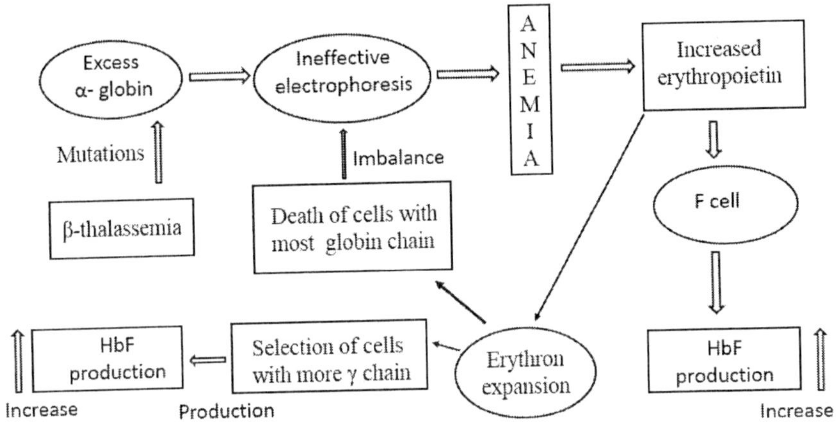

Figure 3. Proposed mechanism of HbF production in thalassemia.

Presence of an increased level of HbF by analyzing the globin genes is one of the primary methods (with a minimum set of other tests) utilized in the diagnosis of hemoglobinopathies such as HPFH and δβ-thalassemia syndromes.

Hemoglobin F and Hereditary Persistence of Fetal Hemoglobin (HPFH)

Molecular and functional abnormalities of the different types of hemoglobin usually manifest as anemic conditions like β-hemoglobinopathies (Kohne 2011). However, in rare circumstances like hereditary persistence of fetal hemoglobin, the presence of an unusual trigger may mask other errors. HPFH is a rare medical condition that has raised hopes of using HbF as a therapeutic alternative for β-hemoglobinopathies (Wienert et al. 2018). The search for agents that can trigger HbF reactivation has led to the discovery of many drug-repurposing candidates although with limited success. Hydroxyurea is one such drug that has been found to increase HbF levels in adolescents (Papadopoulou et al. 2015).

HPFH is a genetic abnormality wherein adult hemoglobin synthesis fails to become the dominant hemoglobin synthesis route. Persistently elevated numbers of cells containing HbF in the mother after delivery may be a symptom of HPFH and it is commonly observed as a co-inheritance in asymptomatic patients of sickle-cell disease (Rochette et.al. 1994; Borg et.al. 2010). Two types of mutation lead to the development of HPFH:

a) **Loss of function mutation**- It is caused by large deletions of the β-globin gene cluster. The first chromosomal region revealed to carry DNA sequence variants that affect the concentration of fetal hemoglobin was the β-globin gene cluster. It was discovered that large deletions within the β-globin locus were associated with higher fetal hemoglobin concentrations (deletional HPFH) while smaller deletions were associated with more modest increases in fetal hemoglobin concentration (δβ0-thalassaemia) (Rochette et.al. 1994).
b) **Gain of function mutation**- Fetal stage specific γ genes are normally silenced in adulthood because elevated levels of γ-globin among adults help in ameliorating the disease severity of β-thalassemia and sickle cell anemia (Berry et.al. 2012). Point mutations in the promoters of the γ-globin genes lead to the haplotype 19p13.12–13 which has an identical chromosome segment (Borg et.al. 2010; Rochette et.al. 1994).

Sickle Cell Anemia

Sickle-cell anemia is a monogenic disorder caused by the autosomal recessive inheritance of an adenine to thiamine transversion at the sixth codon of the β-globin gene which results in Glutamic acid to Valine amino acid substitution.

Fetal hemoglobin (HbF) has been known to modulate the phenotype of sickle cell anemia by inhibiting deoxy sickle hemoglobin (HbS) polymerization (Steinberg et al. 2014).

Janet Watson was the first to co-relate the role of fetal hemoglobin with sickle cell disease by confirming that infants with sickle cell disease had few symptoms and that their deoxygenated erythrocytes took longer to sickle and did not deform as extensively as did their sickle cell trait-carrying mother's cells.

This led to the discovery that fetal hemoglobin is excluded from the sickle hemoglobin polymer. Since the level of HbF and its distribution among sickle erythrocytes is highly variable; it may be concluded that it is the major genetic modulator of the hematologic and clinical features of sickle cell disease (Akinsheye et al. 2011).

HbF is used in the treatment of sickle-cell disease because it has been observed that if fetal hemoglobin remains the predominant form of hemoglobin after birth, the number of painful episodes decreases in patients with sickle-cell disease. Thus, hydroxyurea promotes the production of fetal hemoglobin and has been utilized in the treatment of sickle-cell disease (Akinsheye et al. 2011).

Thalassemia

High levels of HbF in the β thalassemia syndromes have been observed to the extent that it has been speculated that the highest adult levels of HbF are seen in β and δβ thalassemia where levels of HbF range between 2.1-10% or even higher from the normal level of <1.5% (Mosca et.al. 2009; Olaniyi et al. 2010).

Marrow Hypoplasia

Bone marrow hypoplasia is predominantly caused by marrow suppression which in turn is defined by a decrease in bone marrow hematopoietic cells leading to peripheral cytopenias. Patients may exhibit fatigue, pallor, dyspnea on exertion, tachycardia due to anemia, and easy bruising or mucocutaneous bleeding due to thrombocytopenia. Patients may also present with infections due to leukopenia. Congenital bone marrow failures are traditionally complex with varying combinations of cytopenias, macrocytosis, elevated levels of HbF, hypocellular BM, immunologic deficiency, physical malformations, and predisposition to leukemia.

Aplastic Anemia

Aplastic anemia is a condition that occurs when the body stops producing enough new blood cells. Its symptoms include symptoms may include: fatigue, shortness of breath with exertion, rapid or irregular heart rate, pale skin, frequent or prolonged infections, unexplained or easy bruising, nosebleeds and bleeding gums, prolonged bleeding from cuts, skin rashes, dizziness and even headache. Fetal hemoglobin levels are important for their prognostic value in determining the treatment of aplastic anemia (Bloom and Diamond, 1968; Dover et.al. 1978).

Leukemia

A reversion to a fetal form of erythropoiesis occurs in chronic myeloid leukemia and in erythroleukemia, thereby leading to elevated levels of hemoglobin F and simultaneous reduction in the production of HbA2 and the carbonic-anhydrase isozymes during the course of the illness (Sheridan et.al. 1976).

Thyrotoxicosis

Thyrotoxicosis is a condition in which there is an excess of thyroid hormones in the body wherein patients suffer from a low level of thyroid stimulating hormone, TSH, in their bloodstream, because of pituitary gland malfunction (Kuhn et.al. 1983). This hyperthyroidism affects erythrocyte size and is associated with elevated levels of 2,3-diphosphoglycerate which is known to be associated with reduced oxygen-hemoglobin affinity. It has been known to be associated with an increased production of HbF and with increased production of δ-globin with a simultaneous decrease in the production of β-globin. Due to the common symptoms of microcytosis, elevated HbF, it has been described as similar to β-thalassemia.

Acquired Increases in HbF Levels

The level of fetal hemoglobin (HbF) in adults declines to less than 1% of the total hemoglobin under normal circumstances. However, in acquired disorders that result in increased HbF production, normal erythropoiesis is disturbed, thus resulting in an increased demand for erythrocytes. This in turn leads to 'stress' erythropoiesis with preferential production of F-cells (Wood, 1993). As a sporadic finding, elevated HbF has been reported in cases of hydatidiform mole, choriocarcinoma, pernicious anemia, thyrotoxicosis, trophoblastic testicular tumor, bronchogenic carcinoma and hepatoma (Wood et al. 1977). Increased HbF production is thus often observed in situations in which there is acute erythroid expansion as in severe iron deficiency anemia (Dover, 1979). Additionally, acquired elevation in HbF levels in the neonatal period have been observed in infants of diabetic mothers (Bard and Prosmanne, 1985; Perrine et al, 1985).

THERAPEUTICS

Therapeutics involved in hemoglobinopathies differ extensively on the specific disease forms (Ricerca et al. 2009). For β-thalassemia, the standard management plans depend on regular and frequent blood transfusions and iron-chelation therapy (Saliba et al. 2015). Increased understandings of the molecular and pathogenic variables that regulate the mechanisms of the disease have suggested novel and unconventional modalities to treat hemoglobinopathies (Lee et al. 2019). One of the methods demonstrating exciting prospects involves reactivation of the fetal globin gene by pharmacological compounds. Understanding of previously unknown mechanisms of fetal hemoglobin gene silencing has led to the development of new therapies. Researchers are aiming not only to improve treatment but also to possibly cure the disease.

REFERENCES

Ager JAM and Lehmann H. 1958. Observations on some fast hemoglobins: K, J, N, and "Bart's." *Br. Med. J.* 1: 929–931.

Akinsheye I, Alsultan A, Nadia S, Ngo D, Clinton BT, Paola S, Chui HKD and Martin SH. 2011. Fetal hemoglobin in sickle cell anemia. *Blood.* 118(1): 19–27.

Angastiniotis M and Lobitz S. 2019. Thalassemias: An Overview. *Int. J. Neonatal Screen.* 5:16. doi:10.3390/ijns501001

Babin DR, Schroeder WA, Shelton JR, Shelton JB, and Robberson B. 1963. The amino acid sequence of the y chain of bovine fetal hemoglobin. *Biochemistry.* 5: 1297-1309.

Bard H and Prosmanne J. 1985. Relative rates of fetal hemoglobin and adult hemoglobin synthesis in cord blood of infants of insulin-dependent diabetic mothers. *Pediatrics.* 75(6): 1143-1147.

Beaven GH and Gratzer WB. 1959. Foetal hemoglobin in the monkey. *Nature*. 184: 1730.

Berry M, Grosveld F and Dillon N. 1992. A single point mutation is the cause of the Greek form of hereditary persistence of fetal hemoglobin. *Nature.* 358 (6386): 499.

Bertles JF and Dobler J. 1969. Reversible and Irreversible Sickling: a Distinctionby Electron Microscopy. *Blood* 33(6): 884-898.

Bloom GE and Diamond LK. 1968. Prognostic value of fetal hemoglobin levels in acquired aplastic anemia. *New England Journal of Medicine* 278(6): 304-307.

Borg J, Papadopoulos P, Georgitsi M, Gutiérrez L, Grech G, Fanis P, Marios P et al. 2010. Haploinsufficiency for the erythroid transcription factor KLF1 causes hereditary persistence of fetal hemoglobin. *Nature genetics.* 42(9): 801.

Boyer SH, Belding TK, Margolet L and Noyes AN. 1975. Fetal hemoglobin restriction to a few erythrocytes (F cells) in normal human adults. *Science.* 188:361.

Buettner-Janusch J and Twichell JB. 1961. Alkali-resistant hemoglobin in prosimian primates. *Nature.* 192: 669.

Cao Antonio and Galanello Renzo. 2010. Beta-thalassemia. *Genetics in Medicine.* 12(2) :61-76.

Chung Jacky, Caiyong Chen and Barry H Paw. 2012. Heme Metabolism and Erythropoiesis. *Curr Opin Hematol.* 19(3): 156–162.

Dailey A Harry, Dailey A Tamara, Gerdes Svetlana, Jahn Dieter, Jahn Martina, O'Brian R Mark and Warren J Martin. 2017. Prokaryotic Heme Biosynthesis: Multiple Pathways to a Common Essential Product. *Microbiol Mol Biol Rev.* 81(1): e00048-16.

David SR, Sawal NS, Bin Hamzah MN and Rajabalaya R. 2018. The blood blues: A review on methemoglobinemia. *J Pharmacol Pharmacother* 9:1-5.

Davies J and Bull GM. 1971. Drugs affecting foetal hemoglobin synthesis. *Transactions of the Royal Society of Tropical Medicine and Hygiene.* 65: 405-406.

Dover GJ, Boyer SH, Charache S and Heintzelman K. 1978. Individual variation in the production and survival of F cells in sickle-cell disease. *N Engl J Med.* 299:1428.

Dover GJ, Boyer SH, Zinkham WH. 1979. Production of erythrocytes that contain fetal hemoglobin in anemia. *J Clin Invest.* 63: 173-176.

deJong WWW. 1971. Structure of the chain of chimpanzee hemoglobin A2. *Nature New Biology.* 234: 176-177.

deWitt W and Ingram VM. 1967. Acetylated peptide chains in Bullfrog hemoglobins. *Biochemical and Biophysical Research Communications*. 27. 236-241.

Fessas P, Mastrokalos N and Phostiropoulos G. 1959. New Variant of Human Fœtal Hæmoglobin. *Nature*. 183: 30–31.

Galanello R. 2013. *Prevention of thalassemia and other Hemoglobin disorders*, vol 1. TIF publication. Nicosia.

Gibney GT, Panhuysen CIM, So JCC, Ma ESK, Ha SY, Li CK, Lee ACW, Li CK, Yuen HL, Lau YL, Johnson DM, Farrell JJ, Bisbee AB, Farrer LA, Steinberg MH, Chan LC and Chui DHK. 2008. Variation and heritability of Hb F and F-cells among b-thalassemia heterozygotes in Hong Kong. *Am. J. Hematol.* 83. 458–464.

Huehns ER and Beaven GH. 1971. Developmental changes in human hemoglobins. *The biochemistry of Development.* Edited by P. F. BcIlsol1, pp. 175-203. Heineman Mcdical Books, London.

Huisman THJ, Schroeder WA, Dozy AM, Shelton JR, Shelton JB, Boyd EM, and Apell G. 1969. *Evidence for multiple structural genes for the gamma chains of human fetal hemoglobin in hereditary persistence of fetal hemoglobins*. Annals of the New York Academy of Sciences. 165: 320-331.

Jwa SC, Fujiwara T, Yamanobe Y, Kozuka K and Jwa HS. 2015. Changes in maternal hemoglobin during pregnancy and birth outcomes. *BMC Pregnancy and Childbirth.* 15:80 doi 10.1186/s12884-015-0516-1.

Kaufman DP and Lappin SL. 2019. Physiology, Fetal Hemoglobin. In: StatPearls [Internet]. *Treasure Island* (FL): StatPearls Publishing. https://www.ncbi.nlm.nih.gov/books/NBK500011/

Kessler J, Rasmussen S, Godfrey K. 2008. Longitudinal Study of Umbilical and Portal Venous Blood flow to the Fetal Liver: Low Pregnancy Weight Gain is Associated With Preferential Supply to the Fetal Left Liver Lobe. *Pediatr Res.* 63, 315–320.

Kitchen H, Eaton JW and Stenger VG. 1968. Hemoglobin types of adult, fetal and newborn subhuman primates: Macaca speciosu. *Archives of Biochemistrv and Biophysics,* 123, 227-234.

Kleihauer E and Stoffler G.1968. Embryonic hemoglobins of different animal species. Quantitative and qualitative data about the production of and properties of hemoglobin during early developmental stages of pig, cattle, and sheep. *Molecular and General Genetics.* 101: 59-69.

Korber E. 1866. Ueber Differenzen des Blutfarbstoffes. *Inaugural dissertation*, Dorpat. (Cited in Bertles, 1969).

Krywko DM and Shunkwiler SM. 2019. Kleihauer Betke Test. In: StatPearls [Internet]. *Treasure Island* (FL): StatPearls Publishing. .https://www.ncbi.nlm.nih.gov/books/NBK430876

Kohne E. 2011. Hemoglobinopathies Clinical Manifestations, Diagnosis, and Treatment. *Dtsch Arztebl Int.* 108(31-32): 532–540.

Kuhn JM. 1983. *Influence of thyroid status on hemoglobin A2 expression* [Internet].

Lee KY, Kim HJ, Lee K, Park HS, Song SH, Seong MW, Kim M and Jin HY. 2019. Recent progress in laboratory diagnosis of thalassemia and hemoglobinopathy: a study by the Korean Red Blood Cell Disorder Working Party of the Korean Society of Hematology. *Blood Res.* 54(1): 17–22.

Manning RL, Russell JE, Padovan CJ, Chait TB, Popowicz A, Manning SR and Manning JM. 2007. Human embryonic, fetal, and adult hemoglobins have different subunit interface strengths. Correlation with lifespan in the red cell. *Protein Sci.* 16(8): 1641–1658.

Marcus SJ, Kinney TR, Schultz WH, O'Branski EE and Ware RE. 1997. Quantitative Analysis of Erythrocytes Containing Fetal Hemoglobin (F Cells) in Children with Sickle Cell Disease. *American Journal of Hematology*. 54:40–46.

Mehdi Syed Riaz and Dahmash Badr Abdullah Al. 2013. Analysis of hemoglobin electrophoresis results and physicians investigative practices in Saudi Arabia. *Indian J Hum Genet.* 19(3): 337–341.

Mohan R, Bajaj A, Mohan G. 2017. Human amnion membrane: Potential applications in oral and periodontal field. *Journal of International Society of Preventive and Community Dentistry.* 7(1): 15-21.

Mosca A, Paleari R, Leone D and Ivaldi G. 2009. The relevance of hemoglobin F measurement in the diagnosis of thalassemias and related hemoglobinopathies. *Clinical biochemistry*. 42(18): 1797-1801.

Musallam KM, Taher AT, Cappellini MD and Sankaran VG. 2013. Clinical experience with fetal hemoglobin induction therapy in patients with b-thalassemia. *Blood*. 121(12). 2199-2212.

Nute PE and Stamatoyannopoulos G. 1971. Evidence for duplication of the hemoglobin y chain locus in Macaca nemestrina. *Nature New Biology*. 229:145-147.

Nye GA, Ingram E, Johnstone ED, Jensen OE, Schneider H, Lewis RM, Chernyavsky IL and Paul B. 2018. Human placental oxygenation in late gestation: experimental and theoretical approaches. *J Physiol* 596(23): 5523–5534.

Ogun AS and Valentine M. 2019. Biochemistry, Heme Synthesis. *Treasure Island* (FL): StatPearls Publishing. Available from: https://www.ncbi.nlm.nih.gov/books/NBK-537329.

Olaniyi JA, Arinola OG and Odetunde AB. 2010. Fetal hemoglobin (HbF) status in adult sickle cell anemia patients in Ibadan, Nigeria. *Annals of Ibadan postgraduate medicine*. 8(1): 30-33.

Papadopoulou E, Teli A, Theodoridou S, Gompakis N and Economou M. 2015. Safety and efficacy of hydroxyurea in children and adolescents with sickle/beta-thalassemia: two-year experience. *Hippokratia*. 19(2): 172–175.

Perrine SP, Greene MF, and Faller DV. 1985. Delay in the fetal globin switch in infants of diabetic mothers. *New England Journal of Medicine*. 312(6): 334-338.

Perutz MF. 1978. Hemoglobin structure and respiratory transport. *Scientific American* 239(6):92-125.

Ricerca MB, Girolamo DA and Rund D. 2009. Infections in Thalassemia and Hemoglobinopathies: Focus on Therapy-Related Complications. *Mediterr J Hematol Infect Dis*. 1(1): e2009028.

Rochette J, Craig JE, and Thein SL. 1994. Fetal hemoglobin levels in adults. *Blood reviews*. 8(4): 213-224.

Saliba NA, Harb RA and Ali TT. 2015. Iron chelation therapy in transfusion-dependent thalassemia patients: current strategies and future directions. *J Blood Med.* 6: 197–209.

Sankaran Vijay G and Orkin Stuart H.2013. The Switch from Fetal to Adult Hemoglobin. *Cold Spring Harb Perspect Med.* 3(1): a011643.

Sen NN, Das KC, and Aikat BK. 1960. Foetal hemoglobin in the monkey. *Nature*. 186: 977-978.

Sheridan BL, Weatherall DJ et.al. 1976. The Patterns of Fetal Hemoglobin Production in Leukaemia. *British journal of Huematofogy.* 32 :487.

Singer K, Chernoff A I and Singer L. 1951. Studies on abnormal hemoglobins. II. Their identification by means of the method of fractional denaturation. *Blood* 6: 429-435.

Steinberg MH, Chui DHK, Dover GJ, Sebastiani P and Alsultan A. 2014. Fetal hemoglobin in sickle cell anemia: a glass half full? *Blood.* 123(4): 481-485.

Synodinos JT, Harteveld CL, Old JM, Petrou M, Galanello R, Giordano P, Angastioniotis M, Salle BDL, Henderson S and Alison M. 2015. EMQN Best Practice Guidelines for molecular and haematology methods for carrier identification and prenatal diagnosis of the hemoglobinopathies *European Journal of Human Genetics.* 23: 426–437.

Tabrizi FNM and Barjasteh S. 2015. Maternal Hemoglobin Levels during Pregnancy and their Association with Birth Weight of Neonates. *Iranian Journal of Pediatric Hematology Oncology* 5(4): 211-217.

Thom CS, Dickson CF, Gell DA and Weiss MJ. 2013 Hemoglobin variants: biochemical properties and clinical correlates. *Cold Spring Harb Perspect Med.* 3(3): a011858.

Wienert B, Gabriella ME, Funnell PWA, Quinlan GRK and Merlin C. 2018. Wake-up Sleepy Gene: Reactivating Fetal Globin for β-Hemoglobinopathies. *Trends in Genetics.* 34(12): 927-940.

Wilson J B, Edwards WC, McDaniel M, Dobbs M and Huisman T HJ. 1966. The structure of sheep hemoglobins II. The amino acid composition of the tryptic peptides of the non a chains of hemoglobins A, B, C, and F. *Archives of Biochemistry and Biophysics.* 115: 385-400.

Wood WG. 1993. Increased HbF in adult life. *Baillière's clinical haematology*. 6(1): 177-213.

Wood WG, Clegg JB and Weatherall DJ. 1977. Developmental biology of human hemoglobins. *Progress in hematology*. 10: 43-90.

In: Fetal Hemoglobin
Editor: Nishant Chakravorty

ISBN: 978-1-53617-955-2

Chapter 4

REGULATION OF THE GAMMA GLOBIN GENE: CURRENT CONCEPTS AND PERSPECTIVES

Anindita Basak[1,*] and Suman Kumar Ray[2]

[1]Boston Children's Hospital, Harvard Medical School, Boston, MA, USA

[2]School of Medical Science and Technology, Indian Institute of Technology Kharagpur, Kharagpur, West Bengal, India

ABSTRACT

Fetal-to-adult hemoglobin switch and fetal hemoglobin (HbF) silencing mechanisms have long been topics of interest amongst hematologists and hematobiologists. This interest particularly stems from the promise that HbF holds in alleviating signs and symptoms of β-hemoglobinopathies like sickle cell disease (SCD) and β-thalassemia. Insights into HbF regulation and its reactivation have identified many trans-acting variables in addition to cis-regulatory components of the β-globin gene cluster. Furthermore, researchers have also identified several

* Corresponding Author's Email: abasak@broadinstitute.org.

epigenetic mechanisms influencing HbF functions and silencing. Although the transition from the fetal to the adult hemoglobin form is based on the suppression or silencing of the gamma globin gene, it has been challenging to target the direct transcriptional repressors of the process for therapy. Scientists have therefore additionally focused their attention on epigenetics in order to develop effective and acceptable targets for HbF reactivation. This chapter primarily aims to bring the different theories and hypotheses on HbF silencing under one umbrella, with an expectation that this will enable scientists to investigate newer therapeutic modalities either singly or in combination of different approaches.

Keywords: fetal hemoglobin, globin gene, silencing, fetal to adult hemoglobin switch

INTRODUCTION

Red blood cells (RBCs) or erythrocytes are responsible for oxygen transport to tissues. Erythrocytes contain hemoglobin that binds oxygen. The hemoglobin molecule is a tetramer composed of two subunits each of α-globin and β-globin polypeptides. The heme groups bind to each subunit, allowing for effective oxygen transport in the blood. Out of these two polypeptides, the human β-globin locus, comprising of five paralogous globin genes, is developmentally regulated, such that humans express multiple distinct β-like globin molecules – embryonic, fetal and adult – at different developmental stages (Sankaran and Nathan 2010).

The α- and β-globin loci, located respectively on chromosomes 16 and 11, consist of genes that are sequentially activated and silenced during human ontogeny. The α-globin locus contains three α-like globin genes: HBZ (ζ-globin), HBA2 (α2-globin) and HBA1 (α1-globin). The β-globin locus consists of five β-like globin genes: HBE1 (ε-globin), HBG2 ($^{G}\gamma$-globin), HBG1 ($^{A}\gamma$-globin), HBD (δ-globin) and HBB (β-globin) (Figure 1). In the α-globin locus, the ζ-globin is expressed in the early stages of embryonic life and is subject to silencing within the first trimester. The expression of α-globin (HBA1/HBA2) increases from this stage onwards and continues through adulthood. In contrast, two switches occur on the β-globin locus during human development: the primitive to definitive, and the fetal to adult switches. The major embryonic β-like hemoglobin expressed within the primitive erythrocyte lineage is the ε-globin (HBE1) produced in the first trimester of gestation (Peschle et al. 1985). The primitive to definitive hemoglobin switch, present in all mammals, involves a transition from the embryonic primitive erythrocytes to the earliest definitive erythrocytes arising from the fetal liver (McGrath and Palis 2008). As ε-globin expression is silenced, and throughout the rest of the gestational period, γ-globin, encoded by two duplicated genes (HBG1 and HBG2) within the β-globin gene locus, takes over as the major β-like globin. γ-globin polypeptides associate with α-globin chains to form the fetal hemoglobin (HbF) tetramer ($\alpha_2\gamma_2$), the predominant form of hemoglobin present in fetal life. During birth and in the immediately following months, the second switch involving a transition from fetal to adult hemoglobin is completed (Stamatoyannopoulos 2005). γ-globin expression is effectively silenced, and adult β-globin (HBB) becomes the major β-like globin, so as to form adult hemoglobin (HbA, $\alpha_2\beta_2$). Simultaneous with these switches, the site of erythropoiesis also changes during ontogeny. While early hemoglobin is produced in the blood islands of the embryonic yolk sac, the site of synthesis shifts to the fetal liver and subsequently to the adult bone marrow around the time of birth.

Both types of hemoglobin switching have been historically studied as models for the control of gene regulatory processes. The mammalian β-globin loci were cloned early (Fritsch, Lawn, and Maniatis 1980, Leder et al. 1980), facilitating the understanding of control of gene expression from the β-globin loci. The locus control region (LCR) was identified as a powerful upstream enhancer cluster of the β- globin gene locus that drives strong expression of these genes (Tuan et al. 1985, Forrester et al. 1986, Grosveld et al. 1987, Bender et al. 2000). This LCR, with DNAseI hypersensitive elements in erythroid cells, functions as a super-enhancer and loops to downstream genes to selectively activate appropriate globin genes during human ontogeny. Multiple regulators of this locus were subsequently identified (Cantor and Orkin 2002). LIM domain-binding protein 1 (LDB1) mediated chromosomal looping provided evidence for models for chromosomal looping (Deng et al. 2014). The expression of the globin genes at specific stages during ontogeny are controlled by the binding of transcription factors at the promoters to engage the LCR. Krüppel-like Factor 1 (KLF1) was identified as the erythroid specific transcription factor that binds to the β-globin promoter and controls adult β-globin expression (Miller and Bieker 1993).

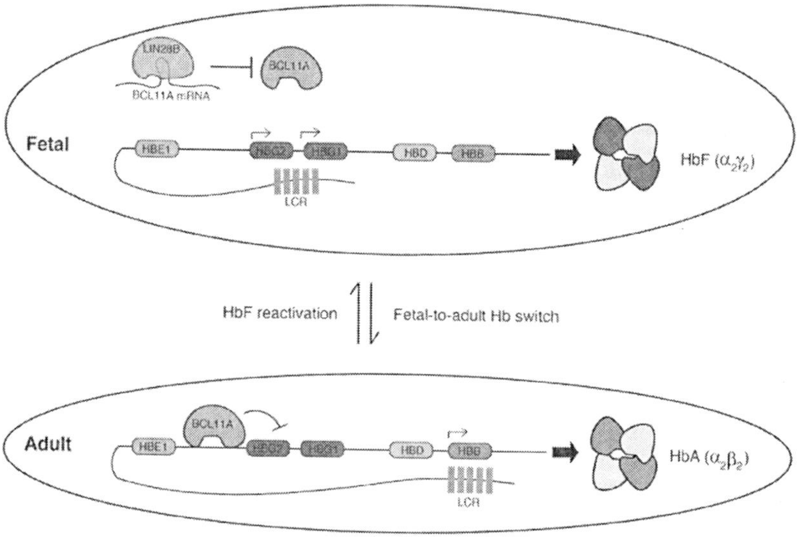

Figure 1. The Human β-globin locus and its Developmental Regulation by BCL11A and LIN28B. The β-globin locus, regulated by the upstream enhancer element Locus Control Region (LCR) is shown. Erythroid cells produced in the fetal liver express LIN28B, which suppresses translation of the key fetal hemoglobin repressor, BCL11A. Expression from the γ-globin genes therefore results in the formation of fetal hemoglobin (HbF). Adult erythroid cells lack LIN28B, and therefore, expression of BCL11A results in the suppression of γ-globin. Adult hemoglobin (HbA) is expressed instead.

Other key erythroid transcription factors involved in transcriptional activation are GATA1, Friend of GATA 1 (FOG1) and SCL/TAL1. However, the transcriptional control and molecular basis of the fetal hemoglobin switch had remained elusive until recent years. After decades of research, the factors required for fetal γ-globin silencing were identified through Genome Wide Association Studies (GWAS). Three regions of the genome were identified to be significant: the β-globin locus of chromosome 11p, the intergenic region between HBS1L and MYB on chromosome 6q, and the zinc-finger transcription factor B Cell Lymphoma 11A (BCL11A) on chromosome 2p (Lettre et al. 2008, Menzel et al. 2007, Uda et al. 2008). While the first two regions had previously been implicated to result in elevated HbF through linkage studies and naturally occurring deletions (Thein et al. 2007), BCL11A was a new candidate for a potential role in fetal hemoglobin regulation. Follow up studies over the past decade have validated it as a major repressor of γ-globin genes (Sankaran et al. 2008, Xu et al. 2010, Bauer et al. 2013, Sankaran et al. 2009, Funnell et al. 2015).

Knockout studies in HUDEP-2 cells and transgenic mice identified the Zinc Finger and BTB Domain Containing 7A (ZBTB7A, also called LRF) as an important fetal globin repressor (Masuda et al. 2016). These two factors contribute as major regulators of the fetal hemoglobin switch. Simultaneous depletion of BCL11A and ZBTB7A results in greater HbF induction than either factor alone, suggesting that they act in independent pathways to repress HbF.

The Role of Molecular Factors in the Regulation of Gamma Globin

BCL11A

The transcription factor BCL11A was initially identified to be required for normal lymphoid development (Liu et al. 2003). Evidence that it may be important in the red blood cell lineage came from the GWAS studies of fetal hemoglobin levels done in humans (Lettre et al. 2008, Menzel et al. 2007, Uda et al. 2008). Multiple studies have now established BCL11A as a major HbF repressor. BCL11A interacts with the NuRD chromatin remodeling complex and the transcription factors FOG1 and GATA1 in erythroid progenitor cells. Knockdown (by short hairpin RNAs, shRNAs) or knockout (by clustered regularly interspaced short palindromic repeats, CRISPR/Cas9) of BCL11A results in increased abundance of HbF, without significant perturbation to erythroid differentiation (Sankaran et al. 2008).

BCL11A occupies several distinct regions within the adult erythroid chromatin, including sites within the LCR and downstream of HBG1. These results demonstrate the importance of BCL11A in directly establishing and maintaining HbF silencing (Sankaran et al. 2008). Studies on transgenic mice have shown that BCL11A restricts the β-like embryonic globin expression to the primitive lineage in mouse, and fetal hemoglobin expression to the fetal and newborn lineage in humans (Sankaran et al. 2009). In BCL11A knockout mouse embryos containing the entire human β-globin locus and its cis-regulatory sequences as a yeast artificial chromosome clone (β-YAC mice), HbF silencing is strongly compromised. Erythroid-specific conditional inactivation of BCL11A in genetically engineered SCD mice rescues SCD phenotypes in the disease model due to HbF reactivation (Sankaran et al. 2008, Xu et al. 2011). Haploinsufficiency of BCL11A has been studied in humans who harbor rare microdeletions of 2p15-16.5 (Basak et al. 2015, Funnell et al. 2015). Importantly, HbF levels are elevated without perturbation of other hematological parameters in such instances. These results confirm the importance of targeting BCL11A for selective reactivation of HbF through biochemical or gene therapy interventions. However, BCL11A is required in hematopoietic stem cells and early hematopoietic progenitors (Tsang et al. 2015, Brendel et al. 2016). Hence, erythroid specific regulatory element-driven shRNAs were used to knockdown BCL11A, and this resulted in a significant induction of HbF without an adverse effect on erythroid development (Brendel et al. 2016). These findings have set the stage for ongoing clinical trials to investigate BCL11A shRNAs as potential therapeutic intervention in patients with β-hemoglobinopathies.

The ~12kb region of intron 2 of the BCL11A gene is densely populated with trait-associated single nucleotide polymorphisms (SNPs) identified by GWAS (Bauer et al. 2013, Canver et al. 2015). This is an important erythroid specific enhancer element, with erythroid transcription factor binding sites, histone modifications and DNaseI hypersensitivity hallmarks. This region is able to drive erythroid-specific expression from a minimal promoter in transgenic mice (Bauer et al. 2013). Deletion of this region with Zn-finger TALENS or CRISPR/Cas9 in multiple cell lines reduces the expression of BCL11A. This enhancer region contains three DNAseI hypersensitive sites (DHS): +55, +58 and +62 kilobases of the transcriptional start site (TSS). A causal SNP (rs1427407), lying within a conserved E-box/GATA site, is reported to be most correlated with high HbF, as well as associated with increased HbF in a SCD cohort (Bauer et al. 2013). Allele-specific chromatin immunoprecipitation (ChIP) shows decreased GATA1 and TAL1 binding to the high HbF allele. These results suggest that this enhancer region can be potentially targeted for selective downregulation of BCL11A in erythroid lineages. A CRISPR/Cas9 saturation mutagenesis screen in HUDEP-2 (human erythroblast cell line) determined the susceptible regions to be located in a small region within the +58 DHS, while the +62 DHS showed less potency. In mouse, however, the +62 DHS shows enhanced potency compared to +58 DHS (Canver et al. 2015). This +58 DHS in humans also exemplifies an attractive candidate for therapeutic modulation, since their disruption does not impair hematopoietic stem cell or B-cell development, but is rather restricted to red blood lineage.

BCL11A exhibits stage-specific expression during human ontogeny: it is sparsely expressed in fetal and embryonic cells that robustly express HbF, while BCL11A expression peaks in adult cells that have low levels of HbF. The mechanisms underlying this developmental expression have only recently been uncovered. BCL11A has been shown to be regulated at the level of mRNA translation during human hemopoietic development by the RNA binding protein LIN28B (Basak et al. 2020). LIN28B is developmentally expressed in a pattern reciprocal to BCL11A and directly interacts with *BCL11A* mRNA to suppress its translation. Such LIN28B-mediated suppression of BCL11A protein expression, leading to HbF induction, was also demonstrated to be independent of the well-documented role of LIN28B in regulating the let-7 family of miRNAs. Additionally, a LIN28B-binding site was identified in the 3' coding region of the *BCL11A* mRNA, which is necessary for suppression of BCL11A translation and subsequent HbF reactivation. These recent studies have therefore illuminated additional pathways and molecular targets for potential therapeutic interventions.

LRF/ZBTB7A/Pokemon

Leukemia/Lymphoma-Related Factor (LRF) is a C2H2 zinc finger transcription factor with identified roles in T cell, B cell, osteoclast and erythroid development (Maeda 2016). LRF has been identified as an effective repressor of embryonic and fetal hemoglobin from studies in HUDEP-2, cultured $CD34^{+}$ HSPCs and mice (Masuda et al. 2016). LRF directly represses the chromatin accessibility of the γ-globin genes, explaining why knockdown of LRF in cultured erythroblasts results in elevated HbF. Ablation of LRF results in delayed erythroid differentiation in cultured erythroblasts, while LRF knockout mice show embryonic lethal phenotypes due to severe anemia (Maeda et al. 2009). Of note, no rare variants from either patients or GWAS studies have been reported to be associated with LRF. There is significant interest in determining if a partial depletion of LRF could result in elevated HbF without adversely affecting normal erythropoiesis.

GATA1

The dual zinc finger motif harboring master transcription factor GATA1 is a critical determinant of erythrocyte development and globin gene expression. *Cis* regulatory elements of globin promoters and enhancers contain the consensus GATA motif (A/TGATAA/G) (Evans, Reitman, and Felsenfeld 1988). GATA1 binds to the β-globin locus (Evans and Felsenfeld 1989, Martin, Tsai, and Orkin 1989, Tsai et al. 1989) to establish the erythrocyte lineage, as well as regulates the expression of multiple genes that control erythrocyte function (Fujiwara et al. 2009, Cheng et al. 2009). Cells depleted of GATA1 fail to initiate erythroid-specific lineage commitment and lose the capacity to survive (Pevny et al. 1995, Weiss and Orkin 1995, Blobel and Orkin 1996, Fujiwara et al. 1996), and GATA1 mutations in humans result in dyserythropoietic anemia and thrombocytopenia (Nichols et al. 2000). GATA1 has been shown to activate mouse *Hbb-b1* expression through the recruitment RNA polymerase II to the LCR and the *Hbb-b1* promoter (Johnson et al. 2001, Johnson et al. 2002, Johnson et al. 2003). Out of the two zinc fingers of GATA1, the one closer to the carboxyl terminus determines the DNA sequence specificity (Martin and Orkin 1990, Merika and Orkin 1993, Ko and Engel 1993), while the other one closer to the amino terminus of GATA1 mediates binding to the transcriptional co-regulator FOG1 (Friend of GATA1) (Tsang et al. 1997, Crispino et al. 1999). FOG1 interacts with the NuRD remodeling complex (Hong et al. 2005, Gregory et al. 2010, Miccio and Blobel 2010, Miccio et al. 2010) to mediate globin gene expression. GATA1, mediated by FOG1, binds regions upstream of *HBG1* (-566 position) and *HBG2* (-567 position) promoter to initiate HbD silencing via the NuRD complex (Harju-Baker et al. 2008).

KLF1

The zinc finger transcription factor KLF1 was discovered as one of the earliest activators of adult HBB (Miller and Bieker 1993). Studies on *klf1*-null mice suggest that it is active in both in the adult and in the primitive stages of erythropoiesis (Donze, Townes, and Bieker 1995, Perkins, Gaensler, and Orkin 1996, Wijgerde et al. 1996, Gillemans et al. 1998). Naturally occurring KLF1 genetic variants have been associated with elevation in HbF (Borg et al. 2010) while human KLF1 variants have a broad spectrum of erythroid abnormalities (Perkins et al. 2016). KLF1 has a widespread role in regulating both primitive and definitive erythropoiesis (Nuez et al. 1995, Perkins, Sharpe, and Orkin 1995, Hodge et al. 2006). It occupies the HS1-HS3 of the human HBB LCR and the *Hbb-b1* promoter in mouse cell lines (Im et al. 2005) such that DNAseI hypersensitivity at these sites are lost with *Klf1* disruption (Tewari et al. 1998). KLF1 is postulated to regulate the chromatin structure at the HBB locus through the formation of a higher-order chromatin loop (Drissen et al. 2004). However, a direct role of KLF1 in human hemoglobin switching remains to be addressed.

MYB

Multiple variants at the *HBS1L-MYB* intergenic region on chromosome 6q23 that are associated with HbF levels have been identified through GWAS studies (Lettre et al. 2008, Uda et al. 2008). Within this locus, MYB is the causal gene for HbF induction, since fine mapped *MYB* coding variants (Galarneau et al. 2010) as well as *in vitro* MYB knockdown in primary erythroid cells result in increased HbF expression. However, MYB is also associated with multiple other erythroid parameters (Ganesh et al. 2009, Soranzo et al. 2009) and thus has a comprehensive role as a master regulator of erythropoiesis and hematopoiesis.

CIS-ACTING ELEMENTS

As mentioned earlier, high level expression of the globin genes is facilitated by the LCR region located 6-20 kb upstream of the HBE1. Mediated by transcription factors KLF1 and GATA1 (Drissen et al. 2004, Vakoc et al. 2005) and multiple protein complexes, the LCR interacts with the globin gene promoters, sequentially and one at a time, to regulate expression (Tolhuis et al. 2002). Looping of the LCR with the promoter is facilitated by cofactors such as LDB1 (Song, Hou, and Dean 2007) and LMO2 (Wadman et al. 1997) via interactions with erythroid transcription factors GATA1 and TAL1 (Yun et al. 2014), and the chromatin-structure regulating transcription factor CTCF (Lee et al. 2017). Introduction of a fusion construct containing the LDB1 dimerization domain and a zinc-finger that binds strictly to HBG promoter sequences resulted in reduced sickling and increased HbF expression in SCD-patient derived erythroblasts, through the forced recruitment of the LCR to the HBG promoter and a simultaneous decrease of HBB promoter interaction (Deng et al. 2012, Deng et al. 2014).

Early studies on naturally occurring mutations and deletions in β-thalassemia patients identified deletions related to δβ-thalassemia and Hereditary Persistence of Fetal Hemoglobin (HPFH) (Forget 1998). Cas9-mediated 13kb deletion mimicking a naturally occurring HPFH deletion resulted in elevated HbF (Ye et al. 2016).

Mutations in HBG promoters result in elevated HbF in non-deletional forms of HPFH (Huisman, Miller, and Schroeder 1975, Weatherall et al. 1975). *In vitro* gene editing at -115 of the HBG promoter, involving the naturally-occurring HPFH mutations, results in increased HbF expression in HUDEP-2 cells and primary cell culture (Traxler et al. 2016), likely through the *in vitro* disruption of CCAAT box elements in this region which are likely involved in fetal hemoglobin regulation (Martyn, Quinlan, and Crossley 2017). Similarly, TALEN-mediated -175T>C mutation in HBG2 promoter and the -198T>C mutation in the HBG1 promoter results in elevated HbF through increased binding of the transcription factors TAL1 and KLF1, respectively (Wienert et al. 2015, Wienert et al. 2017).

EPIGENETIC CONTROL

Epigenetic mechanisms (Such as methylation, acetylation etc.) have long been recognized as an important regulatory mediator of HbF silencing. This interest primarily stems from usage of some drugs, like hydroxyurea, azacytidine, decitabine etc. that can induce HbF production by epigenetic modulation and thus hold promise to improve the clinical symptoms of β-hemoglobinopathies like sickle cell disease and β-thalassemia (Mettananda et al. 2015).

Methylation

DNA Methylation

DNA methylation is a process in which methyl groups are added to the DNA molecule without changing the DNA sequence. It is one of the best known epigenetic processes for human globin gene regulation (Kiefer et al. 2009). DNA methylation is one of the most studied molecular processes that are known to repress gene transcription (Jin et al. 2011). Various DNA methyltransferases (DNMTs) such as DNMT3A, DNMT3B, DNMT1 etc. which are involved in this process are widely studied and well characterized (Dean 2014). Ginder (2015) studied DNMT1 as a target for reversing globin gene expression and used 5-azacytidine as an inhibitor of DNMT1. Concerns about the safety issues with this type of DNMT1 inhibitors have restricted its clinical applications for the treatment of β-hemoglobinopathies (Figure 2). Experimental evidences have suggested that DNA methylation and another epigenetic process known as histone modification are closely associated with each other (Cheng 2014). However, a different study has revealed that DNA methylation played a pivotal role in activation of methylcytosine-binding proteins and subsequently recruits co-repressor complexes containing histone modifying enzymes which are responsible for the histone methylation phenomenon (Javaid et al. 2017).

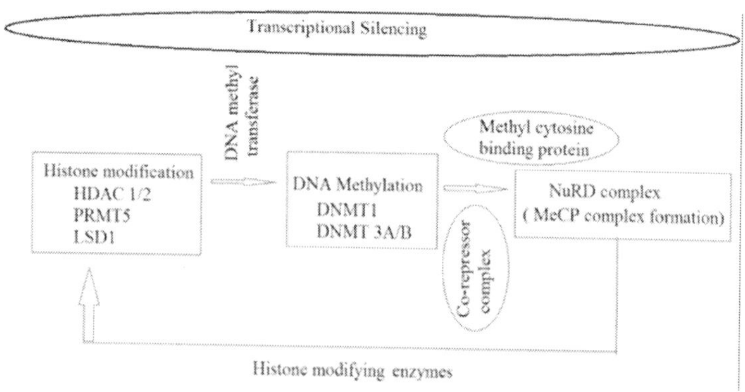

Figure 2. Epigenetic gene silencing: an interdependent phenomenon.

Methylated CpGs and HbF

CpG islands are specific regions of the genome which are distinguished by the high prevalence of CpG sites ("C" nucleotide is followed by a "G" nucleotide) and widely distributed throughout the genome. Methylation of CpGs by DNMTs can lead to changes in gene expression patterns, leading to various patho-physiological conditions (Long et al. 2017). Consensus binding sites of different transcription factors near the CpGs tend to be more methylated when a transcription factor is less expressed or not expressed at all. Researchers have also explored different transcription factor binding sites within different loci which are important for erythroid development (Zhu et al. 2016). A comparative study of the hypomethylated CpGs between fetal and adult erythroblast has identified some of the most relevant transcription factor binding sites (TFBS) for erythroid development (Lessard et al. 2015). Various differentially methylated CpGs in TFBS are mentioned below (Table 1).

Table 1. List of TFBS near differentially methylated CpGs. Motif name and their relevant consensus sequences are mentioned

Motif name	Consensus sequence	Remarks
SOX2	BCCATTGTTC	Transcription factor that forms a complex on DNA and controls the expression of a number of genes involved in embryonic development. Critical for early embryogenesis and for embryonic stem cell pluripotency.
TCF3	ASWTCAAAGG	Transcription factor which enhances Hes1 expression that is involved in Notch signaling.
SOX6	CCATTGTTNY	SOX is involved in developmental programs
GATA1	AGATGKDGAGATAAG	GATA-1 regulates the transcription of erythroid genes and is involved in HbF to HbA switching.
MAZ	GGGGGGGG	MAZ is involved in *chromatin binding* phenomenon.
TCF4	ASATCAAAGGVA	TCF4 is involved in neural differentiation by binding to DNA during pregnancy.
FOXA1	WAAGTAAACA	It is a transcriptional activator and involved in embryonic development, tissue-specific gene expression and regulation

Methylcytosine Binding Domain Proteins

Another group of proteins that are critical for the methylation of β globin gene is commonly known as Methyl-cytosine Binding Domain (MBD). MBD proteins are usually categorized as four groups: MBD1, MBD2, MECP2 and MBD4 which are responsible for globin gene silencing in various disease condition (Hendrich et al. 1998). Rupon et al. (2006) have documented higher expression of γ-globin gene after knockdown of MBD2 in experiments with transgenic mice.

γ-Globin Gene Expression and Promoter Methylation

Experimental evidences indicate towards a progressive demethylation of the γ-globin promoter region in Hematopoietic Stem and Progenitor cells (HSPCs) during their differentiation (Singh et al. 2007). This indicates that demethylation of the promoter site is essential for the development of the γ-globin gene. Recent finding shows that the methylation by DNMT1 in the promoter region of the γ-globin gene is very important for its fate determination. Consequently, DNMT inhibitors have been successfully investigated as therapeutic molecules for HbF induction. However, their usage as HbF inducers has been limited due to their side-effects.

Histone Modification

Histone Acetylation

Histone acetylation is an important process in gene regulation wherein lysine in the N-terminal part of the histone octamer becomes acetylated. A dynamic action played by two functionally contrasting enzymes, Histone acetyltransferase (HAT) and histone deacetylase (HDAC) helps in maintaining stability of the genome and normal physiological process (Kuo et al. 1998). In 2007 Fathallah and his group found a positive correlation between histone acetylation and HbF expression. Histone deacetylase (HDAC) inhibitors prevent the deacetylation process in the promoter region (Bassett et al. 2014) and are therefore helpful in HbF induction. Regulation of the γ-globin genes by HDACs is not entirely explored till now. However, it has been found that the HDAC inhibitors like butyrate and trichostatin-A initiate γ-globin expression through a p38/MAPK pathway (Witt et al. 2003). Further, Sangerman et al. (2006) found that Activating Transcription Factor-2 (ATF-2) of p38/MAPK pathway and cyclic AMP reaction component restricting protein (CREB), are associated with HbF induction which leads to the conclusion that HbF induction by butyrate and trichostatin-A requires activation of p38/MAPK pathway that can be targeted in various therapeutic as well as clinical intervention.

Histone Methylation

In contrast to histone acetylation, histone methylation can either activate or silence various genes of physiological importance. Histone H3K4me3 (tri-methylation at the 4th Lysine residue of the 3rd Histone protein) methylation is typically associated with regulation of gene expression while histone H3K9 (methylation at the 9th Lysine residue of the 3rd Histone protein) and H3K27me3 (tri-methylation at the 27th Lysine residue of the 3rd Histone protein) methylation are related with gene silencing (Miller et al. 2013). A different study has showed protein arginine methyltransferases (PRMT) and lysine-specific methylases also play vital roles in histone methylation. Researchers have identified a PRMT5-dependant protein complex which is intricately associated in regulation of human γ-globin gene expression and fetal hemoglobin silencing (Rank et al. 2010). Similar study involving Adenosine-2′, 3′ dialdehyde (Adox), a methyltransferase inhibitor has shown the potentiality to enhance γ-globin gene expression by silencing the PRMT5 enzymatic activity (Ginder 2015). Another methyltransferase PRMT1, in association with FoP (friend of PRMT1) protein regulates γ-globin gene silencing and subsequent knockdown of FoP has been shown to increase the γ-globin gene expression in *in vitro* cell culture of erythroid cells (Ginder 2015).

Recently researchers have discovered specific lysine demethylases which are associated with γ-globin gene silencing in human as well as murine erythroid cells (Xu et al. 2013). LSD1, a Lysine-explicit demethylase 1 has been found to associate with BCL11A through a complex containing the CoREST (component 1 quieting variable corepressor-1) repressor (Xu et al. 2013). Cui and his colleagues (2015) found that LSD1 is associated with the TR2/TR4/DRED complex, along with other co-repressor complexes. Inhibition of LSD1 in β-globin locus-inserted transgenic mice results in increased γ -globin gene expression (Shi et al. 2013).

Together with DNA methyltransferase 3A (DNMT3A), the arginine methyltransferase PRMT5 is involved in repression of γ-globin (Zhao et al. 2009). Histone methyltransferases, EHMT1 (GLP) and EHMT2 (G9a) carry out the demethylation of histone H3 lysine 9 (H3K9me2) and mediate epigenetic silencing of gene expression. Through large-scale proteomic screens, they were identified as potent γ-globin silencers (Chaturvedi et al. 2009, Krivega et al. 2015, Renneville et al. 2015).

Nucleosome Remodeling and Deacetylase Complex

The Nucleosome Remodeling and Deacetylase (NuRD) complex has been found to be involved in Methyl-cytosine Binding Domain-2 (MBD2) mediated gene silencing process (Feng et al. 2001). The MBD2 protein binds to DNA containing methylated CpG rich sequences with high affinity and recruits other members of NuRD co-repressor complex through specific protein-protein interactions (Wood et al. 2016). Five different types of protein subunits, forming the complete NuRD complex (Zhang et al. 2016) are (i) histone deacetylase core proteins (e.g., HDAC1 and HDAC2) (ii) histone-binding proteins (e.g., RbAp46 and RbAp48) (iii) metastasis-associated proteins (e.g., MTA1) (iv) methyl CpG binding domain protein (e.g., MBD2 or MBD3) and (v) chromodomain helicase DNA binding protein (e.g., CHD3 or CHD4). A close association of transcription factors like GATA1 with the MBD-NuRD complex is shown in the Figure 3.

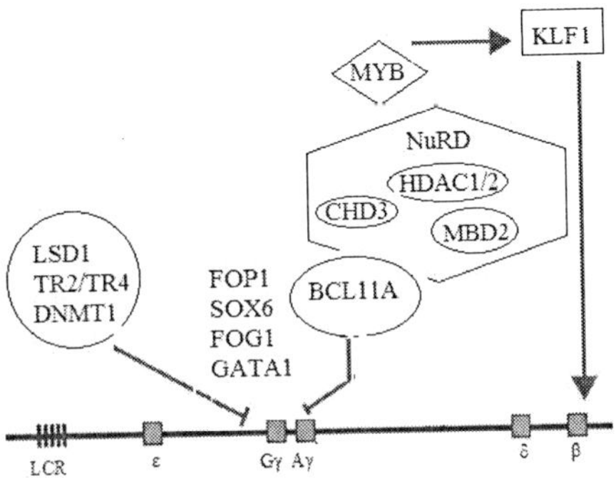

Figure 3. Schematic network diagram of γ -globin regulators in adult life.

NuRD has been implicated in HbF repression both in chicken and mammalian erythroid cells (Amaya et al. 2013, Gnanapragasam et al. 2011), and shRNA mediated knockdown of NuRD components in mice and primary human erythroid cultures results in HbF reactivation. The deletion of CHD4 (Costa et al. 2012, Amaya et al. 2013, Xu et al. 2013) as well as HDAC1/2 (Shearstone et al. 2016) result in upregulated γ-globin. Both the transcription factors BCL11A and LRF interact with components of the NuRD complex, suggesting that NuRD may be the converging point for both BCL11A- and LRF-pathways to result in HbF repression (Xu et al. 2013, Masuda et al. 2016). However, BCL11A and LRF are not found to co-occupy the same complex, suggesting they may associate with distinct NuRD sub-complexes.

MBD2 contains a conserved methyl-CpG binding domain (MBD), which binds to methylated DNA (Nan, Meehan, and Bird 1993). MBD2, as a part of the methyl-CpG binding complex 1 (MeCP1) (Feng and Zhang 2001), recognizes DNA methylation at the γ-globin locus, and recruits the nucleosome remodeling deacetylase (NuRD) silencing complex (Gnanapragasam et al. 2011). *Mbd2*-null mice with β-YAC have elevated HBG1/G2, reiterating the importance of MBD2 and DNA methylation in silencing HBG1/G2 (Rupon et al. 2006). However, while both increased DNA methylation and the HbF induction observed in these mice is only moderate, MBD2 may not directly bind to the *HBG1/G2* promoter (Rupon et al. 2006).

HbF regulation involves BCL11A, KLF1, MYB, nuclear receptors TR2/TR4 (Suzuki et. al. 2014). These are associated with DNMT1 and LSD1 (Suzuki et. al. 2014).

CHROMATIN REMODELING

Ikaros-PYR Complex

The chromatin remodeling PYR complex (Figure 4), present in adult mouse and human HSCs, binds to a 250 bp polypyrimidine stretch upstream of the human HBD gene (O'Neill et al. 1999, Bank 2006). It consists of the SWI/SNF and NuRD complexes and has Ikaros as its DNA-binding subunit. Studies have reported that Ikaros null mice exhibit delayed fetal to adult hemoglobin switching and thereby highlight its importance in regulation of γ-globin (Lopez et al. 2002, Keys et al. 2008, Bottardi et al. 2009). These mice, however, also exhibit multiple other erythropoietic defects and thus direct effects on erythropoiesis are unclear (Narlikar, Fan, and Kingston 2002)

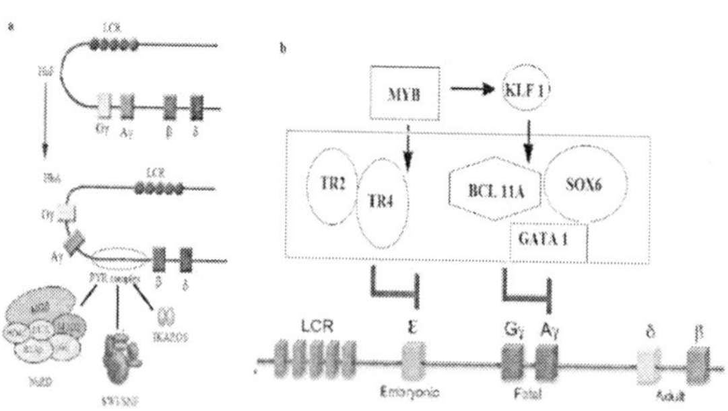

Figure 4. (a). Human hemoglobin switching and PYR complex. The SWI/SNF complex subunits, the NuRD subunits, and DNA-binding subunit Ikaros of the PYR complex are shown in the figure. (b). A current interacting model of HbF regulation is shown in the figure. TR2/TR4 and GATA1 repress HbF genes directly and MYB represses indirectly. KLF1 activates BCL11A and represses HbF genes after interacting with SOX6.

BRG1 (SWI/SNF)

The ATP dependent chromatin remodeling complexes, such as the SWI/SNF, are specialized machineries that utilize ATP to restructure the nucleosome to make the DNA accessible. Although the exact mechanisms are unclear, this complex is implicated to regulate globin gene expression (Lopez et al. 2002, O'Neill et al. 1999). The ATPase catalytic subunit of this complex, BRG1 (Brahma-related Gene 1) occupies HS2-4 and *Hbb-b1* promoter in mice and GATA1-mediated increase in BRG1 occupancy at some of these sites has been reported (Im et al. 2005). BRG1 interacts with key erythroid factors such as KLF1 (Armstrong et al. 1998; Kadam et al. 2000), the MafK complex in murine erythroleukemia (MEL) cells and the β-globin LCR associated chromatin remodeling complex LARC (Brand et al. 2004; Mahajan et al. 2005).

Conclusion

Fetal hemoglobin silencing and the switch to adult hemoglobin is indeed a complex process involving multiple layers of complexities. This chapter has attempted to cover the most important players. Recent discoveries in the field have been able to provide us with multiple insights into this complex process. This chapter discusses the key players that have been found to play an instrumental role in HbF and γ globin silencing. While we have unearthed a lot of information on this, we must acknowledge the fact that many more unknowns are yet to be discovered.

References

Amaya, M., M. Desai, M. N. Gnanapragasam, S. Z. Wang, S. Zu Zhu, D. C. Williams, Jr., and G. D. Ginder. 2013. "Mi2beta-mediated silencing of the fetal gamma-globin gene in adult erythroid cells." *Blood* 121 (17):3493-501. doi: 10.1182/blood-2012-11-466227.

Bank, A. 2006. "Regulation of human fetal hemoglobin: new players, new complexities." *Blood* 107 (2):435-43. doi: 10.1182/blood-2005-05-2113.

Basak, A., M. Hancarova, J. C. Ulirsch, T. B. Balci, M. Trkova, M. Pelisek, M. Vlckova, K. Muzikova, J. Cermak, J. Trka, D. A. Dyment, S. H. Orkin, M. J. Daly, Z. Sedlacek, and V. G. Sankaran. 2015. "BCL11A deletions result in fetal hemoglobin persistence and neurodevelopmental alterations." *J Clin Invest* 125 (6):2363-8. doi: 10.1172/JCI81163.

Basak, A., M. Munschauer, C. A. Lareau, K. E. Montbleau, J. C. Ulirsch, C. R. Hartigan, M. Schenone, J. Lian, Y. Wang, Y. Huang, X. Wu, L. Gehrke, C. M. Rice, X. An, H. A. Christou, N. Mohandas, S. A. Carr, J. J. Chen, S. H. Orkin, E. S. Lander, and V. G. Sankaran. 2020. "Control of human hemoglobin switching by LIN28B-mediated regulation of BCL11A translation." *Nat Genet* 52 (2):138-145. doi: 10.1038/s41588-019-0568.

Bassett AS and Matthew BPG. 2014. The Role of Dietary Histone Deacetylases (HDACs) Inhibitors in Health and Disease. *Nutrients.* 6(10): 4273–4301.

Bauer, D. E., S. C. Kamran, S. Lessard, J. Xu, Y. Fujiwara, C. Lin, Z. Shao, M. C. Canver, E. C. Smith, L. Pinello, P. J. Sabo, J. Vierstra, R. A. Voit, G. C. Yuan, M. H. Porteus, J. A. Stamatoyannopoulos, G. Lettre, and S. H. Orkin. 2013. "An erythroid enhancer of BCL11A subject to genetic variation determines fetal hemoglobin level." *Science* 342 (6155):253-7. doi: 10.1126/science.1242088.

Bender, M. A., M. Bulger, J. Close, and M. Groudine. 2000. "Beta-globin gene switching and DNase I sensitivity of the endogenous beta-globin locus in mice do not require the locus control region." *Mol Cell* 5 (2):387-93. doi: 10.1016/s1097-2765(00)80433-5.

Blobel, G. A., and S. H. Orkin. 1996. "Estrogen-induced apoptosis by inhibition of the erythroid transcription factor GATA-1." *Mol Cell Biol* 16 (4):1687-94. doi: 10.1128/mcb.16.4.1687.

Borg, J., P. Papadopoulos, M. Georgitsi, L. Gutierrez, G. Grech, P. Fanis, M. Phylactides, A. J. Verkerk, P. J. van der Spek, C. A. Scerri, W. Cassar, R. Galdies, W. van Ijcken, Z. Ozgur, N. Gillemans, J. Hou, M. Bugeja, F. G. Grosveld, M. von Lindern, A. E. Felice, G. P. Patrinos, and S. Philipsen. 2010. "Haploinsufficiency for the erythroid transcription factor KLF1 causes hereditary persistence of fetal hemoglobin." *Nat Genet* 42 (9):801-5. doi: 10.1038/ng.630.

Bottardi, S., J. Ross, V. Bourgoin, N. Fotouhi-Ardakani, B. Affar el, M. Trudel, and E. Milot. 2009. "Ikaros and GATA-1 combinatorial effect is required for silencing of human gamma-globin genes." *Mol Cell Biol* 29 (6):1526-37. doi: 10.1128/MCB.01523-08.

Brendel, C., S. Guda, R. Renella, D. E. Bauer, M. C. Canver, Y. J. Kim, M. M. Heeney, D. Klatt, J. Fogel, M. D. Milsom, S. H. Orkin, R. I. Gregory, and D. A. Williams. 2016. "Lineage-specific BCL11A knockdown circumvents toxicities and reverses sickle phenotype." *J Clin Invest* 126 (10):3868-3878. doi: 10.1172/JCI87885.

Cantor, A. B., and S. H. Orkin. 2002. "Transcriptional regulation of erythropoiesis: an affair involving multiple partners." *Oncogene* 21 (21):3368-76. doi: 10.1038/sj.onc.1205326.

Canver, M. C., E. C. Smith, F. Sher, L. Pinello, N. E. Sanjana, O. Shalem, D. D. Chen, P. G. Schupp, D. S. Vinjamur, S. P. Garcia, S. Luc, R. Kurita, Y. Nakamura, Y. Fujiwara, T. Maeda, G. C. Yuan, F. Zhang, S. H. Orkin, and D. E. Bauer. 2015. "BCL11A enhancer dissection by Cas9-mediated in situ saturating mutagenesis." *Nature* 527 (7577):192-7. doi: 10.1038/nature15521.

Chaturvedi, C. P., A. M. Hosey, C. Palii, C. Perez-Iratxeta, Y. Nakatani, J. A. Ranish, F. J. Dilworth, and M. Brand. 2009. "Dual role for the methyltransferase G9a in the maintenance of beta-globin gene transcription in adult erythroid cells." *Proc Natl Acad Sci U S A* 106 (43):18303-8. doi: 10.1073/pnas.0906769106.

Cheng X. 2014. Structural and Functional Coordination of DNA and Histone Methylation. *Cold Spring Harb Perspect Biol.* 8(8): 196.

Cheng, Y., W. Wu, S. A. Kumar, D. Yu, W. Deng, T. Tripic, D. C. King, K. B. Chen, Y. Zhang, D. Drautz, B. Giardine, S. C. Schuster, W. Miller,

F. Chiaromonte, Y. Zhang, G. A. Blobel, M. J. Weiss, and R. C. Hardison. 2009. "Erythroid GATA1 function revealed by genome-wide analysis of transcription factor occupancy, histone modifications, and mRNA expression." *Genome Res* 19 (12):2172-84. doi: 10.1101/gr.098921.109.

Costa, F. C., H. Fedosyuk, A. M. Chazelle, R. Y. Neades, and K. R. Peterson. 2012. "Mi2beta is required for gamma-globin gene silencing: temporal assembly of a GATA-1-FOG-1-Mi2 repressor complex in beta-YAC transgenic mice." *PLoS Genet* 8 (12):e1003155. doi: 10.1371/journal.pgen.1003155.

Crispino, J. D., M. B. Lodish, J. P. MacKay, and S. H. Orkin. 1999. "Use of altered specificity mutants to probe a specific protein-protein interaction in differentiation: the GATA-1:FOG complex." *Mol Cell* 3 (2):219-28. doi: 10.1016/s1097-2765(00)80312-3.

Cui, S., K. C. Lim, L. Shi, M. Lee, N. Jearawiriyapaisarn, G. Myers, A. Campbell, D. Harro, S. Iwase, R. C. Trievel, A. Rivers, J. DeSimone, D. Lavelle, Y. Saunthararajah, and J. D. Engel. 2015. "The LSD1 inhibitor RN-1 induces fetal hemoglobin synthesis and reduces disease pathology in sickle cell mice." *Blood* 126 (3):386-96. doi: 10.1182/blood-2015-02-626259.

Dean W. 2014. DNA Methylation and Demethylation: A Pathway to Gametogenesis and Development. *Molecular Reproduction & Development.* 81:113–125.

Deng, W., J. Lee, H. Wang, J. Miller, A. Reik, P. D. Gregory, A. Dean, and G. A. Blobel. 2012. "Controlling long-range genomic interactions at a native locus by targeted tethering of a looping factor." *Cell* 149 (6):1233-44. doi: 10.1016/j.cell.2012.03.051.

Deng, W., J. W. Rupon, I. Krivega, L. Breda, I. Motta, K. S. Jahn, A. Reik, P. D. Gregory, S. Rivella, A. Dean, and G. A. Blobel. 2014. "Reactivation of developmentally silenced globin genes by forced chromatin looping." *Cell* 158 (4):849-860. doi: 10.1016/j.cell.2014.05.050.

Donze, D., T. M. Townes, and J. J. Bieker. 1995. "Role of erythroid Kruppel-like factor in human gamma- to beta-globin gene switching." *J Biol Chem* 270 (4):1955-9. doi: 10.1074/jbc.270.4.1955.

Drissen, R., R. J. Palstra, N. Gillemans, E. Splinter, F. Grosveld, S. Philipsen, and W. de Laat. 2004. "The active spatial organization of the beta-globin locus requires the transcription factor EKLF." *Genes Dev* 18 (20):2485-90. doi: 10.1101/gad.317004.

Evans, T., and G. Felsenfeld. 1989. "The erythroid-specific transcription factor Eryf1: a new finger protein." *Cell* 58 (5):877-85. doi: 10.1016/0092-8674(89)90940-9.

Evans, T., M. Reitman, and G. Felsenfeld. 1988. "An erythrocyte-specific DNA-binding factor recognizes a regulatory sequence common to all chicken globin genes." *Proc Natl Acad Sci U S A* 85 (16):5976-80. doi: 10.1073/pnas.85.16.5976.

Fathallah H., Weinberg S. R., Yelena G., Millicent S. and George A. F. 2007. Role of epigenetic modifications in normal globin gene regulation and butyrate-mediated induction of fetal hemoglobin. *Blood* 110 (110), 9: 3391- 3397.

Feng, Q., and Y. Zhang. 2001. "The MeCP1 complex represses transcription through preferential binding, remodeling, and deacetylating methylated nucleosomes." *Genes Dev* 15 (7):827-32. doi: 10.1101/gad.876201.

Forget, B. G. 1998. "Molecular basis of hereditary persistence of fetal hemoglobin." *Ann N Y Acad Sci* 850:38-44. doi: 10.1111/j.1749-6632.1998.tb10460.x.

Forrester, W. C., C. Thompson, J. T. Elder, and M. Groudine. 1986. "A developmentally stable chromatin structure in the human beta-globin gene cluster." *Proc Natl Acad Sci U S A* 83 (5):1359-63. doi: 10.1073/pnas.83.5.1359.

Fritsch, E. F., R. M. Lawn, and T. Maniatis. 1980. "Molecular cloning and characterization of the human beta-like globin gene cluster." *Cell* 19 (4):959-72. doi: 10.1016/0092-8674(80)90087-2.

Fujiwara, T., H. O'Geen, S. Keles, K. Blahnik, A. K. Linnemann, Y. A. Kang, K. Choi, P. J. Farnham, and E. H. Bresnick. 2009. "Discovering hematopoietic mechanisms through genome-wide analysis of GATA factor chromatin occupancy." *Mol Cell* 36 (4):667-81. doi: 10.1016/j.molcel.2009.11.001.

Fujiwara, Y., C. P. Browne, K. Cunniff, S. C. Goff, and S. H. Orkin. 1996. "Arrested development of embryonic red cell precursors in mouse embryos lacking transcription factor GATA-1." *Proc Natl Acad Sci U S A* 93 (22):12355-8. doi: 10.1073/pnas.93.22.12355.

Funnell, A. P., P. Prontera, V. Ottaviani, M. Piccione, A. Giambona, A. Maggio, F. Ciaffoni, S. Stehling-Sun, M. Marra, F. Masiello, L. Varricchio, J. A. Stamatoyannopoulos, A. R. Migliaccio, and T. Papayannopoulou. 2015. "2p15-p16.1 microdeletions encompassing and proximal to BCL11A are associated with elevated HbF in addition to neurologic impairment." *Blood* 126 (1):89-93. doi: 10.1182/blood-2015-04-638528.

Galarneau, G., C. D. Palmer, V. G. Sankaran, S. H. Orkin, J. N. Hirschhorn, and G. Lettre. 2010. "Fine-mapping at three loci known to affect fetal hemoglobin levels explains additional genetic variation." *Nat Genet* 42 (12):1049-51. doi: 10.1038/ng.707.

Ganesh, S. K., N. A. Zakai, F. J. van Rooij, N. Soranzo, A. V. Smith, M. A. Nalls, M. H. Chen, A. Kottgen, N. L. Glazer, A. Dehghan, B. Kuhnel, T. Aspelund, Q. Yang, T. Tanaka, A. Jaffe, J. C. Bis, G. C. Verwoert, A. Teumer, C. S. Fox, J. M. Guralnik, G. B. Ehret, K. Rice, J. F. Felix, A. Rendon, G. Eiriksdottir, D. Levy, K. V. Patel, E. Boerwinkle, J. I. Rotter, A. Hofman, J. G. Sambrook, D. G. Hernandez, G. Zheng, S. Bandinelli, A. B. Singleton, J. Coresh, T. Lumley, A. G. Uitterlinden, J. M. Vangils, L. J. Launer, L. A. Cupples, B. A. Oostra, J. J. Zwaginga, W. H. Ouwehand, S. L. Thein, C. Meisinger, P. Deloukas, M. Nauck, T. D. Spector, C. Gieger, V. Gudnason, C. M. van Duijn, B. M. Psaty, L. Ferrucci, A. Chakravarti, A. Greinacher, C. J. O'Donnell, J. C. Witteman, S. Furth, M. Cushman, T. B. Harris, and J. P. Lin. 2009. "Multiple loci influence erythrocyte phenotypes in the CHARGE Consortium." *Nat Genet* 41 (11):1191-8. doi: 10.1038/ng.466.

Gillemans, N., R. Tewari, F. Lindeboom, R. Rottier, T. de Wit, M. Wijgerde, F. Grosveld, and S. Philipsen. 1998. "Altered DNA-binding specificity mutants of EKLF and Sp1 show that EKLF is an activator of the beta-globin locus control region in vivo." *Genes Dev* 12 (18):2863-73. doi: 10.1101/gad.12.18.2863.

Ginder D. G. 2015. Epigenetic Regulation of Fetal Globin Gene Expression in Adult Erythroid Cells. *Transl Res.* 165(1): 115–125.

Gnanapragasam, M. N., J. N. Scarsdale, M. L. Amaya, H. D. Webb, M. A. Desai, N. M. Walavalkar, S. Z. Wang, S. Zu Zhu, G. D. Ginder, and D. C. Williams, Jr. 2011. "p66Alpha-MBD2 coiled-coil interaction and recruitment of Mi-2 are critical for globin gene silencing by the MBD2-NuRD complex." *Proc Natl Acad Sci U S A* 108 (18):7487-92. doi: 10.1073/pnas.1015341108.

Gregory, G. D., A. Miccio, A. Bersenev, Y. Wang, W. Hong, Z. Zhang, M. Poncz, W. Tong, and G. A. Blobel. 2010. "FOG1 requires NuRD to promote hematopoiesis and maintain lineage fidelity within the megakaryocytic-erythroid compartment." *Blood* 115 (11):2156-66. doi: 10.1182/blood-2009-10-251280.

Grosveld, F., G. B. van Assendelft, D. R. Greaves, and G. Kollias. 1987. "Position-independent, high-level expression of the human beta-globin gene in transgenic mice." *Cell* 51 (6):975-85. doi: 10.1016/0092-8674(87)90584-8.

Harju-Baker, S., F. C. Costa, H. Fedosyuk, R. Neades, and K. R. Peterson. 2008. "Silencing of Agamma-globin gene expression during adult definitive erythropoiesis mediated by GATA-1-FOG-1-Mi2 complex binding at the -566 GATA site." *Mol Cell Biol* 28 (10):3101-13. doi: 10.1128/MCB.01858-07.

Hendrich B. and Bird A. 1998. Identification and Characterization of a Family of Mammalian Methyl-CpG Binding Proteins. *Mol Cell Biol.* 18(11): 6538–6547.

Hodge, D., E. Coghill, J. Keys, T. Maguire, B. Hartmann, A. McDowall, M. Weiss, S. Grimmond, and A. Perkins. 2006. "A global role for EKLF in definitive and primitive erythropoiesis." *Blood* 107 (8):3359-70. doi: 10.1182/blood-2005-07-2888.

Hong, W., M. Nakazawa, Y. Y. Chen, R. Kori, C. R. Vakoc, C. Rakowski, and G. A. Blobel. 2005. "FOG-1 recruits the NuRD repressor complex to mediate transcriptional repression by GATA-1." *EMBO J* 24 (13):2367-78. doi: 10.1038/sj.emboj.7600703.

Huisman, T. H., A. Miller, and W. A. Schroeder. 1975. "A G gamma type of the hereditary persistence of fetal hemoglobin with beta chain production in cis." *Am J Hum Genet* 27 (6):765-77.

Im, H., J. A. Grass, K. D. Johnson, S. I. Kim, M. E. Boyer, A. N. Imbalzano, J. J. Bieker, and E. H. Bresnick. 2005. "Chromatin domain activation via GATA-1 utilization of a small subset of dispersed GATA motifs within a broad chromosomal region." *Proc Natl Acad Sci U S A* 102 (47):17065-70. doi: 10.1073/pnas.0506164102.

Javaid N. and Sangdun C. 2017. Acetylation- and Methylation-Related Epigenetic Proteins in the Context of Their Targets. *Genes.* 8(196)1-37. doi:10.3390/genes8080196.

Jin B., Yajun L. and Robertson D. 2011. Keith DNA Methylation Superior or Subordinate in the Epigenetic Hierarchy? *Genes Cancer*. 2(6): 607–617.

Johnson, K. D., H. M. Christensen, B. Zhao, and E. H. Bresnick. 2001. "Distinct mechanisms control RNA polymerase II recruitment to a tissue-specific locus control region and a downstream promoter." *Mol Cell* 8 (2):465-71. doi: 10.1016/s1097-2765(01)00309-4.

Johnson, K. D., J. A. Grass, M. E. Boyer, C. M. Kiekhaefer, G. A. Blobel, M. J. Weiss, and E. H. Bresnick. 2002. "Cooperative activities of hematopoietic regulators recruit RNA polymerase II to a tissue-specific chromatin domain." *Proc Natl Acad Sci U S A* 99 (18):11760-5. doi: 10.1073/pnas.192285999.

Johnson, K. D., J. A. Grass, C. Park, H. Im, K. Choi, and E. H. Bresnick. 2003. "Highly restricted localization of RNA polymerase II within a locus control region of a tissue-specific chromatin domain." *Mol Cell Biol* 23 (18):6484-93. doi: 10.1128/mcb.23.18.6484-6493.2003.

Keys, J. R., M. R. Tallack, Y. Zhan, P. Papathanasiou, C. C. Goodnow, K. M. Gaensler, M. Crossley, J. Dekker, and A. C. Perkins. 2008. "A mechanism for Ikaros regulation of human globin gene switching." *Br J Haematol* 141 (3):398-406. doi: 10.1111/j.1365-2141.2008.07065.x.

Kiefer C. M., Hou C., Little J., and Dean A. 2008. Epigenetics of β-globin gene regulation. *Mutat Res.* 647(1-2): 68–76. doi:10.1016/j.mrfmmm.2008.07.014.

Ko, L. J., and J. D. Engel. 1993. "DNA-binding specificities of the GATA transcription factor family." *Mol Cell Biol* 13 (7):4011-22. doi: 10.1128/mcb.13.7.4011.

Krivega, I., C. Byrnes, J. F. de Vasconcellos, Y. T. Lee, M. Kaushal, A. Dean, and J. L. Miller. 2015. "Inhibition of G9a methyltransferase stimulates fetal hemoglobin production by facilitating LCR/gamma-globin looping." *Blood* 126 (5):665-72. doi: 10.1182/blood-2015-02-629972.

Kuo M. H. and David C. 1998. Allis Roles of histone acetyltransferases and deacetylases in gene regulation. *Bioessays.* https://doi.org/10.1002/(SICI)1521-1878(199808)20:8.

Leder, P., J. N. Hansen, D. Konkel, A. Leder, Y. Nishioka, and C. Talkington. 1980. "Mouse globin system: a functional and evolutionary analysis." *Science* 209 (4463):1336-42. doi: 10.1126/science.7414319.

Lee, J., I. Krivega, R. K. Dale, and A. Dean. 2017. "The LDB1 Complex Co-opts CTCF for Erythroid Lineage-Specific Long-Range Enhancer Interactions." *Cell Rep* 19 (12):2490-2502. doi: 10.1016/j.celrep.2017.05.072.

Lessard S., Beaudoin M., Benkirane K. and Lettre G. 2015. Comparison of DNA methylation profiles in human fetal and adult red blood cell progenitors. *Genome Med.* 7(1): 1.

Lettre, G., V. G. Sankaran, M. A. Bezerra, A. S. Araujo, M. Uda, S. Sanna, A. Cao, D. Schlessinger, F. F. Costa, J. N. Hirschhorn, and S. H. Orkin. 2008. "DNA polymorphisms at the BCL11A, HBS1L-MYB, and beta-globin loci associate with fetal hemoglobin levels and pain crises in sickle cell disease." *Proc Natl Acad Sci U S A* 105 (33):11869-74. doi: 10.1073/pnas.0804799105.

Liu, P., J. R. Keller, M. Ortiz, L. Tessarollo, R. A. Rachel, T. Nakamura, N. A. Jenkins, and N. G. Copeland. 2003. "Bcl11a is essential for normal lymphoid development." *Nat Immunol* 4 (6):525-32. doi: 10.1038/ni925.

Long D. M., Dominic S. J. and Campbell M. J. 2017. The Genomic Impact of DNA CpG Methylation on Gene Expression; Relationships in Prostate Cancer. *Biomolecules.* 7(1): 15.

Lopez, R. A., S. Schoetz, K. DeAngelis, D. O'Neill, and A. Bank. 2002. "Multiple hematopoietic defects and delayed globin switching in Ikaros null mice." *Proc Natl Acad Sci U S A* 99 (2):602-7. doi: 10.1073/pnas.022412699.

Maeda, T. 2016. "Regulation of hematopoietic development by ZBTB transcription factors." *Int J Hematol* 104 (3):310-23. doi: 10.1007/s12185-016-2035-x.

Maeda, T., K. Ito, T. Merghoub, L. Poliseno, R. M. Hobbs, G. Wang, L. Dong, M. Maeda, L. C. Dore, A. Zelent, L. Luzzatto, J. Teruya-Feldstein, M. J. Weiss, and P. P. Pandolfi. 2009. "LRF is an essential downstream target of GATA1 in erythroid development and regulates BIM-dependent apoptosis." *Dev Cell* 17 (4):527-40. doi: 10.1016/j.devcel.2009.09.005.

Martin, D. I., and S. H. Orkin. 1990. "Transcriptional activation and DNA binding by the erythroid factor GF-1/NF-E1/Eryf 1." *Genes Dev* 4 (11):1886-98. doi: 10.1101/gad.4.11.1886.

Martin, D. I., S. F. Tsai, and S. H. Orkin. 1989. "Increased gamma-globin expression in a nondeletion HPFH mediated by an erythroid-specific DNA-binding factor." *Nature* 338 (6214):435-8. doi: 10.1038/338435a0.

Martyn, G. E., K. G. R. Quinlan, and M. Crossley. 2017. "The regulation of human globin promoters by CCAAT box elements and the recruitment of NF-Y." *Biochim Biophys Acta Gene Regul Mech* 1860 (5):525-536. doi: 10.1016/j.bbagrm.2016.10.002.

Masuda, T., X. Wang, M. Maeda, M. C. Canver, F. Sher, A. P. Funnell, C. Fisher, M. Suciu, G. E. Martyn, L. J. Norton, C. Zhu, R. Kurita, Y. Nakamura, J. Xu, D. R. Higgs, M. Crossley, D. E. Bauer, S. H. Orkin, P. V. Kharchenko, and T. Maeda. 2016. "Transcription factors LRF and BCL11A independently repress expression of fetal hemoglobin." *Science* 351 (6270):285-9. doi: 10.1126/science.aad3312.

McGrath, K., and J. Palis. 2008. "Ontogeny of erythropoiesis in the mammalian embryo." *Curr Top Dev Biol* 82:1-22. doi: 10.1016/S0070-2153(07)00001-4.

Menzel, S., C. Garner, I. Gut, F. Matsuda, M. Yamaguchi, S. Heath, M. Foglio, D. Zelenika, A. Boland, H. Rooks, S. Best, T. D. Spector, M. Farrall, M. Lathrop, and S. L. Thein. 2007. "A QTL influencing F cell production maps to a gene encoding a zinc-finger protein on chromosome 2p15." *Nat Genet* 39 (10):1197-9. doi: 10.1038/ng2108.

Merika, M., and S. H. Orkin. 1993. "DNA-binding specificity of GATA family transcription factors." *Mol Cell Biol* 13 (7):3999-4010. doi: 10.1128/mcb.13.7.3999.

Mettananda Sachith, Gibbons J. Richard and Higgs R. Douglas. 2015. α-Globin as a molecular target in the treatment of β-thalassemia. *Blood.* 125(24): 3694–3701.

Miccio, A., and G. A. Blobel. 2010. "Role of the GATA-1/FOG-1/NuRD pathway in the expression of human beta-like globin genes." *Mol Cell Biol* 30 (14):3460-70. doi: 10.1128/MCB.00001-10.

Miccio, A., Y. Wang, W. Hong, G. D. Gregory, H. Wang, X. Yu, J. K. Choi, S. Shelat, W. Tong, M. Poncz, and G. A. Blobel. 2010. "NuRD mediates activating and repressive functions of GATA-1 and FOG-1 during blood development." *EMBO J* 29 (2):442-56. doi: 10.1038/emboj.2009.336.

Miller, I. J., and J. J. Bieker. 1993. "A novel, erythroid cell-specific murine transcription factor that binds to the CACCC element and is related to the Kruppel family of nuclear proteins." *Mol Cell Biol* 13 (5):2776-86. doi: 10.1128/mcb.13.5.2776.

Miller J. L. and Grant P. A. 2013. The role of DNA methylation and histone modifications in transcriptional regulation in humans. *Subcell Biochem.* 61: 289–317.

Nan, X., R. R. Meehan, and A. Bird. 1993. "Dissection of the methyl-CpG binding domain from the chromosomal protein MeCP2." *Nucleic Acids Res* 21 (21):4886-92. doi: 10.1093/nar/21.21.4886.

Narlikar, G. J., H. Y. Fan, and R. E. Kingston. 2002. "Cooperation between complexes that regulate chromatin structure and transcription." *Cell* 108 (4):475-87. doi: 10.1016/s0092-8674(02)00654-2.

Nichols, K. E., J. D. Crispino, M. Poncz, J. G. White, S. H. Orkin, J. M. Maris, and M. J. Weiss. 2000. "Familial dyserythropoietic anaemia and thrombocytopenia due to an inherited mutation in GATA1." *Nat Genet* 24 (3):266-70. doi: 10.1038/73480.

Nuez, B., D. Michalovich, A. Bygrave, R. Ploemacher, and F. Grosveld. 1995. "Defective haematopoiesis in fetal liver resulting from inactivation of the EKLF gene." *Nature* 375 (6529):316-8. doi: 10.1038/375316a0.

O'Neill, D., J. Yang, H. Erdjument-Bromage, K. Bornschlegel, P. Tempst, and A. Bank. 1999. "Tissue-specific and developmental stage-specific DNA binding by a mammalian SWI/SNF complex associated with human fetal-to-adult globin gene switching." *Proc Natl Acad Sci U S A* 96 (2):349-54. doi: 10.1073/pnas.96.2.349.

Perkins, A. C., K. M. Gaensler, and S. H. Orkin. 1996. "Silencing of human fetal globin expression is impaired in the absence of the adult beta-globin gene activator protein EKLF." *Proc Natl Acad Sci U S A* 93 (22):12267-71. doi: 10.1073/pnas.93.22.12267.

Perkins, A. C., A. H. Sharpe, and S. H. Orkin. 1995. "Lethal beta-thalassaemia in mice lacking the erythroid CACCC-transcription factor EKLF." *Nature* 375 (6529):318-22. doi: 10.1038/375318a0.

Perkins, A., X. Xu, D. R. Higgs, G. P. Patrinos, L. Arnaud, J. J. Bieker, S. Philipsen, and K. L. F. Consensus Workgroup. 2016. "Kruppeling erythropoiesis: an unexpected broad spectrum of human red blood cell disorders due to KLF1 variants." *Blood* 127 (15):1856-62. doi: 10.1182/blood-2016-01-694331.

Peschle, C., F. Mavilio, A. Care, G. Migliaccio, A. R. Migliaccio, G. Salvo, P. Samoggia, S. Petti, R. Guerriero, M. Marinucci, and et al. 1985. "Haemoglobin switching in human embryos: asynchrony of zeta-alpha and epsilon-gamma-globin switches in primitive and definite erythropoietic lineage." *Nature* 313 (5999):235-8. doi: 10.1038/313235a0.

Pevny, L., C. S. Lin, V. D'Agati, M. C. Simon, S. H. Orkin, and F. Costantini. 1995. "Development of hematopoietic cells lacking transcription factor GATA-1." *Development* 121 (1):163-72.

Rank Gerhard, Cerruti Loretta, Simpson J. Richard., Moritz L. Robert, Jane M. Stephen and Zhao Quan. 2010. Identification of a PRMT5-dependent repressor complex linked to silencing of human fetal globin gene expression. *Blood.* 116(9): 1585–1592.

Renneville, A., P. Van Galen, M. C. Canver, M. McConkey, J. M. Krill-Burger, D. M. Dorfman, E. B. Holson, B. E. Bernstein, S. H. Orkin, D. E. Bauer, and B. L. Ebert. 2015. "EHMT1 and EHMT2 inhibition induces fetal hemoglobin expression." *Blood* 126 (16):1930-9. doi: 10.1182/blood-2015-06-649087.

Rupon, J. W., S. Z. Wang, K. Gaensler, J. Lloyd, and G. D. Ginder. 2006. "Methyl binding domain protein 2 mediates gamma-globin gene silencing in adult human betaYAC transgenic mice." *Proc Natl Acad Sci U S A* 103 (17):6617-22. doi: 10.1073/pnas.0509322103.

Sangerman J., Seung L. M., Xiao Y., Eugene O., Cheng-Hui H., Li W., Sima Z., Solomon O. A. F. and Betty S. 2006. Pace Mechanism for fetal hemoglobin induction by histone deacetylase inhibitors involves γ-globin activation by CREB1 and ATF-2. *Blood.* 108(10): 3590–3599.

Sankaran, V. G., T. F. Menne, J. Xu, T. E. Akie, G. Lettre, B. Van Handel, H. K. Mikkola, J. N. Hirschhorn, A. B. Cantor, and S. H. Orkin. 2008. "Human fetal hemoglobin expression is regulated by the developmental stage-specific repressor BCL11A." *Science* 322 (5909):1839-42. doi: 10.1126/science.1165409.

Sankaran, V. G., and D. G. Nathan. 2010. "Reversing the hemoglobin switch." *N Engl J Med* 363 (23):2258-60. doi: 10.1056/NEJMcibr1010767.

Sankaran, V. G., J. Xu, T. Ragoczy, G. C. Ippolito, C. R. Walkley, S. D. Maika, Y. Fujiwara, M. Ito, M. Groudine, M. A. Bender, P. W. Tucker, and S. H. Orkin. 2009. "Developmental and species-divergent globin switching are driven by BCL11A." *Nature* 460 (7259):1093-7. doi: 10.1038/nature08243.

Shearstone, J. R., O. Golonzhka, A. Chonkar, D. Tamang, J. H. van Duzer, S. S. Jones, and M. B. Jarpe. 2016. "Chemical Inhibition of Histone Deacetylases 1 and 2 Induces Fetal Hemoglobin through Activation of

GATA2." *PLoS One* 11 (4):e0153767. doi: 10.1371/journal.pone.0153767.

Shi Lihong, Shuaiying Cui, James D Engel and Osamu Tanabe. 2013. Lysine-specific demethylase 1 is a therapeutic target for fetal hemoglobin induction. *Nat Med.* 19(3): 291-294. doi: 10.1038/nm.3101.

Singh M., Lavelle D., Vaitkus K., Mahmud N., Hankewych M. and DeSimone J. 2007. The gamma-globin gene promoter progressively demethylates as the hematopoietic stem progenitor cells differentiate along the erythroid lineage in baboon fetal liver and adult bone marrow. *Exp Hematol.* 35(1): 48-55.

Song, S. H., C. Hou, and A. Dean. 2007. "A positive role for NLI/Ldb1 in long-range beta-globin locus control region function." *Mol Cell* 28 (5):810-22. doi: 10.1016/j.molcel.2007.09.025.

Soranzo, N., T. D. Spector, M. Mangino, B. Kuhnel, A. Rendon, A. Teumer, C. Willenborg, B. Wright, L. Chen, M. Li, P. Salo, B. F. Voight, P. Burns, R. A. Laskowski, Y. Xue, S. Menzel, D. Altshuler, J. R. Bradley, S. Bumpstead, M. S. Burnett, J. Devaney, A. Doring, R. Elosua, S. E. Epstein, W. Erber, M. Falchi, S. F. Garner, M. J. Ghori, A. H. Goodall, R. Gwilliam, H. H. Hakonarson, A. S. Hall, N. Hammond, C. Hengstenberg, T. Illig, I. R. Konig, C. W. Knouff, R. McPherson, O. Melander, V. Mooser, M. Nauck, M. S. Nieminen, C. J. O'Donnell, L. Peltonen, S. C. Potter, H. Prokisch, D. J. Rader, C. M. Rice, R. Roberts, V. Salomaa, J. Sambrook, S. Schreiber, H. Schunkert, S. M. Schwartz, J. Serbanovic-Canic, J. Sinisalo, D. S. Siscovick, K. Stark, I. Surakka, J. Stephens, J. R. Thompson, U. Volker, H. Volzke, N. A. Watkins, G. A. Wells, H. E. Wichmann, D. A. Van Heel, C. Tyler-Smith, S. L. Thein, S. Kathiresan, M. Perola, M. P. Reilly, A. F. Stewart, J. Erdmann, N. J. Samani, C. Meisinger, A. Greinacher, P. Deloukas, W. H. Ouwehand, and C. Gieger. 2009. "A genome-wide meta-analysis identifies 22 loci associated with eight hematological parameters in the HaemGen consortium." *Nat Genet* 41 (11):1182-90. doi: 10.1038/ng.467.

Stamatoyannopoulos, G. 2005. "Control of globin gene expression during development and erythroid differentiation." *Exp Hematol* 33 (3):259-71. doi: 10.1016/j.exphem.2004.11.007.

Suzuki M., Yamamoto M. and Engel J. D. 2014. Fetal Globin Gene Repressors as Drug Targets for Molecular Therapies to treat the β-Globinopathies. *Mol Cell Biol.* 34(19): 3560–3569.

Tewari, R., N. Gillemans, M. Wijgerde, B. Nuez, M. von Lindern, F. Grosveld, and S. Philipsen. 1998. "Erythroid Kruppel-like factor (EKLF) is active in primitive and definitive erythroid cells and is required for the function of 5'HS3 of the beta-globin locus control region." *EMBO J* 17 (8):2334-41. doi: 10.1093/emboj/17.8.2334.

Thein, S. L., S. Menzel, X. Peng, S. Best, J. Jiang, J. Close, N. Silver, A. Gerovasilli, C. Ping, M. Yamaguchi, K. Wahlberg, P. Ulug, T. D. Spector, C. Garner, F. Matsuda, M. Farrall, and M. Lathrop. 2007. "Intergenic variants of HBS1L-MYB are responsible for a major quantitative trait locus on chromosome 6q23 influencing fetal hemoglobin levels in adults." *Proc Natl Acad Sci U S A* 104 (27):11346-51. doi: 10.1073/pnas.0611393104.

Tolhuis, B., R. J. Palstra, E. Splinter, F. Grosveld, and W. de Laat. 2002. "Looping and interaction between hypersensitive sites in the active beta-globin locus." *Mol Cell* 10 (6):1453-65. doi: 10.1016/s1097-2765(02) 00781-5.

Traxler, E. A., Y. Yao, Y. D. Wang, K. J. Woodard, R. Kurita, Y. Nakamura, J. R. Hughes, R. C. Hardison, G. A. Blobel, C. Li, and M. J. Weiss. 2016. "A genome-editing strategy to treat beta-hemoglobinopathies that recapitulates a mutation associated with a benign genetic condition." *Nat Med* 22 (9):987-90. doi: 10.1038/nm.4170.

Tsai, S. F., D. I. Martin, L. I. Zon, A. D. D'Andrea, G. G. Wong, and S. H. Orkin. 1989. "Cloning of cDNA for the major DNA-binding protein of the erythroid lineage through expression in mammalian cells." *Nature* 339 (6224):446-51. doi: 10.1038/339446a0.

Tsang, A. P., J. E. Visvader, C. A. Turner, Y. Fujiwara, C. Yu, M. J. Weiss, M. Crossley, and S. H. Orkin. 1997. "FOG, a multitype zinc finger protein, acts as a cofactor for transcription factor GATA-1 in erythroid and megakaryocytic differentiation." *Cell* 90 (1):109-19. doi: 10.1016/ s0092-8674(00)80318-9.

Tsang, J. C., Y. Yu, S. Burke, F. Buettner, C. Wang, A. A. Kolodziejczyk, S. A. Teichmann, L. Lu, and P. Liu. 2015. "Single-cell transcriptomic reconstruction reveals cell cycle and multi-lineage differentiation defects in Bcl11a-deficient hematopoietic stem cells." *Genome Biol* 16:178. doi: 10.1186/s13059-015-0739-5.

Tuan, D., W. Solomon, Q. Li, and I. M. London. 1985. "The "beta-like-globin" gene domain in human erythroid cells." *Proc Natl Acad Sci U S A* 82 (19):6384-8. doi: 10.1073/pnas.82.19.6384.

Uda, M., R. Galanello, S. Sanna, G. Lettre, V. G. Sankaran, W. Chen, G. Usala, F. Busonero, A. Maschio, G. Albai, M. G. Piras, N. Sestu, S. Lai, M. Dei, A. Mulas, L. Crisponi, S. Naitza, I. Asunis, M. Deiana, R. Nagaraja, L. Perseu, S. Satta, M. D. Cipollina, C. Sollaino, P. Moi, J. N. Hirschhorn, S. H. Orkin, G. R. Abecasis, D. Schlessinger, and A. Cao. 2008. "Genome-wide association study shows BCL11A associated with persistent fetal hemoglobin and amelioration of the phenotype of beta-thalassemia." *Proc Natl Acad Sci U S A* 105 (5):1620-5. doi: 10.1073/pnas.0711566105.

Vakoc, C. R., D. L. Letting, N. Gheldof, T. Sawado, M. A. Bender, M. Groudine, M. J. Weiss, J. Dekker, and G. A. Blobel. 2005. "Proximity among distant regulatory elements at the beta-globin locus requires GATA-1 and FOG-1." *Mol Cell* 17 (3):453-62. doi: 10.1016/j.molcel.2004.12.028.

Wadman, I. A., H. Osada, G. G. Grutz, A. D. Agulnick, H. Westphal, A. Forster, and T. H. Rabbitts. 1997. "The LIM-only protein Lmo2 is a bridging molecule assembling an erythroid, DNA-binding complex which includes the TAL1, E47, GATA-1 and Ldb1/NLI proteins." *EMBO J* 16 (11):3145-57. doi: 10.1093/emboj/16.11.3145.

Weatherall, D. J., R. Cartner, J. B. Clegg, W. G. Wood, I. A. Macrae, and A. Mackenzie. 1975. "A form of hereditary persistence of fetal haemoglobin characterized by uneven cellular distribution of haemoglobin F and the production of haemoglobins A and A2 in homozygotes." *Br J Haematol* 29 (2):205-20. doi: 10.1111/j.1365-2141.1975.tb01815.x.

Weiss, M. J., and S. H. Orkin. 1995. "Transcription factor GATA-1 permits survival and maturation of erythroid precursors by preventing apoptosis." *Proc Natl Acad Sci U S A* 92 (21):9623-7. doi: 10.1073/pnas.92.21.9623.

Wijgerde, M., J. Gribnau, T. Trimborn, B. Nuez, S. Philipsen, F. Grosveld, and P. Fraser. 1996. "The role of EKLF in human beta-globin gene competition." *Genes Dev* 10 (22):2894-902. doi: 10.1101/gad.10.22.2894.

Witt O., Monkemeyer S., Rönndahl G., Erdlenbruch B., Reinhardt D., Kanbach K. and Pekrun A. 2003. Induction of fetal hemoglobin expression by the histone deacetylase inhibitor apicidin. *Blood.* 101(5):2001-2007.

Wood H. Kathleen and Zhou Zhaolan. 2016. Emerging Molecular and Biological Functions of MBD2, a Reader of DNA Methylation. *Frontiers in Genetics.* 7(93): 1-14.

Xu, J., D. E. Bauer, M. A. Kerenyi, T. D. Vo, S. Hou, Y. J. Hsu, H. Yao, J. J. Trowbridge, G. Mandel, and S. H. Orkin. 2013. "Corepressor-dependent silencing of fetal hemoglobin expression by BCL11A." *Proc Natl Acad Sci U S A* 110 (16):6518-23. doi: 10.1073/pnas.1303976110.

Xu, J., C. Peng, V. G. Sankaran, Z. Shao, E. B. Esrick, B. G. Chong, G. C. Ippolito, Y. Fujiwara, B. L. Ebert, P. W. Tucker, and S. H. Orkin. 2011. "Correction of sickle cell disease in adult mice by interference with fetal hemoglobin silencing." *Science* 334 (6058):993-6. doi: 10.1126/science.1211053.

Xu, J., V. G. Sankaran, M. Ni, T. F. Menne, R. V. Puram, W. Kim, and S. H. Orkin. 2010. "Transcriptional silencing of {gamma}-globin by BCL11A involves long-range interactions and cooperation with SOX6." *Genes Dev* 24 (8):783-98. doi: 10.1101/gad.1897310.

Ye, L., J. Wang, Y. Tan, A. I. Beyer, F. Xie, M. O. Muench, and Y. W. Kan. 2016. "Genome editing using CRISPR-Cas9 to create the HPFH genotype in HSPCs: An approach for treating sickle cell disease and beta-thalassemia." *Proc Natl Acad Sci U S A* 113 (38):10661-5. doi: 10.1073/pnas.1612075113.

Yun, W. J., Y. W. Kim, Y. Kang, J. Lee, A. Dean, and A. Kim. 2014. "The hematopoietic regulator TAL1 is required for chromatin looping between the beta-globin LCR and human gamma-globin genes to activate transcription." *Nucleic Acids Res* 42 (7):4283-93. doi: 10.1093/nar/gku072.

Zhang W., Aubert A., Gomez de Segura J. M., Karuppasamy M., Basu S., Murthy A. S., Diamante A., Drury T. A., Balmer J., Cramard J., Watson A. A., Lando D., Lee S. F., Palayret M., S. Kloet S. L., Smits A. H., Deery M. J., Vermeulen M., Hendrich B., Klenerman D., Schaffitzel C., Berger I. and Laue E. D. 2016. The Nucleosome Remodeling and Deacetylase Complex NuRD Is Built from Preformed Catalytically Active Sub-modules. *J Mol Biol.* 428: 2931–2942.

Zhao Quan, Rank Gerhard, Tan T. Yuen, Li Haitao, Moritz L. Robert, Simpson J. Richard, Cerruti Loretta, Curtis J. David, Patel J. Dinshaw, Allis C. David, Cunningham M. John and Jane M. Stephen. 2009. PRMT5-mediated methylation of histone H4R3 recruits DNMT3A, coupling histone and DNA methylation in gene silencing. *Nat Struct Mol Biol.* 16(3): 304–311.

Zhu H., Guohua W. and Jiang Q. 2016. Transcription factors as readers and effectors of DNA methylation. *Nat Rev Genet.* 17(9): 551–565.

In: Fetal Hemoglobin
Editor: Nishant Chakravorty

ISBN: 978-1-53617-955-2

Chapter 5

ROLE OF CELL SIGNALING PATHWAYS IN FETAL HEMOGLOBIN REGULATION IN BETA-HEMOGLOBINOPATHIES

Sankha Subhra Das and Nishant Chakravorty*
School of Medical Science and Technology,
Indian Institute of Technology Kharagpur, West Bengal, India

ABSTRACT

Beta-hemoglobinopathies are among the most common hematological disorders encountered by clinicians worldwide. Such disorders are caused by mutations in the beta-globin gene that result in an alteration of hemoglobin synthesis and hence its structure. Fetal hemoglobin, the predominant hemoglobin form in the fetus and newborn, is replaced by the adult form within six months of postnatal life. This hemoglobin is known to be significant modifier of the clinical severity of beta-hemoglobinopathies like beta-thalassemia and sickle cell anemia. Therefore understanding the mechanisms of fetal to adult globin switching and reactivation of fetal hemoglobin in beta-hemoglobinopathies have been the prime focus of hematology research globally. This chapter seeks

---

* Corresponding Author's Email: nishant@smst.iitkgp.ac.in.

to understand the clinical significance of fetal hemoglobin, and tries to summarize the recent understanding on the involvement of cell signaling pathways in erythropoiesis and fetal hemoglobin regulation.

Keywords: cell signaling pathways, beta-hemoglobinopathies, fetal hemoglobin

FETAL HEMOGLOBIN

What Is Fetal Hemoglobin?

Fetal hemoglobin is the predominant form of hemoglobin in the growing fetus and the newborn. Although the level of fetal hemoglobin is around 50-95% at birth, this level falls substantially in around six months of postnatal life (Thomas and Lumb 2012). Fetal hemoglobin comprises of two alpha-globin chains and two gamma-globin chains. Alpha-globin chains are encoded by the alpha-globin genes which are present in the alpha-globin locus on chromosome 16; whereas gamma-globin chains are encoded by two nearly identical gamma-globin genes (HBG1 and HBG2) which are located in beta-globin gene locus on chromosome 11 (Wienert et al. 2018). With its high oxygen affinity, fetal hemoglobin promotes oxygen transfer between fetal and maternal circulation through the placenta.

Elevated Fetal Hemoglobin Level

The fetal hemoglobin level is significantly increased in various pathophysiological conditions. Elevated fetal hemoglobin levels have been observed in individuals with hereditary persistence of fetal hemoglobin (HPFH) (Forget 1998). The level of fetal hemoglobin usually ranges between 10% and 40% in HPFH heterozygotes, whereas the level can reach nearly 100% in HPFH homozygous (Thein and Craig 1998). Individuals with sickle cell anemia and beta-thalassemia also demonstrate higher levels of fetal hemoglobin when compared with normal individuals. Higher levels of HbF have also been noted in cases of HbE-beta-thalassemia (Rees et al. 1999). HbE-beta-thalassemia is caused by coinheritance of a beta-thalassemia allele and a hemoglobin E variant (Glu26Lys). Details about these and other conditions causing raised HbF levels have been discussed in other chapters.

Clinical Significance of Fetal Hemoglobin

Janet Watson was the first person to recognize the clinical significance of fetal hemoglobin (Watson 1948). She observed lower sickling of blood in infants with sickle cell anemia and attributed this observation to high fetal hemoglobin levels in early birth. Further, it has been reported that individuals with mutations leading to "hereditary persistence of fetal hemoglobin" appear to be clinically normal (Akinsheye et al. 2011), indicating the beneficial role of fetal hemoglobin in beta-hemoglobinopathies. In sickle cell anemia, fetal hemoglobin reduces the polymerization of sickle hemoglobin (HbS), and hence ameliorates the clinical symptoms and disease severity (Steinberg et al. 2014). In addition to this, high fetal hemoglobin levels are also known to reduce various complications in sickle cell disease, such as leg ulcers, acute painful episodes, osteonecrosis, and acute chest syndrome, etc. (Akinsheye et al. 2011). In beta-thalassemia, the increased gamma-globin level decreases the accumulation and precipitation of excess alpha-globin chains and thereby reduces the consequences of ineffective erythropoiesis (Fard et al. 2013). All these findings indicate that fetal hemoglobin is a significant modifier that can overcome the clinical severity of various beta-hemoglobinopathies. To develop fetal hemoglobin-based therapies, the primary focus of researchers needs to be directed towards understanding the mechanisms involved in inactivation of the gamma globin gene and its switchover to beta-globin.

MECHANISM OF GLOBIN SWITCHING

Detailed discussion on the mechanisms involved in gamma to beta switch over has been provided in other chapters of the book. However, a concise description of some of the hypotheses and concepts is necessary to understand the role of signaling pathways involved in gamma globin regulation.

The beta globin gene locus consists of five developmentally organized globin genes (epsilon, gamma-A, gamma-B, delta, and beta) and a distal enhancer region. During the developmental stages of the fetus, the epsilon gene is the first one from the beta-globin locus to get associated with one of the genes from the alpha locus (the epsilon) and forms Hemoglobin Gower 1. This is expressed from the yolk sac. Around 6 to 8 weeks of gestation, the expression of epsilon is silenced and is replaced by gamma-globins – thereby allowing synthesis of Hemoglobin Gower 2. Subsequently, the epsilon globin expression is replaced by alpha globin expression, leading to the formation of fetal hemoglobin which remains the predominant form throughout the fetal stage. Following birth, the gamma-globin expression starts decreasing and levels of beta-globin is seen to increase dramatically leading to increased expression of the adult hemoglobin form. Research indicates two globin switching mechanisms: a competitive interaction of beta-globin and gamma-globin genes, and an autonomous silencing mechanism.

Competitive Interaction of Globin Gene Family Members

Experiments on transgenic mice models first explained the competitive mechanism of globin gene switching during the developmental stages of fetus (Behringer et al. 1990). The study suggested that the switch-over of gamma-globin to beta-globin gene results from a competition between gamma-globin and beta-globin genes for their interaction with the DNase I super hypersensitive site sequence. When either the gamma or the beta-globin genes were incorporated individually, down-stream to the hypersensitive site sequence, neither of them showed any developmental specificity. The beta globin gene was able to show a similar expression pattern in embryonic cells as well as in adult cells. While the gamma globin genes did show a higher expression in embryonic cells as expected, they also showed expression in adult cells, albeit at a lower level. Conversely, gamma and beta-globin genes showed developmental specification, when they were placed in the same fragment with hypersensitive sites. Thus the gamma-globin gene showed higher expression in embryonic cells whereas the beta-globin gene showed higher expression in adult cells. Another study by Enver et al., demonstrated that the mutual interaction between the locus activation region and the individual globin genes results in gamma to beta-globin switching during development (Enver et al. 1990). These observations indicate that the gamma-globin gene preferentially interacts with the hyper sensitive site during the fetal stage of development, resulting in silencing of the beta-globin gene; whereas, during the adult stage of development, preferential interaction between hypersensitive sites and the beta-globin gene turn off the gamma-globin gene (Stamatoyannopoulos 2005).

Autonomous Globin Gene Silencing

A study by Dillon et al. was among the first ones to contradict the earlier report of the involvement of globin gene competition in gamma-globin silencing (Dillon and Grosveld 1991). Experiments on transgenic mice models showed that the expression of the gamma-globin genes does not depend on the presence of hypersensitive sites and the beta-globin gene. Hyper sensitive sites linked to gamma-globin genes were silenced in transgenic mice even in the absence of the beta-globin gene. Further research revealed the role of an upstream silencer region in gamma-globin silencing (Stamatoyannopoulos et al. 1993). The study showed that the interaction between the locus control regions and -382 to -730 silencer region participated in gamma-globin silencing. A study by Harju et al., showed that the deletion of -378 to -730 regions resulted in activation of gamma-globin in adult, indicating that this silencing region participates in gamma-globin silencing (Harju-Baker et al. 2008). A subsequent study found that mutation(s) in this region is/are associated with hereditary persistence of fetal hemoglobin (Stamatoyannopoulos 2005). A transgenic mice study on the beta-globin locus carried on yeast artificial chromosome showed that the insertion of gamma-A globin gene between locus control region and delta-globin gene causes silencing of gamma-A globin. Deletion of 352 bp region between -730 and -378 restored the expression gamma-A globin gene (Harju-Baker et al. 2008). Further analysis of the silencer region indicated that -566 GATA site is occupied with various transcription factors such as Mi2, FOG-1, and GATA1 when the gamma-globin gene is silenced. Direct experimental evidence by Yu et al. suggest that gamma-globin gene silencing is mediated by autonomous silencing (Yu et al. 2006). The study found that the replacement of the beta-globin promoter with gamma-globin promoter sequence caused inhibition of beta-globin gene expression during adult stage of development. Although studies indicated that silencing of the gamma-globin gene is controlled by an autonomous silencing mechanism, there are studies that provide evidence of the contribution of both of the mechanisms in gamma-globin silencing.

Pharmacological Induction of Fetal Hemoglobin

The beneficial effects of fetal hemoglobin in beta-hemoglobinopathies have been known for many years. Therefore, identification of fetal hemoglobin inducing therapeutic agents has been one of the prime foci of the researchers. To identify fetal hemoglobin inducing agents, researchers have used a number of experimental systems such as primary cells, immortalized cell lines, genetically engineered mice, and clinical trials on humans.

The first attempt of drug mediated induction of fetal hemoglobin was made on anemic baboons in the early 1980s (DeSimone et al. 1982). The study demonstrated that 5-Azacytidine, a DNA methyltransferase inhibitor, can stimulate the synthesis of fetal hemoglobin. There is a correlation between methylation of cytosine residues and expression of the globin gene

(Mavilio et al. 1983). In general, inactive genes are hyper-methylated by methyltransferase enzymes. The drug 5-Azacytidine has been found to be an effective inhibitor of DNA methyltransferase. Later a study on erythroid marrow cells has demonstrated the toxic effects of 5-Azacytidine in late erythroid cells (Torrealba-de Ron et al. 1984). Many subsequent studies have reported the cytotoxic effects of this drug in humans and thus limited its usage.

Hydroxyurea is another drug that has been found to induce fetal hemoglobin production. Letvin et al. were among the early researchers who demonstrated such an effect on anemic monkeys (Letvin et al. 1984). Later studies have also confirmed that hydroxyurea can increase fetal hemoglobin levels in sickle cell anemia, (Veith et al. 1985, Platt et al. 1984). Veith et al. suggested that the regeneration of reticulocyte is the major reason of fetal hemoglobin stimulation. Although hydroxyurea is not carcinogenic and is extensively used for other diseases like myeloproliferative disorders (Cao 2004), it still has cytotoxic effects limiting its usage to increase fetal hemoglobin levels in beta-hemoglobinopathies.

Butyrate, a histone deacetylase inhibitor, has also been identified as a potential pharmacological agent to increase gamma-globin synthesis (Constantoulakis, Papayannopoulou, and Stamatoyannopoulos 1988). Perrine et al. showed that sodium butyrate induces fetal hemoglobin synthesis in erythroid progenitor cells derived from sickle cell anemia and beta-thalassemia (Perrine et al. 1989).

In addition to the above drugs, there are several other fetal hemoglobin inducing agents (Mabaera et al. 2008) such as valproic acid (Liakopoulou et al. 1995), phenylbutyrate (Dover, Brusilow, and Charache 1994), decitabine (Saunthararajah et al. 2003), vorinostat (Skarpidi et al. 2003), trichostatin A (McCaffrey et al. 1997), thalidomide (Aerbajinai et al. 2007), rapamycin (Mischiati et al. 2004), etc. Detailed discussion on pharmacological induction of fetal hemoglobin has been made elsewhere in the book.

Signaling Pathways Involved in Erythroid Differentiation

Differentiation of hematopoietic stem cells to enucleated erythrocytes is a multistep process which involves activation of a number of cell signaling pathways (Rahim et al. 2013). Interaction between erythropoietin and its receptor plays a major role in the activation of these cell signaling pathways (Pace et al. 2015). Major pathways that are involved in the various stages of erythropoiesis includes the JAK-STAT pathway (Witthuhn et al. 1993), Mitogen-Activated Protein Kinase (MAPK) pathway and Phosphatidyl Inositol 3 Kinase (PI3K) pathway (Nakao et al. 2008).

The mammalian MAPK pathway includes ERK1/ERK2 cascades, SAPK/JNK, and the p38 MAPK pathway. Among these pathways, the ERK1/ERK2 pathway is known to play a significant role in the proliferation and differentiation of erythroid precursor cells. ERK1/2 play an important role in the growth and differentiation of erythroid cells (Woessmann and Mivechi 2001). The p38 MAPK, on the other hand, has been found to be involved in terminal differentiation of erythroid cells (Moosavi, Yazdanparast, and Lotfi 2007). The activation of SAPK/JNK determines the fate of the cell and plays a pivotal role in erythroid differentiation (Nagata and Todokoro 1999).

The JAK-STAT pathway is associated with proliferation and survival of erythroid cells (Cokic et al. 2012). Seven family members of the STAT protein (STAT1, STAT2, STAT3, STAT4, STAT5A, STAT5B, and STAT6) have been identified in mammals (Ward, Touw, and Yoshimura 2000). In hematopoietic progenitor cells, the transcription factor STAT5A is activated by erythropoietin (Ratajczak et al. 2001) and this influences erythroid cell proliferation and differentiation (Schuringa et al. 2004). The phosphorylation of tyrosine residues of the STAT transcription factor is mediated by JAK2 - a cytoplasmic tyrosine kinase (Schuringa et al. 2004).

The PI3K pathway is another important pathway which is involved in erythropoietin induced proliferation and maturation of erythroid progenitor cells (Myklebust et al. 2002). PI3K, the downstream target of AKT, is activated by both erythropoietin and stem cell factor (Cokic et al. 2012). Studies by Bouscary et al., suggested that PI3K mediates erythroid cell proliferation through down-regulation of p27 (kip1) protein (Bouscary et al. 2003). Another study demonstrated the role of PI3K pathway associated genes such as STATs, GRB2, and PIK3 in erythropoietin mediated survival and proliferation of erythroid cells (Cokic et al. 2012).

Table 1. Cell signaling pathways involved in erythropoiesis

Cell signaling pathway	Role in erythropoiesis	References
JAK-STAT pathway	Erythroid cell proliferation	(Cokic et al. 2012)
MAPK pathway (ERK1/2)	Growth and differentiation of erythroid cells	(Woessmann and Mivechi 2001)
MAPK pathway (p38)	Terminal differentiation of erythroid cells	(Moosavi, Yazdanparast, and Lotfi 2007)
MAPK pathway (SAPK/JNK)	Determine the cell fate and erythroid differentiation	(Nagata and Todokoro 1999)
PI3K pathway	Proliferation and maturation of erythroid cells	(Myklebust et al. 2002, Cokic et al. 2012)

SIGNALING PATHWAYS INVOLVED IN FETAL HEMOGLOBIN INDUCTION

Integrated Stress Response Pathway

In response to various stresses and physiological changes, the integrated stress response (ISR) signaling is activated in eukaryotic cells (Pakos-Zebrucka et al. 2016). Activation of this signaling pathway is triggered by phosphorylation of eukaryotic translation initiation factor 2 alpha subunit (eIF2α). Several kinases are known to catalyze the phosphorylation of eIF2α. In erythroid cells, the most predominant kinase is the heme-regulated eIF2α kinase (HRI). This kinase is activated under various stresses and physiological conditions such as oxidative stress, environmental stress, and heme deficiency, and ultimately leads to phosphorylation of eIF2α (Chen and Perrine 2013). The level of HRI has been found to be elevated during the erythroid differentiation process and this contributes more than 90% increase in phosphorylation of the eIF2α (Liu et al. 2008). Under stressed conditions like ER stress, viral infection, heme deprivation and other cellular stress conditions, phosphorylation of eIF2α reduces the synthesis of proteins globally while increasing the synthesis of specific protein such as activating transcription factor 4 (ATF4) (Pakos-Zebrucka et al. 2016). A study on a murine erythroid model demonstrated the role of HRI-eIF2P-ATF4 pathway in erythroid differentiation (Suragani et al. 2012).

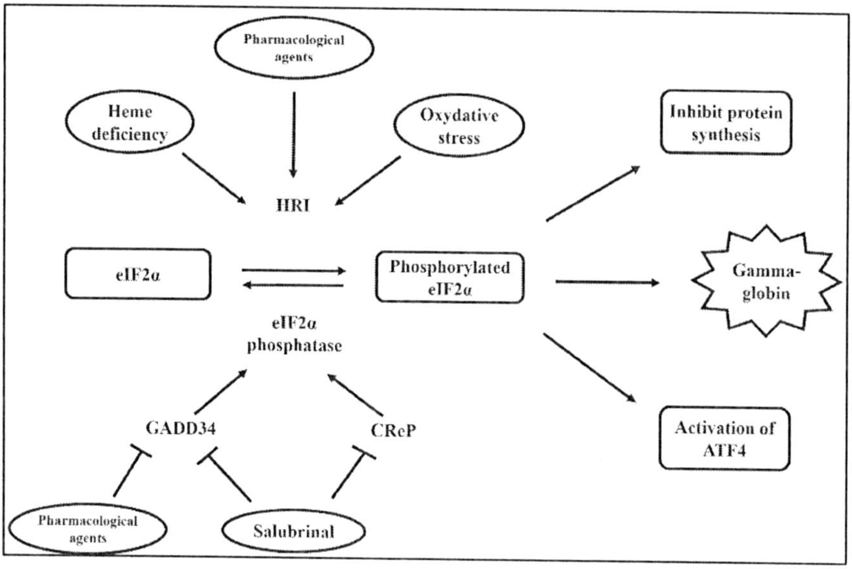

Figure 1. Mechanisms of integrated stress response pathway in fetal hemoglobin regulation.

Dephosphorylation of eIF2αP is facilitated by the enzyme, eIF2αP phosphatase. Two proteins, GADD34 (growth arrest and DNA damage-inducible protein 34) and CReP (constitutive repressor of eIF2α phosphorylation) are involved in the dephosphorylation process (Novoa et al. 2001, Jousse et al. 2003). A study in 2005 found that the drug salubrinal inhibits eIF2αP dephosphorylation by down regulating CReP and GADD34 (Boyce et al. 2005). Most importantly, salubrinal has also been found to enhance the level of phosphorylated eIF2α and increase the level of fetal hemoglobin in hematopoietic progenitor cells by activating the integrated stress response pathway (Hahn and Lowrey 2013). This study further confirmed the involvement of integrated stress response pathway in fetal hemoglobin induction by assessing two pharmacological agents- BTdCPU and GBZ. BTdCPU is known for its role in the activation of HRI and phosphorylation of eIF2α, whereas GBZ inhibits dephosphorylation of eIF2α by disrupting GADD34. Primary erythroid cells treated with BTdCPU and GBZ are seen to have higher levels of absolute fetal hemoglobin level. Therefore, the integrated stress response pathway plays a pivotal role in post-transcriptional activation of fetal hemoglobin.

Cyclic GMP Pathway

Cyclic nucleotides play a pivotal role in HbF induction. Studies using erythroleukemic cell lines have elucidated the role of soluble guanylate cyclase-cGMP dependent protein kinase pathway in fetal hemoglobin regulation (Ikuta, Ausenda, and Cappellini 2001). Soluble guanylate cyclase (sGC) converts GTP to cGMP and regulates various biological processes. The study showed that sGC activators protoporphyrin IX and cGMP increased the expression of the gamma-globin gene. In a subsequent study, Cokic et al., demonstrated the role of hydroxyurea in the induction of gamma-globin gene through activation of the soluble guanylate cyclase-cGMP pathway (Cokic et al. 2003).

It is known that cGMP enhances intracellular cAMP levels by inhibiting cAMP-specific phosphodiesterase 3. A study by Keefer et al. in 2006, demonstrated that inhibition of adenylate cyclase in primary erythroid cells reduces drug-mediated induction of HbF substantially. The study further showed that guanylate cyclase has minimal effect on induction HbF in primary erythroid cells (Keefer et al. 2006). These findings put together indicate the possibility of differential mechanisms of HbF induction in primary erythroid cells and erythroleukemia cell lines. A relevant study on primary erythroblasts showed that activation of the cAMP pathway is associated with induction of gamma-globin genes (Kuroyanagi et al. 2006).

However, another study demonstrated that the gamma-globin level is reduced in K562 cells upon activation of the cAMP pathway (Inoue et al. 2004). This research group further showed that the proto-oncogene cMYB is responsible for decreased gamma-globin expression in K562 cells (Kuroyanagi et al. 2006).

Table 2. Cell signaling pathways that are involved in drug-mediated induction of fetal hemoglobin

Drugs	Pathway	Possible mechanism	References
Hydroxyurea	p38 MAPK pathway	Hydroxyurea increases fetal hemoglobin through activation of p38 MAPK	(Park et al. 2001)
Hydroxyurea	NO-cGMP pathway	Hydroxyurea induces fetal hemoglobin by nitric oxide derived activation of cGMP	(Cokic et al. 2003)
Butyrate	p38 MAPK pathway	Butyrate activates p38 MAPK signaling and induces gamma-globin level	(Witt, Sand, and Pekrun 2000, Pace et al. 2003)
Sodium butyrate	cAMP pathway	The production of cAMP is necessary for fetal hemoglobin induction by sodium butyrate	(Keefer et al. 2006)
5-Azacytidine	cAMP pathway	5-Azacytidine increases fetal hemoglobin by inducing cAMP production	(Keefer et al. 2006)
Valproate	p38 MAPK	p38 MAPK pathway is involved in Valproate mediated fetal hemoglobin induction	(Witt et al. 2002)

Some research groups have demonstrated that activation of soluble guanylate cyclase is mediated by nitric oxide (King 2003). Cokic et al. suggested that nitric oxide actively participates in hydroxyurea mediated induction of fetal hemoglobin (Cokic et al. 2003). The study demonstrated that the levels of nitric oxide derived species are extensively increased after hydroxyurea treatment in sickle cell disease. Nitric oxide donor treated human erythroid progenitor cells and erythroleukemic cells show increased cGMP levels and subsequently induce gamma-globin levels.

Later studies demonstrated that the nitric oxide-mediated cGMP signaling pathway activates various transcription factors (Haby et al. 1994,

Ikuta et al. 2016), which also include, c-Jun and c-Fos (known for their role in gamma-globin activation) (Pace et al. 2015). Therefore, an integration of these studies suggested that there is a link between hydroxyurea, nitric oxide, soluble guanylate cyclase, and fetal hemoglobin expression. The exact mechanism of nitric oxide-cGMP mediated gamma-globin activation however is still unknown.

p38 MAPK Pathway

A study on K562 erythroid cells demonstrated the role of the MAPK pathway in valproate mediated induction of fetal hemoglobin (Witt et al. 2002). Although, the level of ERK of MAPK pathway is reduced upon treatment with sodium valproate, the MAPK pathway protein, p38 shows slight increase in phosphorylation whereas JNK does not show any changes in phosphorylation. These results were further confirmed by using a p38 inhibitor, SB230580. Use of p38 inhibitor blocked sodium valproate mediated induction of fetal hemoglobin.

The role of p38 MAPK pathway has been further demonstrated by Li et al. (2011) in a study that elucidated that the extract of *Fructus trichosanthis* enhances the fetal hemoglobin level in K562 erythroid cells. Pre-treatment with SB203580 (p38 MAPK inhibitor) was shown to block the inducing effect of *Fructus trichosanthis* and thereby inhibited fetal hemoglobin as well as the total hemoglobin production.

Histone deacetylase inhibitors are known for their role in fetal hemoglobin induction. A study by Witt et al., demonstrated the role of apicidin, a potent histone deacetylase inhibitor, in p38 MAPK pathway mediated induction of fetal hemoglobin (Witt et al. 2003). Apicidin can enhance fetal hemoglobin levels by up to 10-fold. The study showed that pre-treatment with apicidin activates p38 signaling in K562 erythroleukemia cells.

Scientists exploring the role of plant-based products for treating anemic disorders have identified *Plastrum testudinis* as a source that can be used to treat various anemic and bleeding disorders. *Plastrum testudinis* has been shown to promote erythroid differentiation and induce gamma-globin production in human erythroid cells (Qian et al. 2013). *Plastrum testudinis* has not been found to have any effect on the expression of alpha-globin and beta-globin, which indicates towards the fact that beneficial effects of this extract may be due to rise in HbF levels. The study found that this traditional medicine promotes acetylation of histone H3 and H4 and is involved in the activation of p38 MAPK signaling. The study further showed that the effect of this medicine on gamma-globin expression is abolished when the cells are treated with p38 MAPK inhibitor (SB203580). Another well-known compound, Piceatannol (an analogue of resveratrol) has been found to induce fetal hemoglobin in erythroid progenitor cells derived from sickle cell anemia patients (Zhang et al. 2017). The study found that the level of p38 MAPK was increased in cells pre-treated with various concentration of Piceatannol (Liu and Chang 2010). These observations indicated that Piceatannol is a potent fetal hemoglobin inducer which might act through the p38 MAPK pathway.

Adrenergic agonists such as phenylephrine and isoproterenol have been found to elevate fetal hemoglobin levels in K562 cells by enhancing the acetylation of histone H3 and H4 (Mei et al. 2013). The study further demonstrated that the phenylephrine and isoproterenol mediated induction of fetal hemoglobin is regulated by p38 MAPK pathway and cAMP pathway.

CONCLUSION

In this review, we have discussed various pharmacological agents which activate cell signaling pathways and induce fetal hemoglobin production. Experimental evidence indicated that the p38 MAPK pathway, cAMP pathway, and integrated stress response pathway are the major pathways which regulate fetal hemoglobin expression. Researchers have identified several pharmacological agents which activate these cell signaling pathways; although their mechanisms of action in gamma-globin regulation has not been proposed yet. There are several questions that need to be addressed, for example: 1) What is the complete mechanism of action of each pathway in gamma-globin regulation? 2) What is the series of events (from the introduction of drugs to induction of fetal hemoglobin) that occur in biological systems? 3) Are the pathway inducing drugs effective in clinical settings? In recent years, researchers have been focusing on other approaches and techniques, such as non-coding RNA mediated activation of fetal hemoglobin regulatory pathways, CRISPR-Cas based techniques, etc. More research needs to be done to identify the complete mechanism of action of globin switching and fetal hemoglobin regulation.

ACKNOWLEDGMENT

The authors are thankful to Indian Institute of Technology Kharagpur for providing the infrastructural support. Mr. Sankha Subhra Das gladly acknowledges Council of Scientific and Industrial Research (CSIR), India for providing financial support.

REFERENCES

Aerbajinai, W., Zhu, J., Gao, Z., Chin, K., and Rodgers, G. P. 2007. "Thalidomide induces gamma-globin gene expression through increased reactive oxygen species-mediated p38 MAPK signaling and histone H4 acetylation in adult erythropoiesis." *Blood.* 110 (8):2864-71. doi: 10.1182/blood-2007-01-065201.

Akinsheye, I., Alsultan, A., Solovieff, N., Ngo, D., Baldwin, C. T., Sebastiani, P., Chui, D. H., and Steinberg, M. H. 2011. "Fetal hemoglobin in sickle cell anemia." *Blood.* 118 (1):19-27. doi: 10. 1182/blood-2011-03-325258.

Behringer, R. R., Ryan, T. M., Palmiter, R. D., Brinster, R. L., and Townes, T. M. 1990. "Human gamma- to beta-globin gene switching in transgenic mice." *Genes Dev.* 4 (3):380-9. doi: 10.1101/gad.4.3.380.

Bouscary, Didier, Pene, Frédéric, Claessens, Yann-Erick, Muller, Odile, Chrétien, Stany, Fontenay-Roupie, Michaëla, Gisselbrecht, Sylvie, Mayeux, Patrick, and Lacombe, Catherine. 2003. "Critical role for PI 3-kinase in the control of erythropoietin-induced erythroid progenitor proliferation." *Blood.* 101 (9):3436-3443. doi: 10.1182/blood-2002-07-2332.

Boyce, M., Bryant, K. F., Jousse, C., Long, K., Harding, H. P., Scheuner, D., Kaufman, R. J., Ma, D., Coen, D. M., Ron, D., and Yuan, J. 2005. "A selective inhibitor of eIF2alpha dephosphorylation protects cells from ER stress." *Science.* 307 (5711):935-9. doi: 10.1126/science.1101902.

Cao, Hua. 2004. "Pharmacological Induction of Fetal Hemoglobin Synthesis using Histone Deacetylase Inhibitors." *Hematology.* 9 (3):223-233. doi: 10.1080/10245330410001701512.

Chen, J. J., and Perrine, S. 2013. "Stressing HbF synthesis: role of translation?" *Blood.* 122 (4):467-8. doi: 10.1182/blood-2013-06-506139.

Cokic, V. P., Bhattacharya, B., Beleslin-Cokic, B. B., Noguchi, C. T., Puri, R. K., and Schechter, A. N. 2012. "JAK-STAT and AKT pathway-coupled genes in erythroid progenitor cells through ontogeny." *J Transl Med.* 10:116. doi: 10.1186/1479-5876-10-116.

Cokic, V. P., Smith, R. D., Beleslin-Cokic, B. B., Njoroge, J. M., Miller, J. L., Gladwin, M. T., and Schechter, A. N. 2003. "Hydroxyurea induces fetal hemoglobin by the nitric oxide-dependent activation of soluble guanylyl cyclase." *J Clin Invest.* 111 (2):231-9. doi: 10.1172/jci16672.

Constantoulakis, P, Papayannopoulou, T, and Stamatoyannopoulos, G. 1988. "alpha-Amino-N-butyric acid stimulates fetal hemoglobin in the adult." *Blood.* 72 (6):1961-1967.

DeSimone, J., Heller, P., Hall, L., and Zwiers, D. 1982. "5-Azacytidine stimulates fetal hemoglobin synthesis in anemic baboons." *Proceedings of the National Academy of Sciences of the United States of America.* 79 (14):4428-4431. doi: 10.1073/pnas.79.14.4428.

Dillon, N., and Grosveld, F. 1991. "Human gamma-globin genes silenced independently of other genes in the beta-globin locus." *Nature.* 350 (6315):252-4. doi: 10.1038/350252a0.

Dover, G. J., Brusilow, S., and Charache, S. 1994. "Induction of fetal hemoglobin production in subjects with sickle cell anemia by oral sodium phenylbutyrate." *Blood.* 84 (1):339-43.

Enver, T., Raich, N., Ebens, A. J., Papayannopoulou, T., Costantini, F., and Stamatoyannopoulos, G. 1990. "Developmental regulation of human fetal-to-adult globin gene switching in transgenic mice." *Nature.* 344 (6264):309-13. doi: 10.1038/344309a0.

Fard, Ali Dehghani, Hosseini, Seyed Ahmad, Shahjahani, Mohammad, Salari, Fatemeh, and Jaseb, Kaveh. 2013. "Evaluation of Novel Fetal Hemoglobin Inducer Drugs in Treatment of β-Hemoglobinopathy Disorders." *International journal of hematology-oncology and stem cell research.* 7 (3):47-54.

Forget, B. G. 1998. "Molecular basis of hereditary persistence of fetal hemoglobin." *Ann N Y Acad Sci.* 850:38-44. doi: 10.1111/j.1749-6632.1998.tb10460.x.

Haby, C., Lisovoski, F., Aunis, D., and Zwiller, J. 1994. "Stimulation of the cyclic GMP pathway by NO induces expression of the immediate early genes c-fos and junB in PC12 cells." *J Neurochem.* 62 (2):496-501. doi: 10.1046/j.1471-4159.1994.62020496.x.

Hahn, Cynthia K., and Lowrey, Christopher H. 2013. "Eukaryotic initiation factor 2α phosphorylation mediates fetal hemoglobin induction through a post-transcriptional mechanism." *Blood.* 122 (4):477-485. doi: 10. 1182/blood-2013-03-491043.

Harju-Baker, Susanna, Costa, Flávia C., Fedosyuk, Halyna, Neades, Renee, and Peterson, Kenneth R. 2008. "Silencing of ^Aγ-Globin Gene Expression during Adult Definitive Erythropoiesis Mediated by GATA-1-FOG-1-Mi2 Complex Binding at the −566 GATA Site." *Molecular and Cellular Biology.* 28 (10):3101-3113. doi: 10.1128/ mcb.01858-07.

Ikuta, T., Ausenda, S., and Cappellini, M. D. 2001. "Mechanism for fetal globin gene expression: role of the soluble guanylate cyclase-cGMP-dependent protein kinase pathway." *Proc Natl Acad Sci U S A.* 98 (4):1847-52. doi: 10.1073/pnas.041599798.

Ikuta, T., Sellak, H., Odo, N., Adekile, A. D., and Gaensler, K. M. 2016. "Nitric Oxide-cGMP Signaling Stimulates Erythropoiesis through Multiple Lineage-Specific Transcription Factors: Clinical Implications and a Novel Target for Erythropoiesis." *PLoS One.* 11 (1):e0144561. doi: 10.1371/journal.pone.0144561.

Inoue, A., Kuroyanagi, Y., Terui, K., Moi, P., and Ikuta, T. 2004. "Negative regulation of gamma-globin gene expression by cyclic AMP-dependent pathway in erythroid cells." *Exp Hematol.* 32 (3):244-53. doi: 10.1016/j.exphem.2003.12.006.

Jousse, C., Oyadomari, S., Novoa, I., Lu, P., Zhang, Y., Harding, H. P., and Ron, D. 2003. "Inhibition of a constitutive translation initiation factor 2alpha phosphatase, CReP, promotes survival of stressed cells." *J Cell Biol.* 163 (4):767-75. doi: 10.1083/jcb.200308075.

Keefer, J. R., Schneidereith, T. A., Mays, A., Purvis, S. H., Dover, G. J., and Smith, K. D. 2006. "Role of cyclic nucleotides in fetal hemoglobin induction in cultured CD34+ cells." *Exp Hematol.* 34 (9):1151-61. doi: 10.1016/j.exphem.2006.03.018.

King, S. Bruce. 2003. "A role for nitric oxide in hydroxyurea-mediated fetal hemoglobin induction." *The Journal of clinical investigation.* 111 (2):171-172. doi: 10.1172/JCI17597.

Kuroyanagi, Y., Kaneko, Y., Muta, K., Park, B. S., Moi, P., Ausenda, S., Cappellini, M. D., and Ikuta, T. 2006. "cAMP differentially regulates gamma-globin gene expression in erythroleukemic cells and primary erythroblasts through c-Myb expression." *Biochem Biophys Res Commun.* 344 (3):1038-47. doi: 10.1016/j.bbrc.2006.03.203.

Letvin, N. L., Linch, D. C., Beardsley, G. P., McIntyre, K. W., and Nathan, D. G. 1984. "Augmentation of fetal-hemoglobin production in anemic monkeys by hydroxyurea." *N Engl J Med.* 310 (14):869-73. doi: 10.1056/nejm198404053101401.

Li, H., Ko, C. H., Tsang, S. Y., Leung, P. C., Fung, M. C., and Fung, K. P. 2011. "The Ethanol Extract of Fructus trichosanthis Promotes Fetal Hemoglobin Production via p38 MAPK Activation and ERK

Inactivation in K562 Cells." *Evid Based Complement Alternat Med.* 2011:657056. doi: 10.1093/ecam/neq022.

Liakopoulou, E., Blau, C. A., Li, Q., Josephson, B., Wolf, J. A., Fournarakis, B., Raisys, V., Dover, G., Papayannopoulou, T., and Stamatoyannopoulos, G. 1995. "Stimulation of fetal hemoglobin production by short chain fatty acids." *Blood.* 86 (8):3227-35.

Liu, S., Bhattacharya, S., Han, A., Suragani, R. N., Zhao, W., Fry, R. C., and Chen, J. J. 2008. "Haem-regulated eIF2alpha kinase is necessary for adaptive gene expression in erythroid precursors under the stress of iron deficiency." *Br J Haematol.* 143 (1):129-37. doi: 10.1111/j.1365-2141.2008.07293.x.

Liu, Wen-Hsin, and Chang, Long-Sen. 2010. "Piceatannol induces Fas and FasL up-regulation in human leukemia U937 cells via Ca2+/p38α MAPK-mediated activation of c-Jun and ATF-2 pathways." *The International Journal of Biochemistry & Cell Biology.* 42 (9):1498-1506. doi: https://doi.org/10.1016/j.biocel.2010.05.007.

Mabaera, R., West, R. J., Conine, S. J., Macari, E. R., Boyd, C. D., Engman, C. A., and Lowrey, C. H. 2008. "A cell stress signaling model of fetal hemoglobin induction: what doesn't kill red blood cells may make them stronger." *Exp Hematol.* 36 (9):1057-72. doi: 10.1016/j.exphem.2008.06.014.

Mavilio, F., Giampaolo, A., Carè, A., Migliaccio, G., Calandrini, M., Russo, G., Pagliardi, G. L., Mastroberardino, G., Marinucci, M., and Peschle, C. 1983. "Molecular mechanisms of human hemoglobin switching: selective undermethylation and expression of globin genes in embryonic, fetal, and adult erythroblasts." *Proceedings of the National Academy of Sciences of the United States of America.* 80 (22):6907-6911. doi: 10.1073/pnas.80.22.6907.

McCaffrey, P. G., Newsome, D. A., Fibach, E., Yoshida, M., and Su, M. S. 1997. "Induction of gamma-globin by histone deacetylase inhibitors." *Blood.* 90 (5):2075-83.

Mei, Yun, Yin, Naida, Jin, Xia, He, Jiangyan, and Yin, Zhan. 2013. "The Regulatory Role of the Adrenergic Agonists Phenylephrine and Isoproterenol on Fetal Hemoglobin Expression and Erythroid

Differentiation." *Endocrinology.* 154 (12):4640-4649. doi: 10.1210/en.2013-1535.

Mischiati, C., Sereni, A., Lampronti, I., Bianchi, N., Borgatti, M., Prus, E., Fibach, E., and Gambari, R. 2004. "Rapamycin-mediated induction of gamma-globin mRNA accumulation in human erythroid cells." *Br J Haematol.* 126 (4):612-21. doi: 10.1111/j.1365-2141.2004.05083.x.

Moosavi, M. A., Yazdanparast, R., and Lotfi, A. 2007. "ERK1/2 inactivation and p38 MAPK-dependent caspase activation during guanosine 5'-triphosphate-mediated terminal erythroid differentiation of K562 cells." *Int J Biochem Cell Biol.* 39 (9):1685-97. doi: 10.1016/j.biocel.2007.04.016.

Myklebust, J. H., Blomhoff, H. K., Rusten, L. S., Stokke, T., and Smeland, E. B. 2002. "Activation of phosphatidylinositol 3-kinase is important for erythropoietin-induced erythropoiesis from CD34(+) hematopoietic progenitor cells." *Exp Hematol.* 30 (9):990-1000.

Nagata, Y., and Todokoro, K. 1999. "Requirement of activation of JNK and p38 for environmental stress-induced erythroid differentiation and apoptosis and of inhibition of ERK for apoptosis." *Blood.* 94 (3):853-63.

Nakao, T., Geddis, A. E., Fox, N. E., and Kaushansky, K. 2008. "PI3K/Akt/FOXO3a pathway contributes to thrombopoietin-induced proliferation of primary megakaryocytes in vitro and in vivo via modulation of p27(Kip1)." *Cell Cycle.* 7 (2):257-66. doi: 10.4161/cc.7.2.5148.

Novoa, I., Zeng, H., Harding, H. P., and Ron, D. 2001. "Feedback inhibition of the unfolded protein response by GADD34-mediated dephosphorylation of eIF2alpha." *The Journal of cell biology.* 153 (5):1011-1022. doi: 10.1083/jcb.153.5.1011.

Pace, B. S., Liu, L., Li, B., and Makala, L. H. 2015. "Cell signaling pathways involved in drug-mediated fetal hemoglobin induction: Strategies to treat sickle cell disease." *Exp Biol Med* (Maywood). 240 (8):1050-64. doi: 10.1177/1535370215596859.

Pace, B. S., Qian, X. H., Sangerman, J., Ofori-Acquah, S. F., Baliga, B. S., Han, J., and Critz, S. D. 2003. "p38 MAP kinase activation mediates gamma-globin gene induction in erythroid progenitors." *Exp Hematol.* 31 (11):1089-96.

Pakos-Zebrucka, Karolina, Koryga, Izabela, Mnich, Katarzyna, Ljujic, Mila, Samali, Afshin, and Gorman, Adrienne M. 2016. "The integrated stress response." *EMBO reports.* 17 (10):1374-1395. doi: 10.15252/embr.201642195.

Park, J. I., Choi, H. S., Jeong, J. S., Han, J. Y., and Kim, I. H. 2001. "Involvement of p38 kinase in hydroxyurea-induced differentiation of K562 cells." *Cell Growth Differ.* 12 (9):481-6.

Perrine, S. P., Miller, B. A., Faller, D. V., Cohen, R. A., Vichinsky, E. P., Hurst, D., Lubin, B. H., and Papayannopoulou, T. 1989. "Sodium butyrate enhances fetal globin gene expression in erythroid progenitors of patients with Hb SS and beta thalassemia." *Blood.* 74 (1):454-9.

Platt, O. S., Orkin, S. H., Dover, G., Beardsley, G. P., Miller, B., and Nathan, D. G. 1984. "Hydroxyurea enhances fetal hemoglobin production in sickle cell anemia." *The Journal of clinical investigation.* 74 (2):652-656. doi: 10.1172/JCI111464.

Qian, X., Chen, J., Zhao, D., Guo, L., and Qian, X. 2013. "Plastrum testudinis induces gamma-globin gene expression through epigenetic histone modifications within the gamma-globin gene promoter via activation of the p38 MAPK signaling pathway." *Int J Mol Med.* 31 (6):1418-28. doi: 10.3892/ijmm.2013.1338.

Rahim, F., Allahmoradi, H., Salari, F., Shahjahani, M., Fard, A. D., Hosseini, S. A., and Mousakhani, H. 2013. "Evaluation of Signaling Pathways Involved in gamma-Globin Gene Induction Using Fetal Hemoglobin Inducer Drugs." *Int J Hematol Oncol Stem Cell Res.* 7 (3):41-6.

Ratajczak, J., Majka, M., Kijowski, J., Baj, M., Pan, Z. K., Marquez, L. A., Janowska-Wieczorek, A., and Ratajczak, M. Z. 2001. "Biological significance of MAPK, AKT and JAK-STAT protein activation by various erythropoietic factors in normal human early erythroid cells." *Br J Haematol.* 115 (1):195-204. doi: 10.1046/j.1365-2141.2001.03058.x.

Rees, D.C., Porter, J.B., Clegg, J.B., and Weatherall, D.J. 1999. "Why Are Hemoglobin F Levels Increased in HbE/β Thalassemia?" *Blood.* 94 (9):3199-3204.

Saunthararajah, Yogen, Hillery, Cheryl A., Lavelle, Don, Molokie, Robert, Dorn, Louise, Bressler, Linda, Gavazova, Stefana, Chen, Yi-Hsiang, Hoffman, Ronald, and DeSimone, Joseph. 2003. "Effects of 5-aza-2′-deoxycytidine on fetal hemoglobin levels, red cell adhesion, and hematopoietic differentiation in patients with sickle cell disease." *Blood.* 102 (12):3865-3870. doi: 10.1182/blood-2003-05-1738.

Schuringa, J. J., Chung, K. Y., Morrone, G., and Moore, M. A. 2004. "Constitutive activation of STAT5A promotes human hematopoietic stem cell self-renewal and erythroid differentiation." *J Exp Med.* 200 (5):623-35. doi: 10.1084/jem.20041024.

Skarpidi, E., Cao, H., Heltweg, B., White, B. F., Marhenke, R. L., Jung, M., and Stamatoyannopoulos, G. 2003. "Hydroxamide derivatives of short-chain fatty acids are potent inducers of human fetal globin gene expression." *Exp Hematol.* 31 (3):197-203.

Stamatoyannopoulos, G., Josephson, B., Zhang, J. W., and Li, Q. 1993. "Developmental regulation of human gamma-globin genes in transgenic mice." *Mol Cell Biol.* 13 (12):7636-44. doi: 10.1128/mcb.13.12.7636.

Stamatoyannopoulos, George. 2005. "Control of globin gene expression during development and erythroid differentiation." *Experimental hematology.* 33 (3):259-271. doi: 10.1016/j.exphem.2004.11.007.

Steinberg, M. H., Chui, D. H., Dover, G. J., Sebastiani, P., and Alsultan, A. 2014. "Fetal hemoglobin in sickle cell anemia: a glass half full?" *Blood.* 123 (4):481-5. doi: 10.1182/blood-2013-09-528067.

Suragani, R. N., Zachariah, R. S., Velazquez, J. G., Liu, S., Sun, C. W., Townes, T. M., and Chen, J. J. 2012. "Heme-regulated eIF2alpha kinase activated Atf4 signaling pathway in oxidative stress and erythropoiesis." *Blood.* 119 (22):5276-84. doi: 10.1182/blood-2011-10-388132.

Thein, S. L., and Craig, J. E. 1998. "Genetics of Hb F/F cell variance in adults and heterocellular hereditary persistence of fetal hemoglobin." *Hemoglobin.* 22 (5-6):401-14.

Thomas, Caroline, and Lumb, Andrew B. 2012. "Physiology of haemoglobin." *BJA Education.* 12 (5):251-256. doi: 10.1093/bjaceaccp/mks025.

Torrealba-de Ron, A. T., Papayannopoulou, T., Knapp, M. S., Fu, M. F., Knitter, G., and Stamatoyannopoulos, G. 1984. "Perturbations in the erythroid marrow progenitor cell pools may play a role in the augmentation of HbF by 5-azacytidine." *Blood.* 63 (1):201-10.

Veith, R., Galanello, R., Papayannopoulou, T., and Stamatoyannopoulos, G. 1985. "Stimulation of F-cell production in patients with sickle-cell anemia treated with cytarabine or hydroxyurea." *N Engl J Med.* 313 (25):1571-5. doi: 10.1056/nejm198512193132503.

Ward, Alister C., Touw, Ivo, and Yoshimura, Akihiko. 2000. "The Jak-Stat pathway in normal and perturbed hematopoiesis." *Blood.* 95 (1):19-29.

Watson, Janet. 1948. "Sickling in negro newborns: Its possible relationship to fetal hemoglobin." *The American Journal of Medicine.* 5 (1):159-160. doi: 10.1016/0002-9343(48)90029-1.

Wienert, B., Martyn, G. E., Funnell, A. P. W., Quinlan, K. G. R., and Crossley, M. 2018. "Wake-up Sleepy Gene: Reactivating Fetal Globin for beta-Hemoglobinopathies." *Trends Genet.* 34 (12):927-940. doi: 10.1016/j.tig.2018.09.004.

Witt, O., Monkemeyer, S., Kanbach, K., and Pekrun, A. 2002. "Induction of fetal hemoglobin synthesis by valproate: modulation of MAP kinase pathways." *Am J Hematol.* 71 (1):45-6. doi: 10.1002/ajh.10161.

Witt, O., Monkemeyer, S., Ronndahl, G., Erdlenbruch, B., Reinhardt, D., Kanbach, K., and Pekrun, A. 2003. "Induction of fetal hemoglobin expression by the histone deacetylase inhibitor apicidin." *Blood.* 101 (5):2001-7. doi: 10.1182/blood-2002-08-2617.

Witt, O., Sand, K., and Pekrun, A. 2000. "Butyrate-induced erythroid differentiation of human K562 leukemia cells involves inhibition of ERK and activation of p38 MAP kinase pathways." *Blood.* 95 (7):2391-6.

Witthuhn, B. A., Quelle, F. W., Silvennoinen, O., Yi, T., Tang, B., Miura, O., and Ihle, J. N. 1993. "JAK2 associates with the erythropoietin receptor and is tyrosine phosphorylated and activated following stimulation with erythropoietin." *Cell.* 74 (2):227-36. doi: 10.1016/0092-8674(93)90414-l.

Woessmann, W., and Mivechi, N. F. 2001. "Role of ERK activation in growth and erythroid differentiation of K562 cells." *Exp Cell Res.* 264 (2):193-200. doi: 10.1006/excr.2000.5124.

Yu, Man, Han, Hemei, Xiang, Ping, Li, Qiliang, and Stamatoyannopoulos, George. 2006. "Autonomous Silencing as Well as Competition Controls γ-Globin Gene Expression during Development." *Molecular and Cellular Biology.* 26 (13):4775-4781. doi: 10.1128/mcb.00406-06.

Zhang, Yankai, Chang, Alicia, Sumazin, Pavel, and Sheehan, Vivien. 2017. "Piceatannol Induces Fetal Hemoglobin in Erythroid Progenitor Cells from Patients with Sickle Cell Disease." *Blood.* 130 (Suppl 1):2221-2221.

In: Fetal Hemoglobin
Editor: Nishant Chakravorty

ISBN: 978-1-53617-955-2

Chapter 6

MOLECULAR TOOLS FOR EDITING GLOBIN GENES

Kshitiz Singh*

Center for Fetal Research, Children's Hospital of Philadelphia, Philadelphia, PA, US

ABSTRACT

The latest, novel developments in the field of gene editing have helped in gaining traction to correct diseases of genetic origin. New DNA analysis techniques and gene editing can help in ameliorating hemoglobinopathies. This chapter will cover methods of genome editing (ZFNs, TALENs, CRISPR/Cas9) with various aspects of gene delivery using viral (adenoviral, adeno-associated viral and lentiviral) vectors and nanovectors. Additionally, mechanisms of Non-Homologous End Joining (NHEJ), Homology-Directed Repair (HDR), base editing and prime editing will be

* Corresponding Author's Email: singhk@email.chop.edu.

described. The above methods will be covered in the context of editing globin genes to achieve the correction of beta-hemoglobinopathies.

Keywords: Gene targeting, gene editing, β-thalassemia, CRISPR/Cas9, ZFNs, TALEN, hemoglobin, double-strand break, homology-directed repair

INTRODUCTION

Hemoglobinopathies are among the commonest forms of inherited diseases. Clinically, hemoglobinopathies can present with a wide range of manifestations – from mild anemia to moderate and severe hematological abnormalities. Moreover, additional co-morbidities due to the disease or its life-threatening secondary manifestations (infections, adverse immune reactions, organ damage) lead to frequent blood transfusions and may necessitate bone marrow transplantation. Lack of matching donors for bone marrow transplantation and involvement of a single gene in this group of diseases, make them potential candidates for treatment by gene therapy and gene editing.

Hemoglobinopathies include the genetic diseases of hemoglobin that result in abnormal production of hemoglobin or its structural variants. Hemoglobin disorders are broadly classified into two categories—structural defect of globin chains (e.g., HbS, HbC, HbE) and absence or reduction in synthesis of globin chains (alpha- and beta-thalassemia) (Forget et al. 2013).

Defect in the production and/or the structure of the beta-globin chains constitute a large subset among the hemoglobinopathies. The beta-hemoglobinopathies are caused by a broad spectrum of chromosomal mutations, including point mutations, insertions, deletions, chromosomal rearrangements and epigenetic alterations. Within the past three decades, the advent of innovative genome engineering technologies has allowed us to study and potentially correct the phenotype and molecular defect of these diseases. In this chapter, the use of genome editing methods will be discussed in the context of editing globin genes to achieve the correction of beta-hemoglobinopathies.

Overview of Beta-Hemoglobinopathies

Beta-thalassemia

Beta-thalassemia is characterized by a reduction in synthesis of beta-globin chains and consequently, the amount of adult hemoglobin is reduced. Depending on the clinical severity, beta-thalassemia is classified as thalassemia major, thalassemia intermedia and thalassemia minor with severity ranging from being transfusion-dependent to asymptomatic.

ClinVar PubMed database (https://www.ncbi.nlm.nih.gov/clinvar) lists more than 100 known pathogenic mutations that can cause beta-thalassemia. The disease has a wide spectrum of mutations, ranging from single nucleotide variants, small/large insertions or small/large deletions and duplications. These mutations can be in the coding and/or non-coding regions of the HBB gene, which can result in a decrease in the expression of beta-globin chains (Thein et al. 2013).

Sickle Cell Disease

Sickle cell disease is a common autosomal recessive familial hemoglobinopathy. It results from a mutation in the beta-globin chain, leading to the formation of the abnormal hemoglobin—HbS. HbS is an insoluble variant of hemoglobin, which causes sickling of RBCs (red blood cells). The disease occurs due to a single nucleotide change (A to T) at the sixth codon of the beta-globin gene, which leads to the conversion of glutamic acid codon to valine codon.

Regulation of Alpha-Like and Beta-Like Globin Chains and Hemoglobin Switching

Hemoglobin consists of two pairs of alpha-like and beta-like globin genes (Table 3 and Table 4). The genes at alpha-like globin cluster genomic locus are located on the short arm of the chromosome 16 (16p13.3) and consist of Hemoglobin zeta (HBZ), hemoglobin mu (HBM), Hemoglobin alpha-2 (HBA2), hemoglobin alpha-1 (HBA1) and hemoglobin theta-1 (HBQ1) (OMIM Gene Map). The genes at beta-like globin cluster genomic locus are located on the short arm of chromosome 11 (11p15.4) and consist of HBB (hemoglobin beta), HBD (hemoglobin delta), noncoding beta-globin locus transcript 3 (BGLT3), hemoglobin gamma A (HBG1), hemoglobin gamma G (HBG2), hemoglobin epsilon (HBE1) and locus control region beta (LCRB). BGLT3 is a long non-coding RNA, which regulated the expression of gamma globin chains, as demonstrated by its disruption by CRISPR/Cas9 system (Ivaldi et al. 2018). Moreover, certain polymorphisms are known to increase the production of fetal hemoglobin (Breveglieri et al. 2017; Chen et. al. 2017).

The embryonic form of hemoglobin is Hb Gower 1. At ~12 weeks of fetal age, fetal hemoglobin (HbF, alpha-2gamma2) replaces embryonic hemoglobin. Fetal to adult hemoglobin switch starts at around birth and most of the fetal hemoglobin is replaced with adult hemoglobin by the age of 6 months. However, this switching is not complete and in normal adults, residual levels (less than 1%) of HbF are persisting. In Table 1, diverse hemoglobin molecules synthesized during different developmental phases in humans are described. The switch from fetal to adult hemoglobin involves activation of beta-chain synthesis and the silencing of gamma-chain synthesis. Several regulatory elements, such as BCL11A, SOX6, KLF1, HBS1L-MYB are involved in hemoglobin switching. BCL11A and ZBTB7A are two known repressors of gamma-globin chain (Martyn et al. 2018). BCL11A functions in association with other known gamma-globin repressors and interacts with beta-globin LCR and other regions of HBB cluster.

Table 1. Variants of human hemoglobin

Type of hemoglobin	Globin chains	Remarks
Hb Gower 1	Zeta-2, epsilon-2	Unstable
Hb Gower 2	Alpha-2, epsilon-2	Stable, possible candidate for correction of hemoglobinopathies, if HbF reactivation is contraindicated.
Hb Portland I	Zeta-2, gamma-2	
Hb Portland II	Zeta-2, beta-2	Very unstable, but still a possible candidate for reactivation in β-thalassemia
HbA	Alpha-2, beta-2	
HbA_2	Alpha-2, delta-2	
HbF	Alpha-2, gamma-2	
HbS	E6V substitution in the beta-globin chain	
HbE	E26K substitution in the beta-globin chain	
HbC	E6K substitution in the beta-globin chain	
Hemoglobin Lepore	Fusion of 5' end of HBD sequence and 3' end of HBB sequence.	

The levels of HbF persisting in adults are genetically regulated. In normal adults, the HbF levels are less than 0.6% of total hemoglobin. In HPFH, there are substantially high levels of HbF due to the high expression of gamma-globin chains. Clinically, HPFH does not display any other manifestation apart from elevated HbF levels. Some individuals with HPFH have fetal hemoglobin with glycine-136 (HbG2), whereas others have alanine-136 (HBG1).

HPFH is caused by deletions or point mutations in the HBB cluster. Mutant beta-globin chain results in sickle cell disease, while the absence of beta-globin chain causes beta^0-thalassemia. Deletion and relocation of enhancer elements near beta- and gamma-globin genes can lead to increased HbF expression. Several quantitative trait loci (QTLs) are known to be associated with elevated levels of HbF (Table 2). HPFH by itself does not result in any symptoms. However, the association of elevated HbF in the patients of beta-thalassemia lessens the severity of symptoms (Thein et al.

1994; Craig et al. 1996; Garner et al. 1998; Thein et al. 2007; Galarneau et al. 2010). Studying the mechanisms for hereditary persistence of fetal hemoglobin and development of advanced genome editing tools is revealing creative and novel ways of curing hemoglobinopathies.

Table 2. HBFQTL locus responsible for HPFH

HPFH locus	Chromosomal location	Remarks/Description
HBFQTL	11p15.4	HPFH can result from deletion in HBB cluster; Point mutations in promoter regions of HBG1 and HBG2
HBFQTL	6q22.3-q23.1	HPFH is associated with mutations in HBS1L and MYB gene
HBFQTL	Xp22.2	X-linked FCP gene locus is major determinant for HbF expression
HBFQTL	8q	This locus interacts with the common XmnI-G-gamma polymorphism in HBG2 to alter levels of HbF
HBFQTL	2p15	Elevated HbF levels are known to be associated with C allele of SNP in intron 2 of BAF CHROMATIN REMODELING COMPLEX SUBUNIT BCL11A gene.
HBFQTL	19p13	Mutation in KLF1 gene located on chromosome 19p13. ZBTB7A binding to the gamma-globin promoter.

Table 3. Genes in the alpha-globin cluster locus 16p13.3

Type of hemoglobin	Synonyms
Hemoglobin zeta	HBZ
Hemoglobin mu	HBM
Hemoglobin alpha-2	HBA2, HBH, ECYT7
Hemoglobin alpha-1	HBA1, METHBA
Hemoglobin theta-1	HBQ1

Gene Therapy and Gene Editing

United States Food & Drug Administration (FDA) defines gene therapy as a technique that modifies a person's genes to treat or cure diseases, which encompasses both gene replacement therapy and gene editing. However, traditionally gene therapy meant gene replacement therapy only and gene editing was considered to be distinct from gene therapy or gene replacement therapy. Gene (or genome) editing is defined as the ability to make specific changes to the genome of an organism. This chapter primarily focuses on the use of gene editing to correct beta-hemoglobinopathies by editing globin genes or inducing γ-globin genes. Activation of fetal hemoglobin and concomitant alpha-thalassemia are known factors that can improve beta-thalassemia (Mettananda et al. 2015).

TOOLS FOR TARGETED GENOME EDITING

Gene editing can be mediated by nucleases and nickases in combination with other DNA, RNA or protein (full-length or partial domains) molecules. There are several types of designer nucleases – Zinc Finger nucleases (ZFNs), Transcription Activator Like Effector Nucleases (TALENs), Clustered Regularly Interspaced-Repeats (CRISPR)/CRISPR-Associated (Cas) system and engineered meganucleases (MNs). These nucleases differ in mechanism of target sequence recognition but act similarly, by inducing breaks in the genomic DNA (Figure 1).

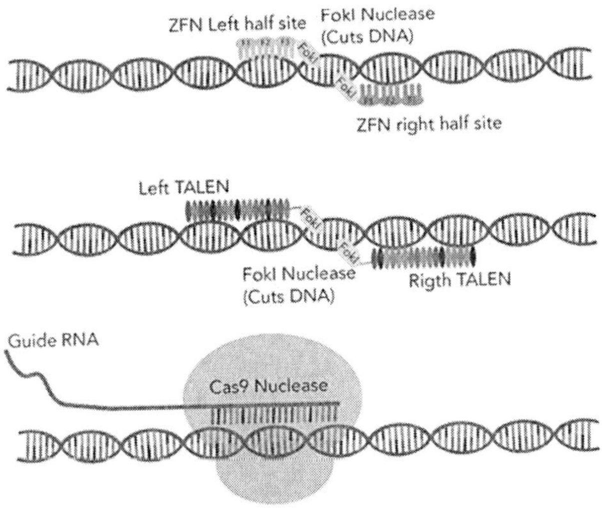

Figure 1. Schematics of genome editing nucleases – ZFNs, TALENs and CRISPR/Cas9 nucleases.

Nucleases and Nickases

Nucleases make double-stranded breaks (DSBs) and in the absence of a repair template, the DSBs get repaired by Non-Homologous End Joining (NHEJ), leading to the formation of indels i.e., insertions and deletions. The nucleases have been altered to create nickases, which generate single-stranded breaks (SSBs). After the breaks in genomic DNA, the DNA repair mechanisms in the cell get activated. The repair of the DNA breaks occurs by error prone end-repair pathways (NHEJ and Microhomology-mediated end joining-MMEJ) or homology-directed repair (HDR).

NHEJ is an imperfect repair mechanism and results in the formation of indels. If the indels are in the coding region of a gene, it can lead to change in the reading frame of the gene or deletions of a few amino acids. Consequently, the functional protein may not be synthesized or the functional region of genomic DNA gets altered. For example, targeting the mouse *F9* gene in exon 1 or exon 6 results in decreased F9 protein activity (Singh et al. 2018). Similarly, the alteration of the promoter region of genes can alter their expression levels. Nucleases are more prone to induce indels, as compared to nickases. Indels formation in BCL11A and ZBTB7A regions can induce the synthesis of fetal hemoglobin (Patsali et al. 2019).

HDR is a precise repair mechanism with lesser efficiency of editing, as compared to NHEJ. For HDR to occur, along with the nucleases or nickases, we provide a DNA repair template, which can help in integrating a specific DNA sequence at the target site. The DNA repair templates can be double-stranded (linear or circular) and single-stranded DNA molecules. The repair template DNA molecules have homology at the cut site. The size of the homology depends on the type of repair template. Typically, integration of long double-stranded templates requires homology of several hundred base pairs at the cut site, while short single-stranded repair templates can also integrate with short homology arms. The approaches such as inhibition of molecules involved in NHEJ (e.g., DNA Ligase-1) or the overexpression of proteins involved in HDR, such as RAD51 are being investigated to increase HDR efficiency. Moreover, for the CRISPR/Cas9 system, the introduction of blocking mutations i.e., alteration of the protospacer adjacent motif

(PAM) site, in the donor template can lead to increased efficiency of HDR, by preventing the repeated excision of the target site (Paquet et al. 2016).

MMEJ occurs as an alternative to or in combination with NHEJ and HDR mechanisms, depending on the active pathways in the cell. In the absence of the Ku protein propensity for MMEJ-based repair pathway is increased (Wang et al. 2016). Even when HDR is occurring MMEJ can happen in appreciable extents.

Meganucleases

Synthetic Meganucleases recognize and cleave 14 to 40 bp long sequences and induce site-specific double-stranded breaks. Among the homing endonuclease family, the eukaryal and archaeal LAGLIDADG family displays maximum specificity. The LAGLIDADG homing endonuclease family uses anti-parallel β-sheets as DNA recognition module, by forming a saddle on the major groove of DNA double helix, catalyzed by divalent metal ions. Engineered meganucleases did not gain favor, due to cumbersome designing procedures and lack of one-to-one correspondence of protein domains with nucleotide residues (Smith et al. 2006).

Fok1-based nucleases — Zinc Finger Nucleases and Transcription Activator-Like Effector Nucleases

ZFNs consist of 3 to 5 zinc finger domains (DNA recognition domain) fused with nuclease domain of fokI restriction enzyme (Bibikova et al. 2001; Kim et al. 1996). These nucleases have been demonstrated to target several genes in the cells of different organisms, including humans. The DNA recognition domain of TALENs is tandem repeats of a 34 amino acid domain, in which amino acids at positions 12 and 13 (NN, NG, NI, HD or NK) are responsible for recognition of nucleotides, A, T, G or C (Miller et al. 2010).

CRISPR/Cas9 Nucleases

The CRISPR/Cas system is the most widely used RNA based genome editing platform. Typically, the CRISPR recognition sequence begins with a guanine (G) and ends with NGG sequence (PAM sequence) and can be 20 to 23 bp long (Mali et al. 2013; Fu et al. 2014). Recently, several different variants of CRISPR nucleases have been developed with a variety of PAM sequences (Kleinstiver et al. 2015). One of the advantages of the CRISPR/Cas system is multiplexing, which enables the targeting of multiple loci in a single cell. Multiplex CRISPR based genome editing can achieve higher rates of gene modifications and help in exploring phenotypes of multigenic diseases (Kabadi et al. 2014; Cong et al. 2013). For specificity enhancement, activity and applicability of the CRISPR/Cas system, several creative strategies have been employed by researchers, e.g., by using truncated guide RNA (Fu et al. 2013; Singh et al. 2018), the fusion of inactivated Cas9 with fokI (Guilinger et al. 2014), paired nickases (Ran et al. 2013) and development of smaller Cas9 (Ran et al. 2015).

Base Editors

The fusion of mutant impaired Cas9 with deaminases led to the development of platforms for base editing. Base editors can perform irreversible single-nucleotide modifications. Cytidine base editors execute the deamination of cytosine and enable the transition from C/G to T/A base pairs. Similarly, adenine base editors deaminate adenine to yield inosine, which is treated as guanine by polymerases (Gaudelli et al. 2017).

Prime Editors

Prime editing can help in targeting those sites which have PAM availability constraints and can be used to induce targeted insertions, deletions and base changes. Prime editing is a nickase based platform, which decreases the possibility of inducing unwanted insertions and deletions. Prime editor consists of the Cas9 H840A nickase fused to the N-terminal of M-MLV reverse transcriptase. The second component of the prime editing system is the prime editing gRNA (pegRNA). The pegRNA encodes for the reverse transcriptase templates and includes the desired editing sequence and homology to the target locus. The complementary sequence to the targeted genomic DNA site serves as the pegRNA or primer binding site (PBS). At this PBS binding site, the reverse transcriptions can be initiated. The pegRNA primer binding site should be at least 8 nucleotides long to induce efficient prime editing in HEK293T cells (Anzalone et al. 2019).

Gene Activation

It is believed that by using the genome editing platforms for altering nucleotide sequences in genomic DNA, we are just scratching the surface of their potential. The human genome possesses not only a genetic code but also a code in epigenetic status (methylation, acetylation, etc.) and histone arrangement. The transcriptional regulation in eukaryotes primarily depends on the chromatin state. In eukaryotes, the core promoters of the genes are inherently inactive because the ground state of eukaryotic promoters is restrictive. Thus, transcription of all eukaryotic genes requires activators.

RuvCI and HNH nuclease domains of Cas9 have been modified to generate an inactive Cas9, which retains the property of binding to the gRNA-DNA complex but does not induce DNA breaks. This inactive Cas9 is referred to as dead Cas9 (dCas9). The dCas9 was then fused to different transcription activation domains (Gilbert et al. 2014). These dCas9 fused activation molecules can be used to induce targeted activation of gamma-chains. Targeted activation of gamma-globin chains can help in inducing HbF synthesis. However, such activation mechanisms are usually transient and will require administration at frequent intervals.

Gene Repression

The fusion of inactive Cas9 with a repressor like KRAB can decrease the expression of target genes by impairing transcription (Gilbert et al. 2014). The targeting guide RNA is designed to target the promoter or enhancer region of the target gene of interest. KRAB domain recruits several histone modifiers, which reversibly repress the gene transcription.

Gene Delivery

The ability to achieve targeted gene editing solves one aspect of correcting genetic diseases. To achieve the effective correction of a disease phenotype, we should be able to deliver the gene editing molecules to target cells and tissues.

An ideal gene delivery vector:

- delivers to a specific cell type
- can package a gene of sufficient size
- expresses the gene for sufficient time to induce editing and fades without a trace after producing the desired effect
- does not induce an immunological reaction
- does not induce an oncogenic alteration
- does not integrate into the genome.

For correcting hemoglobinopathies, the cells involved in hematopoiesis (hematopoietic stem cells; HSCs) - erythroid lineage cells, are the likely targets. HSCs are responsible for the continuous replenishment of blood cells. Hematopoietic Stem cells (HSCs) differentiate into multiple lineage of progenitor cells — lymphoid and myeloid progenitors. Human HSCs are Lin- CD34+ CD38- CD90+ CD45RA- CD49f+ and lack the markers for lineage commitment (Doulatov et al. 2012). The capacity of HSCs to self-renew and the ability to generate progenitor cells help in the adequate lifetime production of the blood cells.

Viral Vectors

For gene delivery applications, it is desirable to distinguish viruses from viral vectors. The viruses are naturally occurring small obligate parasites containing DNA or RNA surrounded by a protein envelope. However, for gene delivery, we harness the machinery of these viruses and modify their genomes such that they exhibit characteristics closer to an ideal delivery system. A multitude of vectors are available as a choice for delivering gene editing molecules — DNA viral vectors (AAV, adenoviral), RNA viral vectors (retroviral, lentiviral, Sendai) and nanovectors (nanoparticles, dendrimers). This chapter briefly covers some viral vector systems, with a focus on application for correcting hemoglobinopathies.

Retroviral Vectors

Retroviral vectors are RNA viruses harboring reverse transcriptase, which can reverse transcribe viral RNA to double-stranded DNA. This DNA can get integrated into the host cell and is capable of producing all viral components. The examples of retroviral vectors are murine leukemia based viral vectors and lentiviral vectors. The key difference between lentiviral and other retroviral vectors is that lentiviral vectors can transduce both dividing and non-dividing cells, while most other retroviral vectors can only transduce dividing cells. Lentiviral vectors are known to transduce HSCs effectively (Magrin et al. 2019).

Adenoviral Vectors

Adenoviral vectors can transduce dividing and non-dividing cells. Double-stranded adenoviral vectors can be modified to insert DNA sequencing up to 30 kilobases (kb) and result in transient expression of genes. However, adenoviral vectors can result in high immune response and possess a tendency to transduce a wide variety of cells.

Adeno-associated Viral Vectors

Similar to adenoviral vectors (AdV), adeno-associated viral vectors (AAVs) can transduce both dividing and non-dividing cells. AAVs induce a low immune response, long-term gene expression and are not pathogenic. Adeno-associated viral vectors can be single-stranded and double-stranded. Single-stranded AAVs (ssAAVs) have packaging capacity of ~4.9 kb. Packaging capacity of self-complementary AAVs (scAAVs) is half of ssAAVs.

Sendai Viral Vectors

Sendai viral vectors (SeV) consist of negative RNA strand surrounded by a spherical envelope (Nakanishi, et al. 2012). Compared to lentiviral and retroviral vectors, SeV does not have any DNA intermediate in its life cycles. Recently, it has been remodeled to deliver CRISPR/Cas9 molecules (Park et al. 2016). SeV has never been known to cause human disease and eliminates the risk of unwanted DNA integration. Moreover, recombinant SeV can target HSCs (Ban et al. 2011; Jin et al. 2003).

Non-Viral Vectors

The non-viral methods of gene delivery encompass liposomes, gold nanoparticles, dendrimers, polymeric micelles, and poly-β-amino esters. Liposomal nanoparticles (LNPs) are non-immunogenic (unless an antigen is packaged in the particle), can co-deliver nucleic acids with other drugs and can be functionalized for targeted delivery. However, the efficiency of delivery is lower, as compared to viral vectors. Similar to LNPs, dendrimers can self-assemble with nucleic acids, are non-immunogenic and can be functionalized for targeted delivery. However, if cationic lipids are used for self-assembly, these may be cytotoxic to the cells. Liposomal nanoparticles made with poly-β-amino esters are biodegradable and lower cytotoxicity than other cationic lipids and may have higher transfection efficiency that other liposomal nanoparticle (Caffery et al. 2019).

The nanoparticle-mediated gene delivery to HSCs can be potentially advantageous for increasing the safety of gene editing. However, the development of chemical combinations that can deliver efficiently to HSCs is a challenging research endeavor. A study using high-throughput methods involving next-generation sequencing can help in studying the chemical and biological combinations, which can result in efficient and safe delivery (Damoiseaux et al. 2011).

Genome Editing Methods Applied to Beta-Hemoglobinopathies

Beta-thalassemia has defective production of beta-globin. Induction of gamma-globin chains can be of therapeutic benefit in beta-thalassemia. These gamma-chains can pair with alpha-chains to synthesize HbF. This also decreases the imbalance of excessive α-chains, which is pathognomic of beta-thalassemia. Further, this can result in the improvement of effective erythropoiesis and less hemolysis. Different drugs like 5-azacytidine, hydroxyurea and short-chain fatty acids, thalidomide (Atweh et al. 2001; Musallam et al. 2013) vorinostat (Mettananda et al. 2019) have been in use for induction of HbF synthesis. Some of these agents act by inducing epigenetic alterations by lesser known mechanisms.

In principle, CRISPR-based genetic and epigenetic editing systems can be used to transiently and permanently induce HbF synthesis, while CRISPR-based repressors can be used to suppress mutant and excessive beta-chains.

Gene disruption, gene integration and gene correction are the primary approaches which can be applied to achieve phenotypic correction in beta-hemoglobinopathies.

Gene Addition and Native Gene Correction *In Situ*

Addition of gamma and beta globin genes using lentiviral vectors containing non-coding and coding regions of beta and gamma-globin genes and anti-sickling variants of beta-globin have been used to correct preclinical models (Pawliuk et al. 2001; Rivella et al. 2003; Miccio et al. 2008; Cavazzana-Calvo et al. 2010).

Additionally, nuclease-mediated HDR can be used for genomic integration or repair of essential globin gene sequences for targeted insertion of gene-expression cassettes, knock-in of crucial gene sequences (promoters, enhancers, coding sequences) and insertion into a safe harbor locus (such as Rosa26 locus).

Methods like HDR and prime editing provide the opportunity to permanently restore the mutated gene. Small or large deletions, point mutations and frameshift mutations in the globin genes can be amenable to correction by these strategies. HDR is an error-free strategy and has less probability of unwanted changes; however, research is needed to get enough efficient editing to achieve therapeutic benefit. Prime editing can help in correcting the mutations which lack the PAM sequence in the targeted regions.

Gene Disruption

Known repressors of the gamma-globin chain (BCL11A and ZBTB7A) can be disrupted to increase the synthesis of fetal hemoglobin. BCL11A targeting is currently under clinical trials to induce fetal hemoglobin. Disruption of a 200 bp genomic region in the second intron of the BCL11A gene and a 5-bp GATAA motif in BCL11A erythroid enhancer are the vital regulatory elements.

Gene Modifications

Certain naturally existing single-nucleotide variants and QTLs (Table 2) are known to affect the levels of fetal hemoglobin. Depending on the variant, a combination of gene editing methods can potentially be used to induce the targeted variation in the genome, to enhance fetal hemoglobin levels.

Ex-Vivo Stem Cell Correction and Transplantation

Allogenic hematopoietic stem cell transplantation (AHSCT) is the only curative treatment of thalassemia. However, this treatment is limited by the availability of HLA-matched donors. Allogenic hematopoietic stem cell transplantation can induce immunotoxicity, therefore autologous HSCT might be preferable. HSCs are low in proportion (less than 0.01%) in the clinically available bone marrow samples. Consequently, *ex vivo* expansion and modification are desirable. Subsequently, the transplantation of corrected HSCs can result in the amelioration of symptoms (Cavazzana et al. 2018; Psatha et al. 2018).

Most clinical trials use autologous CD34+ cells, which are transduced with a viral vector, usually lentiviral vector and then, the transduced cells are infused back into the patient after employing myeloablative regiment of busulfan.

Prenatal Delivery of Gene Editing Therapeutics

Several different systemic and local routes of administration are possible depending on the therapeutic agent and the target tissue or cells. Recently, prenatal gene delivery has been adopted to perform liver-specific therapeutic base editing. Because a fetus has larger stem cell fraction, compared to adults, therapeutic gene editing can be applicable for treating hemoglobinopathies. The prenatal administration provides several advantages — access to larger fraction stem cell pool, low dose requirement and less-developed immune and physical barriers for gene delivery. However, like any new technology, deliberate attention for evaluation of safety is required (Miccio et al. 2008; Khosravi et al. 2019).

Gene Editing in Clinical Trials for Correction of Beta-Hemoglobinopathies

Several clinical trials are underway to test gene editing and gene replacement vectors and molecules to treat hemoglobinopathies. BCL11A gene is one of the frequent targets in several clinical trials. BCL11A is a repressor of fetal hemoglobin and its disruption can induce levels of fetal hemoglobin.

Table 4. Genes in the beta-globin cluster locus 11p15.4

Type of hemoglobin	Synonyms	Remarks
Hemoglobin beta	Beta-globin, ECYT6, CD113t-C,	Responsible for familial autosomal dominant erythrocytosis-6 and results in formation of hemoglobin with high oxygen affinity.
Hemoglobin Delta	HBD	Cause thalassemia due to hemoglobin lepore
Noncoding beta-globin locus transcript 3	BGLT3, LINC01083	BGLT3 is a long noncoding RNA and is involved in the regulation of the pathways responsible for chronic myeloid leukemia (CML).
Hemoglobin gamma A	HBG1	
Hemoglobin gamma G	HBG2, TNCY	Responsible for transient neonatal cyanosis.
Hemoglobin epsilon	HBE1	Responsible for transient neonatal cyanosis.
Locus control region beta	LCRB	Organizes the beta-globin gene cluster into euchromatin and regulates the transcription of genes in beta-globin cluster.

Table 5. Active, recruiting gene editing and gene therapy clinical trials for treating thalassemia and sickle cell disease

Clinicaltrials.gov identifier	Condition	Name of the drug	Technology	Organization/Company	Phase of clinical trial	Route of administration
NCT03432364	beta-thalassemia	ST-400	ZFNs to disrupt BCL11A gene to induce HbF production	Sangamo Therapeutics and Sanofi	Phase 1/2	Transplantation of genetically-modified patient's own cells using ZFNs to disrupt enhancer of the BCL11A gene to increase HbF levels after conditioning the bone marrow with busulfan.
NCT03207009, NCT02906202	beta-thalassemia	LentiGlobin BB305 Lentiviral βA-T87Q-Globin Vector	Autologous CD34+ Stem Cells Transduced Ex Vivo With a Lentiviral βA-T87Q-Globin Vector encoding the human βA-T87Q-globin gene	Bluebird Bio	Phase 3	Transplantation of Autologous CD34+ Stem Cells Transduced Ex Vivo With a Lentiviral βA-T87Q-Globin Vector
NCT03655678, NCT03745287	beta-thalassemia, severe sickle cell disease	CTX001	CRISPR-Cas9 Modified autologous CD34+ Human Hematopoietic Stem and Progenitor Cells (hHSPCs) using CTX001 at erythroid lineage-specific enhancer of the BCL11A gene	Vertex Therapeutics and CRISPR Therapeutics	Phase 1/2	Transfusion of modified autologous CD34+ Human Hematopoietic Stem and Progenitor Cells after conditioning the bone marrow with busulfan.

Table 5. (Continued)

Clinicaltrials.gov identifier	Condition	Name of the drug	Technology	Organization/Company	Phase of clinical trial	Route of administration
NCT02247843	Sickle cell disease	Lenti/G-βAS3-FB lentiviral vector	Expression of an anti-sickling (βAS3) gene in peripheral blood CD34+ cells	University of California and California Institute of Regenerative Medicine	Phase 1/2	Transfusion of patient's own peripheral blood CD34+ positive cells transduced with Lenti/G-βAS3-FB lentiviral vector to express an anti-sickling (βAS3) gene after conditioning the bone marrow with busulfan.

Sangamo Therapeutics is testing ZFNs to induce fetal hemoglobin, by targeting BCL11A in patient's own cells, followed by transplantation. Vertex Pharmaceuticals and CRISPR Therapeutics are testing the efficacy of their CRISPR/Cas9 based product CTX001 (targets erythroid lineage-specific enhancer of the BCL11A gene) for transfusion dependent beta-thalassemia and sickle cell disease. In this study, the CRISPR/Cas9 modified autologous CD34+ hHSPC (Human Hematopoietic Stem and Progenitor Stem cell) are infused by intravenous route after myeloablation by busulfan.

Table 5 summarizes the on-going clinical trials for the use of gene editing and gene therapy to correct hemoglobinopathies.

CONCLUSION

Considering the variety of mutations responsible for beta-thalassemia the use of molecular tools to correct disease needs to be precisely customized for the type of mutation. The technology to manipulate and alter genes has allowed the possibility to reduce the severity or provide the cure for the hemoglobinopathies. After decades of exploration and optimization, gene-editing tools are opening the opportunity for therapy, especially for monogenic diseases. Use of adenine base editors allows for reversing, thus correcting, G-> A or C-> T mutations, and Cytidine base editors can correct T-> C or A-> G mutations.

Off-target editing by gene editors remains a challenge. Optimizations and modifications of the gRNA design strategy can allow for specific gene editing, which can cause less frequent off-target mutations, as compared to the DNA synthesis and repair mechanisms in the cells. Notable advances in gene cloning methods, engineering viral vectors and synthesis of custom oligonucleotides contribute in achieving efficient and specific gene editing.

References

Anzalone, A. V. et al. Search-and-replace genome editing without double-strand breaks or donor DNA. *Nature* 1–9 (2019) doi:10.1038/s41586-019-1711-4.

Atweh, G. F. & Loukopoulos, D. Pharmacological induction of fetal hemoglobin in sickle cell disease and [beta]-thalassemia. *Semin. Hematol.* 38, 367–373 (2001).

Ban, H. et al. Efficient generation of transgene-free human induced pluripotent stem cells (iPSCs) by temperature-sensitive Sendai virus vectors. *Proc. Natl. Acad. Sci. USA.* 108, 14234–14239 (2011).

Bibikova, M. et al. Stimulation of Homologous Recombination through Targeted Cleavage by Chimeric Nucleases. *Mol. Cell. Biol.* 21, 289–297 (2001).

Breveglieri, G. et al. An Aγ-globin G-> A gene polymorphism associated with β039 thalassemia globin gene and high fetal hemoglobin production. *BMC Med. Genet.* 18, 93 (2017).

Caffery, B., Lee, J. S. & Alexander-Bryant, A. A. Vectors for Glioblastoma Gene Therapy: Viral & Non-Viral Delivery Strategies. *Nanomaterials* 9, 105 (2019).

Cavazzana-Calvo, M. et al. Transfusion independence and HMGA2 activation after gene therapy of human β-thalassaemia. *Nature* 467, 318–322 (2010).

Cavazzana, M. & Mavilio, F. Gene Therapy for Hemoglobinopathies. *Hum. Gene Ther.* 29, 1106–1113 (2018).

Chen, D. et al. A Genetic Variant Ameliorates β-Thalassemia Severity by Epigenetic-Mediated Elevation of Human Fetal Hemoglobin Expression. *Am. J. Hum. Genet.* 101, 130–138 (2017).

Cong, L. et al. Multiplex Genome Engineering Using CRISPR/Cas Systems. *Science* 339, 819–823 (2013).

Craig, J. E. et al. Dissecting the loci controlling fetal haemoglobin production on chromosomes 11p and 6q by the regressive approach. *Nat. Genet.* 12, 58–64 (1996).

Damoiseaux, R. et al. No Time To Lose - High Throughput Screening To Assess Nanomaterial Safety. *Nanoscale* 3, 1345–1360 (2011).

Doulatov, S., Notta, F., Laurenti, E. & Dick, J. E. Hematopoiesis: A Human Perspective. *Cell Stem Cell* 10, 120–136 (2012).

Forget, B. G. & Bunn, H. F. Classification of the Disorders of Hemoglobin. *Cold Spring Harb. Perspect. Med.* 3, a011684 (2013).

Fu, Y., Sander, J. D., Reyon, D., Cascio, V. M. & Joung, J. K. Improving CRISPR-Cas nuclease specificity using truncated guide RNAs. *Nat. Biotechnol.* 32, 279–284 (2014).

Galarneau, G. et al. Fine-mapping at three loci known to affect fetal hemoglobin levels explains additional genetic variation. *Nat. Genet.* 42, 1049–1051 (2010).

Garner, C. et al. Haplotype mapping of a major quantitative-trait locus for fetal hemoglobin production, on chromosome 6q23. *Am. J. Hum. Genet.* 62, 1468–1474 (1998).

Gaudelli, N. M. et al. Programmable base editing of A•T to G•C in genomic DNA without DNA cleavage. *Nature* 551, 464–471 (2017).

Gilbert, L. A. et al. Genome-Scale CRISPR-Mediated Control of Gene Repression and Activation. *Cell* 159, 647–661 (2014).

Guilinger, J. P., Thompson, D. B. & Liu, D. R. Fusion of catalytically inactive Cas9 to FokI nuclease improves the specificity of genome modification. *Nat. Biotechnol.* advance online publication, (2014).

Ivaldi, M. S. et al. Fetal γ-globin genes are regulated by the BGLT3 long noncoding RNA locus. *Blood* 132, 1963–1973 (2018).

Jin, C. H. et al. Recombinant Sendai virus provides a highly efficient gene transfer into human cord blood-derived hematopoietic stem cells. *Gene Ther.* 10, 272–277 (2003).7.

Kabadi, A. M., Ousterout, D. G., Hilton, I. B. & Gersbach, C. A. Multiplex CRISPR/Cas9-based genome engineering from a single lentiviral vector. *Nucleic Acids Res.* gku749 (2014) doi:10.1093/nar/gku749.

Khosravi, M. A. et al. Targeted deletion of BCL11A gene by CRISPR-Cas9 system for fetal hemoglobin reactivation: A promising approach for gene therapy of beta thalassemia disease. *Eur. J. Pharmacol.* 854, 398–405 (2019).

Kim, Y. G., Cha, J. & Chandrasegaran, S. Hybrid restriction enzymes: zinc finger fusions to Fok I cleavage domain. *Proc. Natl. Acad. Sci. U. S. A.* 93, 1156 (1996).

Kleinstiver, B. P. et al. Engineered CRISPR-Cas9 nucleases with altered PAM specificities. *Nature* 523, 481–485 (2015).

Mali, P. et al. RNA-Guided Human Genome Engineering via Cas9. *Science* 339, 823–826 (2013).

Magrin, E., Miccio, A. & Cavazzana, M. Lentiviral and genome-editing strategies for the treatment of β-hemoglobinopathies. *Blood* 134, 1203–1213 (2019).

Martyn, G. E. et al. Natural regulatory mutations elevate the fetal globin gene via disruption of BCL11A or ZBTB7A binding. *Nat. Genet.* 50, 498–503 (2018).

Mettananda, S. et al. Synergistic silencing of α-globin and induction of γ-globin by histone deacetylase inhibitor, vorinostat as a potential therapy for β-thalassaemia. *Sci. Rep.* 9, 1–8 (2019).

Mettananda, S., Gibbons, R. J. & Higgs, D. R. α-Globin as a molecular target in the treatment of β-thalassemia. *Blood* 125, 3694–3701 (2015).

Miccio, A. et al. *In vivo* selection of genetically modified erythroblastic progenitors leads to long-term correction of beta-thalassemia. *Proc. Natl. Acad. Sci. U. S. A.* 105, 10547–10552 (2008).

Miller, J. C. et al. A TALE nuclease architecture for efficient genome editing. *Nat. Biotechnol.* 29, 143–148 (2010).

Musallam, K. M., Taher, A. T., Cappellini, M. D. & Sankaran, V. G. Clinical experience with fetal hemoglobin induction therapy in patients with β-thalassemia. *Blood* 121, 2199–2212 (2013).

Nakanishi, M. & Otsu, M. Development of Sendai Virus Vectors and their Potential Applications in Gene Therapy and Regenerative Medicine. *Curr. Gene Ther.* 12, 410–416 (2012).

OMIM Gene Map - Chromosome: 16 https://omim.org/geneMap/16/18?start=-3&limit=10&highlight=18].

Park, A. et al. Sendai virus, an RNA virus with no risk of genomic integration, delivers CRISPR/Cas9 for efficient gene editing. *Mol. Ther. Methods Clin. Dev.* 3, 16057 (2016).

Paquet, D. et al. Efficient introduction of specific homozygous and heterozygous mutations using CRISPR/Cas9. *Nature* 533, 125–129 (2016).

Patsali, P. et al. The Scope for Thalassemia Gene Therapy by Disruption of Aberrant Regulatory Elements. *J. Clin. Med.* 8, 1959 (2019).

Pawliuk, R. et al. Correction of sickle cell disease in transgenic mouse models by gene therapy. *Science* 294, 2368–2371 (2001).

Psatha, N. et al. Disruption of the BCL11A Erythroid Enhancer Reactivates Fetal Hemoglobin in Erythroid Cells of Patients with β-Thalassemia Major. *Mol. Ther. - Methods Clin. Dev.* 10, 313–326 (2018).

Ran, F. A. et al. Double Nicking by RNA-Guided CRISPR Cas9 for Enhanced Genome Editing Specificity. *Cell* 154, 1380–1389 (2013).

Ran, F. A. et al. *In vivo* genome editing using Staphylococcus aureus Cas9. *Nature* 520, 186–191 (2015).

Rivella, S., May, C., Chadburn, A., Rivière, I. & Sadelain, M. A novel murine model of Cooley anemia and its rescue by lentiviral-mediated human beta-globin gene transfer. *Blood* 101, 2932–2939 (2003).

Singh, K. et al. Efficient *In Vivo* Liver-Directed Gene Editing using CRISPR/CAS9. *Mol. Ther.* (2018) doi:10.1016/j.ymthe.2018.02.023.

Smith, J. et al. A combinatorial approach to create artificial homing endonucleases cleaving chosen sequences. *Nucleic Acids Res.* 34, e149 (2006).

Thein, S. L. et al. Detection of a major gene for heterocellular hereditary persistence of fetal hemoglobin after accounting for genetic modifiers. *Am. J. Hum. Genet.* 54, 214–228 (1994).

Thein, S. L. et al. Intergenic variants of HBS1L-MYB are responsible for a major quantitative trait locus on chromosome 6q23 influencing fetal hemoglobin levels in adults. *Proc. Natl. Acad. Sci.* 104, 11346–11351 (2007).

Thein, S. L. The Molecular Basis of β-Thalassemia. *Cold Spring Harb. Perspect. Med.* 3, a011700 (2013).

Wang, H. & Xu, X. Microhomology-mediated end joining: new players join the team. *Cell Biosci.* 7, 6 (2017).

In: Fetal Hemoglobin
Editor: Nishant Chakravorty

ISBN: 978-1-53617-955-2

Chapter 7

GENETIC MODIFIERS OF FETAL HEMOGLOBIN AND THEIR ROLE IN ALLEVIATING THE CLINICAL SEVERITY OF HEMOGLOBINOPATHIES

Anita Nadkarni* and Priya Hariharan
Department of Hematogenetics,
ICMR- National Institute of Immunohematology.
Mumbai – 400012, Maharashtra, India

ABSTRACT

β-thalassemia and sickle cell anemia, although are inherited as single monogenic recessive disorders, display a wide spectrum of clinical heterogeneity. Recent studies have established that elevated levels of fetal

* Corresponding Author's Email: anitahnadkarni@yahoo.com.

hemoglobin may improve the clinical severity of hemoglobinopathies. In both, β-thalassemia and sickle cell anemia, persistent production of fetal hemoglobin interferes with the pathophysiology of the disease, either by correcting the globin chain imbalance or by reducing the intracellular levels of sickle hemoglobin, respectively. The major genetic modifier loci for HbF persistence are: HBS1L-MYB, BCL11A and the β-globin gene cluster-Xmn1-HBG2. This chapter will review the heterogeneity and major genetic factors that participate in regulating the fetal hemoglobin levels and their impact on the clinical severity of hemoglobinopathy patients. Further the chapter will discuss therapies relevant to HbF induction. This will eventually help in understanding the mechanism of fetal hemoglobin regulation in humans and may assist in apt identification of promising therapeutic targets for HbF induction in these patients. Further, genetic association studies in the context of fetal hemoglobin may help in predicting the clinical course of the disease and aid in better clinical management of these diseases.

Keywords: Hemoglobinopathies, BCL11A, HBS1L-MYB, HbF levels

INTRODUCTION

Inherited defects in the globin portion or in the heme group which lead to hereditary anemias are termed as hemoglobinopathies. Hemoglobinopathies are broadly classified into 3 types: the quantitative defects that leads to reduction in the globin chain synthesis (Thalassemia), the qualitative defects which results in the synthesis of structurally abnormal hemoglobin variants (HbS, HbE, HbD Punjab) and the defect in the switch-over process of the globin chain synthesis from fetal to adult hemoglobin that causes hereditary persistence of fetal hemoglobin (HPFH) phenotype (Sarnaik, 2005).

With the carrier frequency varying from 3 to 17% among different population groups in India, β-thalassemia and sickle cell anemia form a major health burden causing high degree of morbidity due to moderate to severe hemolytic anemia (Balgir, 2000). Both the diseases are caused by mutations in the β-globin gene and are inherited as autosomal recessive single gene disorders. However, despite of this apparent genetic simplicity, both the disorders display a remarkable spectrum of phenotypic severity (Thein, 2013). In β-thalassemics, the primary determinant of disease severity is the type of β-globin gene mutation that the patient inherits. Severe β-thalassemia mutation (β^0/β^+) may completely down regulate β-globin chain synthesis by abrupt β-globin chain termination or by affecting the splicing of pre m-RNA. The milder β^{++} mutations are the point mutations present in the conserved 5′ β-globin promoter region and 3′ untranslated region that alter the mRNA expression, leading to moderate synthesis of the β-globin chain.

However, it has been observed that patients with β-hemoglobinopathies having same mutation may show variable disease severity. Thus, the clinical variability observed amongst patients prompted scientists to search for the additional genetic modulators of these diseases.

Several modifier genes have been identified which influence the severity of hemoglobinopathies. The most common secondary modifier is the co-inheritance of α-thalassemia (deletions in the α-globin gene) and the elevated levels of fetal hemoglobin (HbF) (consisting of 2 α and 2 γ globin chains), both of which directly reduce the globin chain imbalance.

The possibility that the clinical severity of hemoglobinopathies could be reduced by induction of fetal hemoglobin (HbF) was realized when it was established that the symptoms of hemoglobinopathies unfold only after 6 months of age when there is gradual reduction in the HbF levels. Besides the physiological switchover of fetal to adult hemoglobin, co-inheritance of genetic determinants can lead to persistence of fetal hemoglobin in adult life. This eventually helps in reducing the α/non α globin chain imbalance and hence ameliorating the severity of hemoglobinopathies. Elevated levels of γ-globin chains combine with the excess free α-globin chains and reduce the detrimental effects of free alpha precipitation on RBC precursors in β-thalassemia (Rund and Fucharoen, 2008). Similarly, in sickle cell anemia, the raised HbF levels compete with β^S globins to bind with alpha globin chains and thus, reduce the intracellular HbS concentration.

With the advent of genome wide association studies, several factors, linked and unlinked to the β-globin gene cluster that can elevate HbF levels have been identified (Badens et al., 2011). Three major quantitative trait loci

(QTLs), namely, the polymorphisms in the HBB locus, BCL11A gene and HBSB1L-cMYB inter-region, along with KLF1 gene variations, bring about 20-50% variation in HbF levels in hemoglobinopathy patients (Badens et al., 2011). Studies have demonstrated that the combined effect of co-inheriting these polymorphisms aid in ameliorating the clinical severity of hemoglobinopathy patients. Predictions based on genetic modifiers like the type of β-globin gene mutations, co-inheritance of α-thalassemia, polymorphisms in the BCL11A and HBS1L-MYB intergenic gene, may foretell the major or intermedia phenotype of β-thalassemia (Galanello et al., 2009). A similar modulating effect can also be observed in sickle cell anemia patients.

Thus, this chapter emphasizes on the complexity of genetic interactions among the fetal hemoglobin modulators that eventually modulate the clinical phenotype of β-thalassemia and sickle cell anemia through HbF induction. Simultaneous inheritance of genetic loci that elevate the γ-globin gene expression along with hemoglobinopathy condition may probably lead to higher basal HbF levels resulting in clinically milder form of the disease. Hence, with this background, we have discussed and analyzed the effects of variations in the γ-globin promoter region, BCL11A, HBS1L-MYB intergenic region and KLF1 on the HbF levels and, have tried to correlate the combined effect of the variant alleles on reducing the disease severity of the hemoglobinopathy patients. The elucidation of the genetic modifiers of fetal hemoglobin may hopefully help in rational and effective management of hemoglobinopathies.

Genetic Modifiers of Fetal Hemoglobin

In the late 1950s, it was observed that patients with Cooley's anemia (thalassemia) survived beyond the period of the fetal to adult hemoglobin switch owing to the prolonged synthesis of fetal hemoglobin in the erythroid cells (Philipsen, 2013). The continued γ- globin gene expression in β-thalassemia may lead to reduction in the globin chain imbalance, (Ngo et al.,

2015). Similarly, the beneficial role of fetal hemoglobin in reducing the sickle cell disease severity was observed by Janet Watson more than 60 years ago. Watson observed that infants with sickle cell disease, having raised HbF levels, presented with milder clinical symptoms and their deoxygenated erythrocytes do not deform as extensively as the sickle cell trait-carrying mother's cells (Watson et al., 1948). The pathophysiology of sickle cell disease is dependent on the polymerization of deoxy sickle hemoglobin under low oxygen conditions, and this phenomenon is retarded due to increased HbF concentration. Elevated HbF levels dilute the intracellular HbS concentration below the threshold level that is required for polymerization. It is hypothesized that both HbF ($\alpha 2\gamma 2$) and its mixed hybrid tetramer ($\alpha 2\beta S\gamma$), interfere with polymerization of the sickle hemoglobin under deoxy state, thus circumventing the cellular damage evoked by HbS polymers (Akinsheye et al., 2011). In both of the hemoglobinopathy conditions, increased γ-globin gene expression acts as a well-known disease modifier. Hence, understanding the molecular mechanism of hemoglobin switching and identification of molecular targets, for reversing fetal to adult hemoglobin switch, is a subject of intense research (Philipsen, 2013). With the dawn of the molecular era, different models for gene switching have suggested that this process involves complex interaction between the cis acting elements (DNA sequences present on the same loci or chromosome) that include the locus control region of the beta globin gene cluster, and trans acting transcription factors which co-coordinately carry out chromatin remodeling activities (Bank, 2006).

Transcription factors form a repressome complex which aid in decreasing the γ-globin expression by blocking the γ-promoter interaction with the locus control region (LCR). Furthermore, the δ-globin gene promoter region is blocked by PYR complex and thus the interaction of LCR with the β-globin gene promoter is enhanced allowing for the adult hemoglobin production (Figure 1).

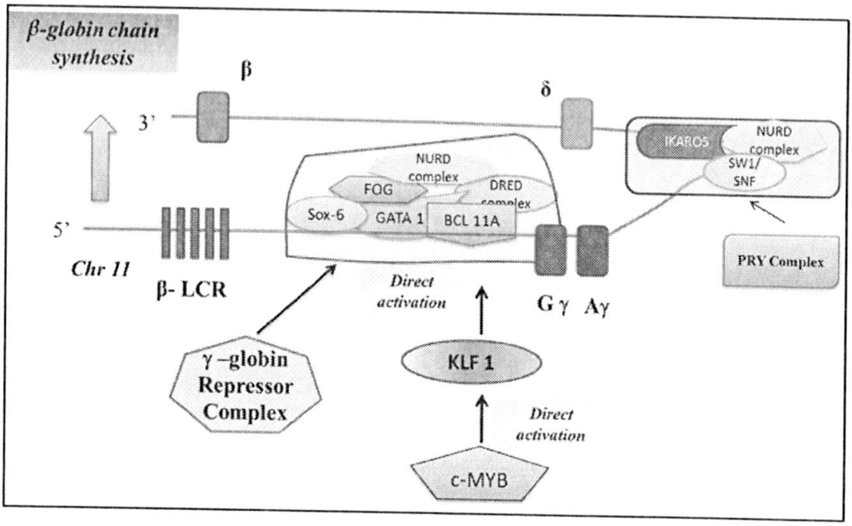

Figure 1. Fetal to adult hemoglobin switch process. The γ-globin to β-globin gene expression switch process is carried out in presence of various transcription factors (repressor complex and PYR complex), which enhance the interaction of LCR with the β-globin gene promoter leading to the β-globin expression instead of γ-globin expression.

The cis- Acting Elements

The β-globin gene cluster contains a 5′ distal regulatory element known as Locus Control Region (LCR), which has 5 active DNAase I hypersensitive sites (HS) 1-5, and these are required for sequential expression of globin genes. Wijgerde et al., proposed a competitive model for regulation of globin genes wherein the LCR participates in the long range looping interaction that leads to only one single productive LCR-globin gene interaction (Bauer and Orkin, 2011). The potent transcriptional activation of the LCR is also due to the clustering of various erythroid transcription factors that bind to the hypersensitivity sites. They act in concert to facilitate the long range looping activity. The duplicated CCAAT box (−84 to −87, −111 to −114), a CACCC box (−140 to −144), and a TATA box (−27 to

–30) are the most crucial regions in the γ–globin gene expression. The TATA box has binding site for RNA-polymerase II.

The CCAAT and the CACCC boxes comprise the proximal promoter regions where regulatory transcription factors bind, and thus influence the transcription rate by RNA polymerase II (Rottgardt et al., 2010). The cis-acting promoter sequences in the individual globin genes and the LCR, thus serve as the templates for binding of various transcription factors (Bauer and Orkin, 2011). It has been observed that the naturally occurring point/ small insertional/ deletional mutations in the γ-globin (HBG2, HBG1) promoter regions, may result in the activation of the γ-globin and hence the synthesis of the Aγ globin or the Gγ globin chains. Natural promoter mutations at -202, -175, -158 of the Gγ globin gene and at -202, -198, -196, -175, -117, in addition to a 4 base deletion in Aγ globin gene (-225 to -222, -AGCA) are associated with prolonged γ globin expression in adult life (Huang et al., 2000). Presence of these mutations ablates the protein binding motifs and eventually the binding of the repressors or enhancers to the site, thus affecting the gene expression. A T→C γ-globin promoter polymorphism at -175 has been found to be associated with raised HbF levels. This polymorphism is known to ablate the binding of GATA-1 and Octamer 1 transcription factor. Thus, these transcription factors perhaps have a repressor role γ globin gene regulation. Similarly, the promoter sequence from -206 to -217 was shown to be the binding site for the repressor complex. Presence of SNPs in this region might be able to disrupt the binding of the repressor complex to this region, thereby inducing the expression of γ-globin gene (Pace and Zien, 2006). Recently, direct binding of BCL11A to both the Gγ and Aγ globin promoter regions has been found to be disrupted in presence of classical HPFH alleles in -117 to -114 region (Liu et al., 2017). This leads to, to increased levels of HbF. Table 1 depicts the γ-globin promoter mutations identified in different population groups and their association with raised HbF levels. In India, the most common polymorphism detected in the γ-globin promoter region is –158 Gγ globin C→T, XmnI polymorphism. The T allele of XmnI polymorphism has been found to be linked to raised HbF levels and milder hemoglobinopathy conditions (Nadkarni et al., 2001). Although, T allele in its homozygous state seems to have little effect on HbF levels in normal individuals, it has been found to be associated with increase in HbF levels, especially during the erythropoietic stress, indicating its role in HbF induction (Mastropietro et al., 2002). The XmnI polymorphism produces a HPFH phenotype by altering the binding of GATA1 and repressor complex to γ-globin promoter region (Pace and Zien, 2006).

In a study on Indian population, the T allele in its homozygous state has been found to be more commonly encountered in thalassemia intermedia patients (12.5%) as compared to thalassemia major patients (5.0%) (Nadkarni et al., 2001). Globally the prevalence of XmnI polymorphism in hemoglobinopathy patients has been found to be extremely diverse (Nadkarni et al., 2001). Nadir et al. 2014 examined the frequency of XmnI polymorphism in Northern Pakistan province among β-thalassemia homozygotes. They observed that 8.0% of β-thalassemia homozygous patients, 8.0% with β-thalassemia trait and 2% of the normal controls were homozygous for the T allele of XmnI polymorphism (Nadir et al., 2014). In a similar study the promoter regions HS-111 and 3′ HS1 regions of Aγ, Gγ - were screened. Among the polymorphisms detected, the T allele of XmnI polymorphism, the A allele at −588 position (A→G) in the Aγ-globin promoter and the G allele of HS111 (-21 A→G), were found to be significantly associated with β-thalassemia intermedia when compared with thalassemia major and this was also linked to elevated HbF levels (Hamid et al., 2009). Said et al. (2015) studied the frequency of XmnI polymorphism in Egyptian β-thalassemia homozygous patients. They observed that 83.3% of the β-thalassemia intermedia cases exhibit heterozygosity for XmnI polymorphism. In contrast to this, 57.6% of β-thalassemia major (57.6%) patients were found to have heterozygous polymorphism for XmnI. The β-thalassemia intermedia patients with single T allele of XmnI showed delayed age of diagnosis, raised HbF levels and milder disease phenotype as compared to patients negative for the XmnI polymorphism (Said et al., 2015). Another study conducted on Iranian population found that 26.7% of β-thalassemia intermedia patients showed homozygosity for the T allele of XmnI polymorphism patients. This was also found to be significantly associated with higher mean HbF levels (85.± 6.8%) as compared to the thalassemia intermedia patients homozygous for XmnI CC genotype (19.5%±29.3) (Miri-Moghaddam et al., 2017). The effect of co-inheritance of γ-globin promoter variations in β-thalassemics in elevating the HbF levels was also observed in β-thalassemia trait individuals. An Indian study has observed that compound heterozygotes for novel γ-globin gene mutation (-ATAAG, -533 to -529) and β-globin gene mutation (619 bp deletion) were healthy, with no history of blood transfusion, and showed exceptionally high level of HbF (HbF: 31.7% and 23.6%) (Hariharan et al., 2017).

Table 1. γ-globin promoter mutations identified in different populations

Mutation	%Hb F (Heterozygotes)	Origin
	$^{G}\gamma$–gene	
-567 (T→G)	5.9-10.2	Iranian
-202 (C→G)	14-21	African
-196 (C→T)	8.6	Greek
-175 (T→C)	17-30	African
-114 (C→G)	8.6	Australian
-114 (C→T)	11-14	Japanese
-110 (A→C)	3.1	Czech
-109 (G→T)	4.1	Greek
	$^{A}\gamma$-gene	
-202 (C→T)	1.6-3.4	African
-201 (C→T)	10.2	Greek
-198 (T→C)	3.5-12	British
-196 (C→T)	14-21	Italian, Chinese
-195 (C→G)	4.5-7.0	Brazilian
-158 (C→T)	2.9-5.1	Greek
-117 (G→A)	7.1-19	Greek, Italian, African

The β^{S} gene which causes sickle cell anemia (SCA) has been found to be associated with five different restriction haplotypes and have been named according to the association of the specific region and the specific β-cluster haplotypes, namely: Senegal, Benin, Bantu, Cameroon and Arab Indian. It has been suggested that the severity of the SCA correlates with the β^S haplotypes (da Silva et al., 2014). The Senegal and the Arab Indian haplotypes are associated with raised HbF levels and milder disease course; as compared to the Bantu haplotype, which is associated with lower levels of HbF and a severe clinical course (Adekile, 2005). Among Indian sickle cell anemia patients, the Arab-Indian haplotype, which is linked to XmnI polymorphism, is found to be the most prevalent haplotype. In 2013, Bhagat et al. studied the correlation of β-globin cluster haplotypes with the HbF levels in SCA patients from Chhattisgarh region. They observed that 65% of the patients showed homozygosity for XmnI polymorphism and Arab Indian Haplotype (HbF: 22.7% ± 5.2), 27% showed heterozygosity for Arab Indian Haplotype (HbF: 20.8% ± 3.02) and 8% showed complete negativity to Arab Indian Haplotype (HbF: 8.7% ± 0.29). This observation indicates to the fact that the presence of Arab-Indian Haplotype in Indian SCA patients may inherently increase the mean HbF levels. However, it has been observed that the patients with same γ-globin promoter genotype, showed varied HbF levels, and this could possibly be due to inheritance of other HbF modifying alleles. A 2010 study on Brazilian sickle cell anemia patients determined that the -396 to -391 deletion, -369 C→G and -309 A→G SNPs in the HBG2 promoter and -271 C→T SNP in the Aγ globin promoter were significantly associated with HbF levels >5.0% (Barbosa et al., 2010). These data suggest that the γ-globin promoter variations play a pivotal role in elevating the HbF levels under stressed erythropoietic conditions.

Trans Acting Factors

Recently, recognition of variability in expression of HbF in non-anemic individuals and patients with β-hemoglobin disorders have stimulated efforts to identify the trans acting genetic factors (the genes encoding transcription factors that regulate the expression of other genes) that may lead to this deviation. Genome wide association mapping studies have identified quantitative trait loci (QTLs) in the transcription factor encoding genes, which play critical roles in the globin gene regulation and hemoglobin switching mechanism. It has been observed that polymorphisms in the genes coding for the negative regulators of γ-globin gene expression have been associated with raised HbF levels. Thus, in this section, various genes and its polymorphisms have been discussed to understand their association with the HbF levels.

B-Cell CLL/Lymphoma 11A [BCL11A]

B-cell CLL/lymphoma 11A is a zinc finger transcription protein which is encoded by BCL11A gene located in chromosome 2p16. The BCL11A gene is expressed in hematopoietic progenitors and plays a pivotal role in lymphocyte development and neurogenesis (Saiki et al., 2000). BCL11A binds to the GC rich motifs of the genes and mediates transcriptional repression (Sankaran et al., 2008). This protein is expressed in 4 different iso-forms (XS, S, L, and XL) and the mutations have also been associated with B-cell malignancies (Liu et al., 2003).

The first genome-wide association study (GWAS) for F cell numbers and HbF level provided the major breakthrough that identified BCL11A as a regulator of γ-globin gene. The role of BCL11A polymorphisms in intron 2 region in modulation of fetal hemoglobin levels was first demonstrated by Uda et al. in 2008 by genome wide association studies in the Sardinian population. The variants of BCL11A were found to be significantly associated with increase in F-cell numbers in hemoglobinopathy patients (Uda et al., 2008). This finding suggests that the second intronic region of BCL11A has regulatory elements or functional motifs responsible for modulating the γ globin gene expression. Eventually, the association studies of BCL11A in diverse ethnic groups around the world showed a strong linkage of BCL11A polymorphisms with the raised HbF levels and in ameliorating the clinical severity in hemoglobinopathy patients. Further, Sankaran et al. demonstrated that BCL11A may function as a potent repressor of the γ- globin gene. There finding was based on the fact that lower levels of BCL11A mRNA is expressed in variants leading to higher levels of HbF (Sankaran et al., 2010). In an earlier study, Sankaran et al. had also demonstrated that BCL11A acts as a direct repressor for γ-globin gene. Knockdown studies of BCL11A in primary erythroid cells demonstrated elevated γ-globin gene expression and subsequent increase in HbF levels (Sankaran et al., 2008). BCL11A has also been shown to bind to discrete sites in the β-globin gene cluster, indicating a direct role of BCL11A in globin gene regulation (Sankaran et al., 2008; Jawaid et al., 2010). Furthermore, erythroid-specific knockout of BCL11A has been found to rescue the phenotypes of genetically engineered SCD mouse models owing to high-level of pancellular HbF expression. Chromatin Immuno Precipitation (ChIP) technology demonstrated that BCL11A, interacts with NuRD chromatin remodeling and repressor complex, erythroid transcription factors – GATA -1, FOG -1, SOX6 and Matrin-3 to silence the gamma globin gene (Sankaran et al., 2010; Xu et al., 2010). All these suggest the pivotal role of BCL11A in gamma globin silencing. It has been observed that the haplo-deficiency of BCL11A leads to a remarkable increase in HbF levels (~15-30%) without affecting lymphopoiesis. These clinical observations highlight the therapeutic value of BCL11A knockdown, which is expected to clinically benefit the hemoglobinopathy patients by inducing the HbF levels (Orkin, 2016). In India, the influence of QTLs on HbF level was studied by Bhanushali et al., in 2014. It was observed that the minor allele T of rs1427407 G→T BCL11A polymorphism showed statistically significant association with HbF levels in the tribal SCA patients from Chhattisgarh region (Bhanushali et al., 2014). Similarly, in a study by Upadhye et al., they observed an association of minor allele A of rs4671393 A→G with higher HbF levels and milder disease phenotype in SCA patients from central India (Upadhye et al., 2016). This is in contrast to the findings of an earlier study by Roy et al., where they did not observe any significant association of this polymorphism with HbF levels in hemoglobinopathy patients from eastern region of India (Roy et al., 2012).

Table 2. BCL11A SNPs associated with raised HbF levels in different populations

HbF inducing Polymorphism	Minor Allele	Population	Patient group	Reference
rs11886868 (C→T)	C	Sardinian (European)	β-thalassemia	Uda M et al., 2008
rs4671393 (A→G)	A	Brazilian	Sickle cell disease	Lettre G et al., 2008
rs766432 (A→C)	C	Chinese, Thai, African American	β-thalassemia trait, HbE trait, SCA	Sedgewick A et al., 2008
rs4671393 (A→G)	A	China (Guangxi)	HbE-β-thalassemia	He Y et al., 2012
rs4671393 (A→G)	A	Cameroon	Sickle cell anemia (SCA)	Wonkam A et al., 2014
rs1427407 (G→T)	T	Tanzanian	SCA	Mtatiro S et al., 2015
rs11886868 (C→T)	C	Indian	β-thalassemia and SCA	Dadeech S et al., 2016

The pioneering experiments on the association of BCL11A polymorphisms with elevated HbF levels and their effect on amelioration of the disease phenotype, was studied by Uda et al., in Sardinian β-thalassemia homozygous patients (Uda et al., 2008). They showed that the C allele of rs11886868 (C→T) formed the major allele in Sardinian population and was significantly associated with elevated HbF levels in the patients. Further, the allelic frequency of C allele was found to be significantly higher in β-thalassemia intermedia patient group, thus indicating the C allele of rs11886868 may contribute in developing a milder disease phenotype (Uda et al., 2008). Correspondingly, another study exhibited the frequency of the C and T alleles of rs11886868 (C→T) in Indian β-thalassemia major and SCA patients (Dadheech et al., 2016). This study found that the C allele formed the minor allele (40%) and it was significantly associated with raised HbF levels and delayed the age of disease presentation in both thalassemia homozygous and SCA group (Dadheech et al., 2016). In a similar study, it was demonstrated that the Indonesian HbE-β-thalassemia patients inheriting the variant alleles (rs11886868, rs766432) of the BCL11A gene showed higher HbF levels and reduced disease severity as compared to patients with wild type alleles (Rujito et al., 2016). In their population, they indicated that the C allele of rs11886868 C→T polymorphism ameliorated the clinical severity of only HbE-β thalassemia patients and not the β-thalassemia intermedia patients (Rujito et al., 2016). In 2014, Bhanushali et al., showed a similar pattern of allelic frequency of rs1427407 (G→T) in Indian SCA patients. They determined that the patients with T allele showed significantly higher HbF levels (p: 0.045) (with a variance in HbF of 23%). In all the patient groups, the presence of T allele showed higher HbF levels with a significant association seen only in thalassemia intermedia and SCA patients (Bhanushali A et al., 2014). Similarly, Chaouch et al., observed that co-inheritance of the C allele of the rs11886868 and the A allele of the rs46713939, ameliorated the clinical phenotype of SCA patients (Chaouch et al., 2016). The polymorphism, rs7606173 (G→C) has been reported to be located in the core enhancer region within the BCL11A gene. A study in 2015 has reported that the presence of minor allele C of this polymorphism is significantly associated with HbF levels sickle cell anemia patients from Cameroon (Pule et al., 2015). Table 2 enlists the BCL11A SNPs associated with raised HbF levels in different populations.

HBS1L- MYB Intergenic Zone [HBS1L-MYB]

Jiang and co-workers were the first to propose the role of MYB in influencing HbF and F cell levels by altering the erythroid kinetics (Jiang et al., 2006). With the genome wide micromapping studies, it has been observed that a set of SNPs located on chromosome 6q23 region is associated with important erythroid traits, especially persistence of fetal hemoglobin (HbF) in adults and diagnostic erythroid parameters like RBC, MCV, MCH (Menzel et al., 2007; Thein et al., 2007). The functional significance of this region was also supported by studies in murine models, in which a transgene inserted in *HBS1L-MYB* intergenic region completely abolished *Myb* transcription and resulted in severe anemia (Mukai et al.,

2006). With the aim of studying the genetic variants associated with the HbF regulation, genome wide association studies identified loci located within 126-kb intergenic region, between the HBS1L and MYB genes. This region is distributed in 3 disequilibrium blocks which are referred to as HBS1L-MYB intergenic polymorphism (HMIP) block 1, 2, 3. Amongst these, block 2, that spans a 24 kb region, has been found to be associated with reduced disease severity in patients with hemoglobinopathies and this has been attributed to increase HbF levels. The presence of regulatory domain in block 2 of MYB intergenic region has been suggested by the presence of histone acetylation, GATA-1, and RNA polymerase II interactions in block 2 (Wahlberg K et al., 2009). Three polymorphisms, namely, rs28384513, rs9399137 and rs4895441 in *HBS1L-MYB* have been found to be robustly associated with elevated HbF levels (Lettre G et al., 2008). These have been observed to ameliorate the phenotypic painful crisis in SCA patients. Although the function of HBS1L in red blood cell development is uncharacterized; the c-Myb transcription factor is known to play a key role in hematopoiesis and erythropoiesis (Stadhouders et al., 2014). The importance of the intergenic region was further strengthened by an observation that the 2 regulatory segments (-84 and -71) bind to transcription factors like KLF1 and LBD1. Such bindings bring long-range chromatin interactions between the intergenic region and the *MYB* promoter and thus regulate the MYB gene expression (Stadhouders et al., 2014). Presence of SNPs in this region, abolish the binding of these transcriptional complexes to this region.

Another mechanism whereby MYB influences HbF levels is by direct activation of KLF1, followed by BCL11A activation. In one study shRNA-mediated knockdown of MYB demonstrated down regulation of BCL11A and KLF1 mRNA expression, thus, suggesting that MYB could be a critical up stream regulator for these genes (Roosjen et al., 2014). The functional importance of the intergenic region was first observed when transgene insertion within the murine HBS1L-MYB intergenic region completely disrupted the binding of the core complex proteins LDB1, GATA1, TAL1 and ETO2, and abolished MYB transcription which in turn resulted in severe anemia (Stadhouders et al., 2014). The most effective of these, are the variants residing within the core enhancer binding region in HMIP-2A and -2B (Farrell et al., 2011). A 3-bp deletion (rs66650371) at HMIP-2A is suspected to directly cause HbF variability. A transcriptional regulatory network has been recently described within the erythroid cells, where KLF1 and MYB alternatively activate each other by forming an Active Chromatin Hub (ACH) structure which brings intergenic enhancer regions in proximity with the MYB promoter (Farrell et al., 2011).

Krüppel Like Factor 1 [KLF 1]

Ira Miller and James Bieker were the first to identify an erythroid specific transcription factor, Erythroid Krüppel like factor (EKLF) or KLF1 that plays a vital role in erythropoiesis (Miller and Bieker, 1993). This founding member of Krüppel like factor family derived its name from the Drosophila body pattern determining Krüppel–gap gene, with which it showed a zinc finger homology (Miller and Bieker, 1993). KLF1 is present on chromosome 19p13.2 and spans through 3 kb of genomic DNA. It consists of three exons which synthesize the KLF1 protein (362 amino acids). KLF1 is a zinc finger transcription factor and has two functional domains: an amino terminal proline rich trans-activation domain that binds to the transcription factors and a carboxy-terminal zinc finger domain that identifies and binds to the CACCC motif (found in regulatory regions of many erythroid genes). This erythroid gene regulator participates in 3 important processes which include, erythroid lineage commitment and initiation of erythropoiesis, the fetal (γ-globin) to adult (β-globin) hemoglobin switch-over process and activation of erythroid specific genes (encoding for globin chain synthesis, heme synthesis enzymes, membrane and cytoskeleton protein, ion channel protein, erythroid metabolic enzymes and cell cycle regulators) (Perkins et al., 2016). KLF1 is also known to modulate the expression of other transcription factors that work together in regulatory networks controlling gene expression in erythroid cells (Grech and Borg, 2014). Expression of KLF1 gene is observed in megakaryocyte-erythroid progenitor (MEP) cells and has been found to play a major role in establishment of erythroid lineage. Its expression is seen to selectively increase in cells of erythroid lineage and this is simultaneous with its downregulation in megakaryocyte progenitor cells (Siatecka and Bieker, 2011). Similarly, its role as a transcription regulator was shown Drissen R et al., 2004, where they found that KLF1 mediates globin gene regulation by formation of active chromatin hub and increasing the interaction of LCR with the CACCC box of the β-globin promoter. This directly activates the expression of β-globin gene (Drissen et al., 2004). The phenomenon was further validated by knockdown of KLF1 gene in mouse embryo that succumbed to severe anemia (Grech and Borg, 2014).

The downstream effects of KLF1 mutations result in a wide range of hematological phenotypes (Waye and Eng, 2015). Until recently, more than 60 KLF1 gene mutations have been enlisted in the Leiden Open Variation Database (LOVD). The deficiency of KLF1 has been associated with several different hematological phenotypes, ranging from benign conditions like persistent elevation of fetal hemoglobin, borderline HbA2 levels, marginal microcytosis and/or hypochromia and the rare In (Lu) blood type or severe hemolytic anemia (Waye and Eng, 2015).

While Singleton and team, primarily highlighted the association of KLF1 gene mutations with clinically benign In (Lu) blood group (2008), Borg and others., were among the pioneers who identified co-inheritance of 2 KLF1 gene variants, p. K288X and p. M39L in, a Maltese family, and found them to be associated with HPFH phenotype (2010) (Singleton et al., 2008; Borg et al., 2010). A similar observation was made in a Sardinian family, wherein compound heterozygotes for 2 KLF1 mutations, S270X and K332Q, showed raised HbF levels of 30.9% in presence of normal hematological indices (Satta et al., 2011). Another similar study reported that HbE patients with KLF1 mutations showed elevation in the HbF levels when compared to the patients with normal KLF1 gene (Tepakhan et al., 2016). It has been reported that KLF1 gene variations are detected in regions where hemoglobinopathies are prevalent. In 2014, Liu and colleagues examined the correlation of KLF1 mutations with β thalassemia, by screening individuals from thalassemia endemic and non-endemic regions of China. They identified 7 functional variants: p.Gly176AlafsX179, p.His299Asp (common variants accounting for 90.6%), p.Ala298Pro, p.Thr334Arg, and c.91311G→A, p.Cys341Tyr, and p.Glu5Lys, prevailing in thalassemia endemic region. Furthermore, zinc finger mutations in KLF1 were selectively present in milder thalassemia homozygous patients (Liu et al., 2014). This observation potentially implicates the complex role of KLF1 under stressed erythropoietic conditions that lead to increase in HbF levels. Similarly, 14 different mutations in KLF1 gene were detected in patients with HbE disorder (Tepakhan et al., 2016). The study determined that KLF1 mutations were relatively common in HbE homozygous patients with elevated HbF levels (mean HbF levels: 10.9%), thus, affirming that KLF1 mutations can be associated with increased HbF production and may help in ameliorating the clinical and hematological features of β-hemoglobinopathies (Tepakhan et al., 2016). KLF1's participation in repression of gamma globin levels via BCL11A activation was proven by conducting gene knock down assays. These experiments showed that the reduction of KLF1 gene expression was markedly associated with decrease in BCL11A levels and this in turn was associated with increase in gamma globin at mRNA levels (Zhou et al., 2010). Thus, it is important to validate and establish the bipartite role of KLF1 gene in activation of β-globin gene and simultaneous suppression of γ globin gene expression. KLF1 plays a critical role in regulating the developmental switch between fetal and adult hemoglobin expression, by direct activation of β-globin along with an indirect repression of γ-globin gene expression in adult erythroid progenitors through BCL11A activation (Siatecka and Bieker, 2011) (Figure 2). These findings suggest that the KLF1 may play an important role in regulating the fetal hemoglobin levels and variations in the KLF1 gene may lead to raised HbF levels.

Combined Inheritance of Polymorphisms and their Role in Ameliorating the Disease Severity

It is known that the HbF modifying polymorphisms, when co-inherited together, ameliorate the clinical severity of a hemoglobinopathy through HbF induction. Further, it has been observed that polymorphisms in HMIP

2A block, which include, rs66650371 and rs9399137, are more strongly associated with HbF levels than others and a strong linkage is observed amongst the 2 polymorphism (Mtatrio et al., 2015). Additionally, a study on β-thalassemia intermedia patients from the Guangxi Province of China has demonstrated that the variant alleles of rs9376090, rs7776054, rs9399137, rs9389268, rs9402685 in the HbS1L-MYB intergenic region and rs189984760 in the BCL11A locus, showed significant association with high HbF levels (Lai et al., 2017).

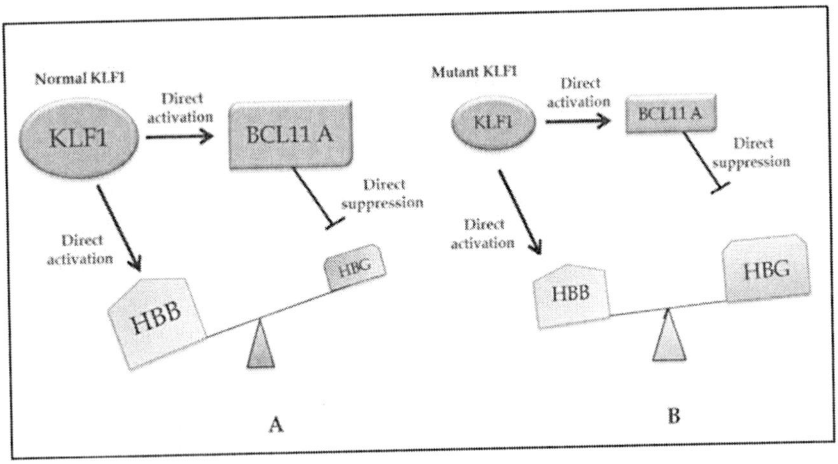

Figure 2. Schematic representation of globin gene regulation by KLF1 and BCL11A. (Reframed from: Philipsen, S. 2013) A: Normal KLF1 directly activates BCL11A and β-globin gene (HBB) and represses the expression of γ-globin gene (HBG). B: Mutant KLF1 partially reverses the hemoglobin switch process by modest elevation γ globin gene expression and with partial reduction in β-globin gene expression.

A study on 5 genetic modifiers of β-thalassemia, which included the type of β-thalassemia mutations: the XmnI Polymorphism, co-inheritance of α-thalassemia, rs11886868 BCL11A polymorphism and HBS1L-MYB intergenic rs9399137 was reported in 2011 (Badens et al., 2011). Regression analysis has revealed that all the 5 types of favorable alleles were significantly found to be associated with thalassemia intermedia phenotype, and that β-globin gene mutations and XmnI polymorphism were the most influential modifiers of the disease severity. A similar observation was reported in 2012, wherein the researchers observed that the phenotypic diversity in disease severity is greatly attributed to the loci modulating the HbF levels (XmnI polymorphisms, HbS1L-MYB and BCL 11A loci) (Danju et al., 2012). They further evaluated the age of first transfusion with respect to the inheritance of HbF boosting alleles and observed that the age of transfusion was found to be delayed in presence of higher number of HbF inducing alleles (Danju et al., 2012). It has been observed that 92% of β-thalassemia intermedia patients inherit more than 2 positive alleles as compared to thalassemia major patients. In β-thalassemia major patients, 65% have been found to carry only 1 or no modifying alleles. These findings lead researchers to speculate that the presence of higher number of disease ameliorating alleles may reduce the disease severity (Galanello et al., 2009). A similar result was shown in a study on Nigerian SCA patients. The study found that the HbF boosting allele T of rs1427407 (G→T), located in the BCL11A gene and the 3 base pair deletional allele in HBS1L intergenic region lead to significantly higher levels of HbF and thus a milder form of the disease with improved hemoglobin levels (Adeyemo et al., 2018). A comparable result was reported in 2015, where the authors determined the genetic modifiers contributing to the β-thalassemia intermedia phenotype in 102 Iraqi Arab β-thalassemia homozygous patients. The main factors leading to milder phenotypes were: the attenuated β-thalassemia alleles, the T allele of XmnI polymorphism and minor allele of BCL11A rs10189857 (Allawi et al., 2015). A study published in 2013 investigated the prevalence of five genetic modifiers (β-globin mutations, coinheritance of α-thalassemia, XmnI polymorphism and single nucleotide polymorphisms (SNPs) in the BCL11A and HBS1L-MYB loci in 306 β-thalassemia patients and predicted the severity of β-thalassemia (Banan et al., 2013). It was observed that the β^{++} mutation and the XmnI variant allele were the strongest predictors of β-thalassemia phenotype. Further, it was reported that the β-thalassemia major patients had inherited less than 2 modifying alleles as opposed to the β-thalassemia intermedia patients who inherited more than 2 modifying alleles.

CONCLUSION

Although β-hemoglobinopathies are well characterized at molecular level, it has been observed that patients with same genotype show differential clinical symptoms. The current treatment strategies for sickle cell anemia (SCA) and β-thalassemia include symptomatic care and regular blood transfusion. Considering the facts that the present therapeutic agents are known to have innumerable side effects and that persistent expression of fetal hemoglobin (HbF) ameliorates the clinical severity of hemoglobinopathies, the study of genetic factors influencing the γ-globin gene expression is highly important. The variability in the clinical severity is associated with various genetic factors amongst which the HbF modifiers have gained importance in the recent genome wide association studies. Thus, this chapter highlights the functional role of the main genetic factors involved in the regulation of γ-globin gene expression. It also expands the knowledge on the effect of polymorphisms present in these loci on elevating the HbF levels. The analysis of the cumulative effect of the HbF modulators may help in identifying the strongest response gene to the HbF level in both β-thalassemia and sickle cell anemia patients. It may also foresee the severity of β-thalassemia and SCA, which may further assist the clinicians, to predict the clinical phenotype of hemoglobinopathy patients at an early stage and thus, may help in the efficient management of the disease.

REFERENCES

Adekile, A. (2005). Mild-phenotype sickle cell disease: molecular basis, clinical presentation and management recommendations. *Current Paediatrics*, *15*, 57-61.

Akinsheye, I., Abdulrahman, A., Solovieff, N., Ngo, D., Baldwin, C., Sebastiani, P., Chui, D. & Steinberg, M. (2011). Fetal hemoglobin in sickle cell anemia. *Blood*, *118*, 19-27.

Al-Allawi, N., Puehringer, H., Raheem, R. & Oberkanins, C. (2015). Genetic modifiers in β-thalassemia intermedia: A study on 102 Iraqi Arab patients. *Genetic testing and molecular biomarkers*, *19*, 242-247.

Ali, N., Ayyub, M., Khan, S. A., Abbas, K., Malik, H. S. & Tashfeen, S. (2015). Frequency of Gγ-globin promoter− 158 (C> T) XmnI polymorphism in patients with homozygous/compound heterozygous beta thalassaemia. *Hematology/oncology and stem cell therapy*, *8*, 10-15.

Badens, C., Joly, P., Agouti, I., Thuret, I., Gonnet, K., Fattoum, S., Francina, A., Simeoni, M. C., Loundou, A. & Pissard, S. (2011). Variants in genetic modifiers of β-thalassemia can help to predict the major or intermedia type of the disease. *Haematologica*, *96*, 1712-1714.

Balgir, R. (2000). The burden of haemoglobinopathies in India and the challenges ahead. *Current Science*, *10*, 1536-1547.

Banan, M., Bayat, H., Namdar, A. P., Azarkeivan, A., Kamali, K., Daneshmand P., Zaker, K. B. & Najmabadi, H. (2013). Utility of the multivariate approach in predicting β-thalassemia intermedia or β-thalassemia major types in Iranian patients. *Hemoglobin*, *37*, 413-422.

Bank, A. (2006). Regulation of human fetal hemoglobin: new players, new complexities. *Blood*, *107*, 435-443.

Barbosa, C. G., Goncalves, S. N., Souza, R. S. B., deMoura, N., José, P., Takahashi, D., Silva, D., Hurtado, G. A., Galvão dos, R. M. & de Souza, G. M. (2010). Promoter region sequence differences in the A and G gamma globin genes of Brazilian sickle cell anemia patients. *Brazilian journal of medical and biological research*, *43*, 705-711.

Bauer, D. & Orkin, S. (2011). Update on fetal hemoglobin gene regulation in hemoglobinopathies. *Current opinion in pediatrics*, *23*, 1.

Bhagat, S., Patra, P. K. & Thakur, A. S. (2013). Fetal haemoglobin and β-globin gene cluster haplotypes among sickle cell patients in Chhattisgarh. *Journal of clinical and diagnostic research*, *7*, 269.

Bhanushali, A. A., Patra, P. K., Nair, D., Verma, H. & Das, B. R. (2015). Genetic variant in the BCL11A (rs1427407), but not HBS1-MYB (rs6934903) loci associate with fetal hemoglobin levels in Indian sickle cell disease patients. *Blood Cells, Molecules, and Disease*, *54*, 4-8.

Borg, J., Papadopoulos, P., Georgitsi, M., Gutiérrez, L., Grech, G., Fanis, P., Phylactides, M., Verkerk, A., vanderSpek, P., Scerri, C., & Cassar, W. (2010). Haploinsufficiency for the erythroid transcription factor KLF1 causes hereditary persistence of fetal hemoglobin. *Nature genetics*, *42*, 801-805.

Chaouch, L., Moumni, I., Ouragini, H., Darragi, I., Kalai, M., Chaouachi, D., Boudrigua, I., Hafsia, R. & Abbes, S. (2016). rs11886868 and rs4671393 of BCL11A associated with HbF level variation and modulate clinical events among sickle cell anemia patients. *Hematology*, *21*, 425-429.

daSilva, M. A. L., Friedrisch, J., Bittar, C., Urnau, M., Merzoni, J., Valim, V., Amorin, B., Pezzi, A, Chies, J. A. & Silla, L. (2014). ß-Globin gene cluster haplotypes and clinical severity in sickle cell anemia patients in Southern Brazil. *Open Journal of Blood Diseases*, *4*, 16.

Dadheech, S., Madhulatha, D., Jain, S., Joseph, J., Jyothy, A. & Munshi, A. (2016). Association of BCL11A genetic variant (rs11886868) with severity in β-thalassaemia major & sickle cell anaemia. *The Indian journal of medical research*, *143*, 449.

Danjou, F., Anni, F., Perseu, L., Satta, S., Dessì, C., Lai, M. E., Fortina, P., Devoto, M. & Galanello, R. (2012). Genetic modifiers of β-thalassemia and clinical severity as assessed by age at first transfusion. *Haematologica*, *97*, 989-993.

Drissen, R. P., Robert, J., Gillemans, N., Splinter, E., Grosveld, F., Philipsen, S. & deLaat, W. (2004). The active spatial organization of the β-globin locus requires the transcription factor EKLF. *Genes & development*, *18*, 2485-2490.

Farrell, J. J., Sherva, R., Chen, Z., Luo, H., Chu, B., Ha, S. Y., Li, C., Lee, A., Li, R., Li, C. & Yuen, H. (2011). A 3-bp deletion in the HBS1L-MYB intergenic region on chromosome 6q23 is associated with HbF expression. *Blood*, *117*, 4935-4945.

Galanello, R., Sanna, S., Perseu, L., Sollaino, M., Satta, S., Lai, M., Barella, S., Uda, M., Usala, G., Abecasis, G. & Cao, A. (2009). Amelioration of Sardinian β0 thalassemia by genetic modifiers. *Blood*, *114*, 3935-3937.

Grech, L. & Borg, J. (2014). Control of globin gene expression by Kruppel-like Factors. *Journal of The Malta Chamber of Scientists*, *2*, 66-77.

Hamid, M., Mahjoubi, F., Akbari, M., Arab, A., Zeinali, S. & Karimipoor, M. (2009). Molecular analysis of γ-globin promoters, HS-111 and 3′ HS1, in β-thalassemia intermedia patients associated with high levels of Hb F. *Hemoglobin*, *33*, 428-438.

Hariharan, P., Sawant, M., Gorivale, M., Manchanda, R., Colah, R., Ghosh, K. & Nadkarni, A. (2017). Synergistic effect of two β globin gene cluster mutations leading to the hereditary persistence of fetal hemoglobin (HPFH) phenotype. *Molecular biology reports*, *44*, 413-417.

He, Y., Chen, P., Li, W. & Luo, J. (2012). Analysis of rs4671393 polymorphism in hemoglobin E/β-thalassemia major in guangxi province of China. *Journal of pediatric hematology/oncology*, *34*, 323-324.

Huang, X. D., Yang, X., Huang, R. B., Zhang, H. Y., Zhao, H. L., Zhao, Y. J., Huang, C. L., Hou, C. J. & Zhang, J. W. (2000). A novel four base-pair deletion within the Aγ-GLOBin gene promoter associated with slight increase of Aγ expression in adult. *American journal of hematology*, *63*, 16-19.

Jawaid, K., Wahlberg, K., Thein, S. & Best, S. (2010). Binding patterns of BCL11A in the globin and GATA1 loci and characterization of the BCL11A fetal hemoglobin locus. *Blood Cells, Molecules, and Diseases*, *45*, 140-146.

Jiang, J., Best, S., Menzel, S., Silver, N., Lai, M., Surdulescu, G., Spector, T. & Thein, S. (2006). cMYB is involved in the regulation of fetal hemoglobin production in adults. *Blood*, *108*, 1077-1083.

Langdon, S. & Kaufman, R. (1998). Gamma-globin gene promoter elements required for interaction with globin enhancers. *Blood*, *91*, 309-318.

Lettre, G., Sankaran, V., Bezerra, M., Araújo, A., Uda, M., Sanna, S., Cao, A., Schlessinger, D., Costa, F., Hirschhorn, J. & Orkin, S. (2008). DNA polymorphisms at the BCL11A, HBS1L-MYB, and β-globin loci associate with fetal hemoglobin levels and pain crises in sickle cell disease. *Proceedings of the National Academy of Sciences*, *105*, 11869-11874.

Liu, N., Hargreaves, V., Zhu, Q., Kurland, J., Hong, J., Kim, W., Sher, F., Macias-Trevino, C., Rogers, J., Kurita, R. &Nakamura, Y. (2018). Direct promoter repression by BCL11A controls the fetal to adult hemoglobin switch. *Cell*, *173*, 430-442.

Liu, D., Zhang, X., Yu, L., Cai, R., Ma, X., Zheng, C., Zhou, Y., Liu, Q., Wei, X., Lin, L. & Yan, T. (2014). KLF1 mutations are relatively more common in a thalassemia endemic region and ameliorate the severity of β-thalassemia. *Blood*, *124*, 803-811.

Liu, P., Jonathan, O., Mariaestela, T. L., Rachel, R., Nakamura, T., Jenkins, N. & Copeland, N. (2003). Bcl11a is essential for normal lymphoid development. *Nature immunology*, *4*, 525.

Mastropietro, F., Modiano, G., Cappabianca, M., Foglietta, E., D'Asero, C., Mezzabotta, M., Ponzini, D., Maffei, L., Amato, A., Lerone, M. & Grisanti, P. (2002). Factors regulating Hb F synthesis in thalassemic diseases. *BMC Hematology*, *2*, 2.

Menzel, S., Jiang, J, Silver, N., Gallagher, J., Cunningham, J., Surdulescu, G., Lathrop, M., Farrall, M., Spector, T. & Thein, S. (2007). The HBS1L-MYB intergenic region on chromosome 6q23. 3 influences erythrocyte, platelet, and monocyte counts in humans. *Blood*. 110: 3624-3626.

Miller, I. & Bieker, J. (1993). A novel, erythroid cell-specific murine transcription factor that binds to the CACCC element and is related to the Krüppel family of nuclear proteins. *Molecular and cellular biology*, *13*, 2776-2786.

Miri-Moghaddam, E., Bahrami, S., Naderi, M., Bazi, A. & Karimipoor, M. (2017). Xmn1-158 Gγ variant in B-thalassemia intermediate patients in South-East of Iran. *International journal of hematology-oncology and stem cell research*, *11*, 165.

Mtatiro, S, Mgaya, J., Singh, T., Mariki, H., Rooks, H., Soka, D., Mmbando, B., Thein, S., Barrett, J., Makani, J. & Cox, S. (2015). Genetic association of fetal-hemoglobin levels in individuals with sickle cell disease in Tanzania maps to conserved regulatory elements within the MYB core enhancer. *BMC medical genetics*, *16*, 4.

Mukai, H., Motohashi, H., Ohneda, O., Suzuki, N., Nagano, M. & Yamamoto, M. (2006). Transgene insertion in proximity to thec-myb gene disrupts erythroid-megakaryocytic lineage bifurcation. *Molecular and cellular biology*, *26*, 7953-7965.

Nadkarni, A., Gorakshakar, A., Lu, C., Krishnamoorthy, R., Ghosh, K., Colah, R. & Mohanty, D. (2001). Molecular pathogenesis and clinical variability of β-thalassemia syndromes among Indians. *American journal of hematology*, *68*, 75-80.

Ngo, D. & Steinberg, M. (2015). Genomic approaches to identifying targets for treating β hemoglobinopathies. *BMC medical genomics*, *8*, 44.

Orkin, S. (2016). Recent advances in globin research using genome-wide association studies and gene editing. *Annals of the New York Academy of Sciences*, *1368*, 5-10.

Pace, B. & Zein, S. (2006). Understanding mechanisms of γ-globin gene regulation to develop strategies for pharmacological fetal hemoglobin induction. *Developmental dynamics: an official publication of the American Association of Anatomists*, *235*, 1727-1737.

Perkins, A., Xu, X., Higgs, D., Patrinos, G., Arnaud, L., Bieker, J. & Philipsen, S. (2016). Krüppeling erythropoiesis: an unexpected broad spectrum of human red blood cell disorders due to KLF1 variants. *Blood*, *127*, 1856-1862.

Philipsen, S. (2013). Molecular control of hemoglobin switching. In *Hematology Education: the education programme for the annual congress of the European Hematology Association*, 7, 1-9.

Pule, G., Ngo, B., Valentina, C., Chemegni, B., Kengne, A., Antonarakis, S. & Wonkam A. (2015). Association between variants at BCL11A erythroid-specific enhancer and fetal hemoglobin levels among sickle cell disease patients in Cameroon: implications for future therapeutic interventions. *Omics: a journal of integrative biology*, *19*, 627-631.

Roosjen, M., McColl, B., Kao, B., Gearing, L., Blewitt, M. & Vadolas, J. (2014). Transcriptional regulators Myb and BCL11A interplay with DNA methyltransferase 1 in developmental silencing of embryonic and fetal β-like globin genes. *The FASEB Journal*, *28*, 1610-1620.

Rottgardt, I., Rothhammer, F. & Dittmar, M. (2010). Native highland and lowland populations differ in γ-globin gene promoter polymorphisms related to altered fetal hemoglobin levels and delayed fetal to adult globin switch after birth. *Anthropological science*, *118*, 41-48.

Roy, P., Bhattacharya, G., Mandal, A., Dasgupta, U., Banerjee, D., Chandra, S. & Das, M. (2012). Influence of BCL11A, HBS1L-MYB, HBBP1 single nucleotide polymorphisms and the HBG2 Xmn I polymorphism on Hb F levels. *Hemoglobin*, *36*, 592-599.

Rujito, L., Basalamah, M., Siswandari, W., Setyono, J., Wulandari G., Mulatsih, S., Sofro, A., Sadewa, A. & Sutaryo, S. (2016). Modifying effect of XmnI, BCL11A, and HBS1L-MYB on clinical appearances: a study on β-thalassemia and hemoglobin E/β-thalassemia patients in Indonesia. *Hematology/oncology and stem cell therapy*, *9*, 55-63.

Rund, D. & Fucharoen, S. (2008). Genetic modifiers in hemoglobinopathies. *Current molecular medicine*, *8*, 600-608.

Said, F. & Abdel, S. A. (2015). XmnI polymorphism: Relation to β-thalassemia phenotype and genotype in Egyptian Children. *Egyptian Journal of Medical Human Genetics*, *16*, 123-127.

Saiki, R., Walsh, S., Levenson, C. & Erlich, H. (1989). Genetic analysis of amplified DNA with immobilized sequence-specific oligonucleotide probes. *Proceedings of the National Academy of Sciences*, *86*, 6230-6234.

Sankaran, V., Xu, J. & Orkin, S. (2010). Transcriptional silencing of fetal hemoglobin by BCL11A. Annals of the New York *Academy of Science*, *1202*, 64-68.

Sankaran, V., Menne, T., Xu, J., Akie, T., Lettre, G., Handel, B. V., Mikkola, H., Hirschhorn, J., Cantor, A. & Orkin, S. (2008). Human fetal hemoglobin expression is regulated by the developmental stage-specific repressor BCL11A. *Science*, *322*, 1839-1842.

Sarnaik, S. (2005). Thalassemia and related hemoglobinopathies. *The Indian Journal of Pediatrics*, *72*, 319-324.

Satta, S., Perseu, L, Moi, P., Asunis, I., Cabriolu, A., Maccioni, L., Demartis, F., Manunza, L., Cao, A. & Galanello, R. (2011). Compound heterozygosity for KLF1 mutations associated with remarkable increase of fetal hemoglobin and red cell protoporphyrin. *Haematologica*, *96*, 767-770.

Sedgewick, A., Timofeev, N., Sebastiani, P., So, J., Ma, E., Chan, L., Fucharoen, G., Fucharoen, S., Barbosa, C., Vardarajan, B. & Farrer, L. (2008). BCL11A is a major HbF quantitative trait locus in three different populations with β-hemoglobinopathies. *Blood Cells, Molecules, and Diseases*, *41*, 255-258.

Siatecka, M. & Bieker, J. (2011). The multifunctional role of EKLF/KLF1 during erythropoiesis. *Blood*, *118*, 2044-2054.

Singleton, B., Burton, N., Green, C., Brady, L. & Anstee, D. (2008). Mutations in EKLF/KLF1 form the molecular basis of the rare blood group In (Lu) phenotype. *Blood*, *112*, 2081-2088.

Stadhouders, R., Aktuna, S., Thongjuea, S., Aghajanirefah, A., Pourfarzad, F., van IJcken, W., Lenhard, B., Rooks, H., Best, S., Menzel, S. & Grosveld, F. (2014). HBS1L-MYB intergenic variants modulate fetal hemoglobin via long-range MYB enhancers. *The Journal of clinical investigation*, *124*, 1699-1710.

Tepakhan, W., Yamsri, S., Sanchaisuriya, K., Fucharoen, G., Xu, X. & Fucharoen, S. (2016). Nine known and five novel mutations in the erythroid transcription factor KLF1 gene and phenotypic expression of fetal hemoglobin in hemoglobin E disorder. *Blood Cells, Molecules, and Diseases*, *59*, 85-91.

Thein, S., Menzel, S., Peng, X., Best, S., Jiang, J., Close, J., Silver, N., Gerovasilli, A., Ping, C., Yamaguchi, M. & Wahlberg, K. (2007). Intergenic variants of HBS1L-MYB are responsible for a major quantitative trait locus on chromosome 6q23 influencing fetal hemoglobin levels in adults. *Proceedings of the National Academy of Sciences*, *104*, 11346-11351.

Thein, S. (2013). The molecular basis of β-thalassemia. *Cold Spring Harbor perspectives in medicine*, *3*, a011700.

Uda, M., Galanello, R., Sanna, S., Lettre, G., Sankaran, V., Chen, W., Usala, G., Busonero, F., Maschio, A., Albai, G. & Piras, M. (2008). Genome-wide association study shows BCL11A associated with persistent fetal hemoglobin and amelioration of the phenotype of β-thalassemia. *Proceedings of the National Academy of Sciences*, *105*, 1620-1625.

Upadhye, D., Jain, D., Trivedi, Y., Nadkarni, A., Ghosh, K. & Colah, R. (2016). Influence of single nucleotide polymorphisms in the BCL11A and HBS1L-MYB gene on the HbF levels and clinical severity of sickle cell anaemia patients. *Annals of hematology*, *95*, 1201-1203.

Wahlberg, K., Jiang, J., Rooks, H., Jawaid, K., Matsuda, F., Yamaguchi, M., Lathrop, M., Thein, S, & Best, S. (2009). The HBS1L-MYB intergenic interval associated with elevated HbF levels shows characteristics of a distal regulatory region in erythroid cells. *Blood*, *114*, 1254-1262.

Watson, J, Stahman, A, Bilello, F. 1948. The significance of the paucity of sickle cells in newborn Negro infants. *Obstetrical & Gynecological Survey*, *3*, 819-820.

Waye, J. & Eng, B. (2015). Krüppel-like factor 1: hematologic phenotypes associated with KLF1 gene mutations. *International journal of laboratory hematology*, *37*, 78-84.

Wonkam, A., Ngo, B., Vorster, A., Ramesar, R., Cooper, R., Tayo, B., Lettre, G. & Ngogang, J. (2014). Association of variants at BCL11A and HBS1L-MYB with hemoglobin F and hospitalization rates among sickle cell patients in Cameroon. *PloS one*, 9, e92506.

Xu, J., Sankaran, V., Ni, M., Menne, T., Puram, R., Kim, W. & Orkin, S. (2010). Transcriptional silencing of γ-globin by BCL11A involves long-range interactions and cooperation with SOX6. *Genes & development*, *24*, 783-798.

Zhou, D., Liu, K., Sun, C. W., Pawlik, K. & Townes, T. (2010). KLF1 regulates BCL11A expression and γ-to β-globin gene switching. *Nature genetics*, *42*, 742.

In: Fetal Hemoglobin
Editor: Nishant Chakravorty

ISBN: 978-1-53617-955-2

Chapter 8

THERAPEUTIC INDUCTION OF FETAL HEMOGLOBIN EXPRESSION: METHODS IN PRACTICE AND NEWER MODALITIES

Rudra Ray and Maitreyee Bhattacharyya*
Institute of Haematology and Transfusion Medicine
Medical College Kolkata, Kolkata, West Bengal, India

ABSTRACT

The underlying pathophysiology of thalassemia is alteration in the balance between the alpha globin and non alpha globin chains leading to precipitation of free alpha hemoglobin causing damage to RBCs. Besides the beta globin, human RBCs also contain delta and gamma globin chains as the non-alpha globin chain content. Fetal hemoglobin (also known as HbF), is a product of gamma globin gene which is the most predominant form of hemoglobin during the fetal form of life and is subsequently switched to the adult hemoglobin (HbA) later on. In beta-thalassemia, the beta globin chain is either absent or synthesized at a reduced rate. Induction

* Corresponding Author's Email: mbhattyacharyya@yahoo.co.in.

of HbF synthesis becomes beneficial in such cases as an increase in gamma globin chains contributes to maintenance of the balance between 'alpha-non alpha' globin chains, and thus prevents the accumulation of excess alpha globin chains. The silencing of gamma globin genes is controlled by a variety of complex genetic mechanisms. Genome wide association studies (GWAS) have identified several loci and polymorphisms within the beta globin gene cluster on chromosome 11 and outside (BCL11A, KLF1, MYB, βLCR etc.) which are associated with the level of HbF production. Additionally, the participation of various epigenetic mechanism and chromatin modifying activities (DNMT3, PRMT5, NuRD etc.) also help in controlling the event. Several agents (HU, 5-Azacytidine, Butyrate, Decitabine, Tranylcypromine and Thalidomide etc.) have been demonstrated to increase the production of HbF in patients with hemoglobinopathies. These agents differ in their mechanisms and their targets of action which vary from HDAC inhibition, DREAD complex inhibition to over expression of TR2/TR4 and de-methylation of gamma globin promoters. Multiple modalities with agents capable of more effective targets and the use of combination of agents having dual or multiple targets to stimulate HbF synthesis are being investigated for use in future.

Keywords: thalassemia, HbF, gamma globin, globin gene

INTRODUCTION

Thalassemia is an inherited autosomal recessive genetic disorder associated with abnormal globin chain synthesis. Adult hemoglobin is a tetramer which consists of two dimers formed between alpha globin and beta globin (non alpha globin) chains. In a patient with thalassemia, the synthesis of globin chain(s) is abnormal due to mutation(s) in the respective globin gene(s) and this leads to faulty production of hemoglobin (Hb). In the case of beta thalassemia, mutations in the beta globin gene cause abnormal production of beta globin chains and thus their dimerization with alpha globin chains is impaired. This leads to the subsequent pathophysiology and phenotypic outcome associated with the disease.

Pathophysiology in beta thalassemia includes ineffective erythropoiesis, hemolysis and reduced hemoglobin synthesis (Thein, 2005). The phenotypic expression in beta thalassemia varies to a large extent from mild anemia requiring occasional blood transfusion to severe anemia requiring regular blood transfusions for survival (Thein, 2005). Although, beta thalassemia is caused by the reduction in the normal beta globin chain synthesis, it is the altered ratio of the globin chains that gives rise to its characteristic pathophysiology (Thein, 2013). As the production of beta globin chain is reduced, there remains less amount of beta globin chains which can bind to alpha globin chains to form the tetramer required in the process of Hb synthesis. Since alpha globin genes are unaffected in beta thalassemia and synthesis of alpha globin chains remains unaffected, the excess amount of free or unbound alpha globin chains accumulate within the erythrocyte precursors. The excess free alpha globin chains, which are insoluble, can form precipitates and lead to oxidative damage of the RBC membranes. Destruction of these defective erythrocyte precursors in the bone marrow ultimately leads to ineffective erythropoiesis. Thus, in beta thalassemia it is the increased amount of alpha globin chains which primarily determines the severity of the diseased phenotype.

Since the severity in thalassemia is a result of excess alpha globin chain accumulation due to alteration of globin chain ratio (alpha: non alpha), induction of fetal hemoglobin (HbF) synthesis becomes a promising therapeutic approach to ameliorate the globin chain imbalance and reduce the disease severity. Fetal hemoglobin (HbF) inducing drugs thus have been considered as an important part of therapy for hemoglobinopathies in recent years (Fard et al. 2013). The high level of HbF has been observed to have a direct relationship with the phenotypic severity in thalassemia (Olivieri, 1999).

Hemoglobin Switch

The pattern of globin chain synthesis varies from embryo to early infancy stage of development. Different globin genes are expressed at different stages of human development (Sankaran et al. 2013) and their relative amount of expression varies at different stages of life. The embryonic switch involves silencing of ε-globin synthesis and upregulation of γ-globins expression. The fetal switch on the other hand, involves progressive silencing of γ globin synthesis and switching on the expression of β globin and δ globin genes. The globin genes are organized in two different clusters known as the alpha globin gene cluster (also known as alpha like gene cluster) and the beta globin gene cluster which are located at chromosome 16 and chromosome 11 respectively. There are three functional genes (zeta or ζ, alpha1 or α1, alpha2 or α2) present in the segment of alpha globin cluster. Zeta, alpha2 and alpha1 are arranged in 5′ to 3′ direction among which alpha 2 and alpha 1 are situated in close proximity. Beta globin gene cluster (or beta like gene cluster) comprises of five functional globin genes. The five functional genes - one Epsilon (ε), Two Gamma (Gγ and Aγ), one Delta (δ) and one Beta (β) globin genes are arranged in 5′ to 3′ direction and they produce epsilon, delta, gamma and beta globin chains respectively. The genes are expressed in the order of their arrangement during different stage of development and switching of the respective gene synthesis occurs to produce different types of hemoglobins at different stage of development (Sankaran et al. 2013).

The embryonic hemoglobins (Hb Gower and Hb Portland) are synthesized in the yolk sac before 8 weeks of gestation (Sankaran et al. 2010a). At the end of seven to eight weeks, the synthesis of ε- and ζ-chain is not detectable any more, and the main globin chains being produced then are alpha and gamma (α and Gγ, Aγ). The gamma globin forms the major part of non-alpha globin chains at this stage (seven to eight weeks) and beta globin chain production is barely measurable at this point. The synthesis of gamma globin and small amounts of beta globin starts after 8 weeks of gestation and it takes place in the fetal liver. The amount gradually increases to make up to 10% of the total non-alpha globin chains by the end of ten weeks (Sankaran et al. 2010b). From the tenth week to nearly the thirty-third week of gestation the major globin chains that are produced are- alpha, gamma and beta (α, Gγ, Aγ, and β). During this phase, the synthesis of beta chains increase and production of gamma chains reduce at a rate such that the synthesis of beta globin constitutes up to around 30% to 50% of the non-alpha globin chains at birth. Following birth, the production of gamma globin chains continues to reduce steadily and is almost replaced by beta globin chains in the adult life. The types of hemoglobin prevailing in adult life and their significance are given in Table 1.

Table 1. Types of Hemoglobins, Their Composition and Function

Types of Hemoglobin	Composition	Significance
HbF	Alpha Globin – Gamma Globin	Major Hb during Fetal stage
HbA0	Alpha Globin – Beta Globin	Major Hb during Adult life
HbA2	Alpha Globin – Delta Globin	Increased in beta thalassemia

Regulation of HbF Expression and Switching

Numerous genetic interplays are involved in the process of fetal hemoglobin (HbF) production. The regulation of HbF expression in beta thalassemia is a complex phenomenon and not yet thoroughly understood. The genetic factors involved in the process of HbF production show heterogeneous features and their frequency also varies from one population to other. Recently genome wide association studies (GWAS) have established the relation between various polymorphisms and SNPs with HbF induction (Jackson et al. 2003; Fan et al. 2015; Paikari et al. 2018; Wilber et al. 2011; Yien et al. 2012; Miri-Moghaddam et al. 2017; vanDijk et al. 2010; Xu et al. 2010; Yien et al. 2012). The LCR region which is situated upstream to beta globin cluster is also known to regulate the promoters of gamma globin gene expression (Jackson et al. 2003). Factors like GATA1, TAL1, E2A, LMO2 and LDB1, MYB, and BCL11A (B-cell lymphoma/leukemia 11A) have been associated with the induction of HbF directly or indirectly (Jackson et al. 2003; Fan et al. 2015; Paikari et al. 2018; Wilber et al. 2011; Yien et al. 2012). A polymorphism commonly found at -158 position of the Gγ gene, the XmnI restriction site has been co-related with induction of HbF (Paikari et al. 2018). The XmnI polymorphism is situated in a region with close proximity to the LCR within the beta globin gene cluster which impacts on the regulation of different globin genes at different stage of development (Miri-Moghaddam et al. 2017). A protein known as Friend of protein arginine methyltransferase I (FOP) controls HbF induction by its interaction with chromatin (vanDijk et al. 2010). Studies have shown that knockdown of FOP remarkably impacts the induction of gamma globin exprcssion and increases the production of HbF in mice model (vanDijk et al. 2010). The epigenetic regulatory machinery also holds a major control on the switching of hemoglobin during different stages of development (Ginder, 2015). The regulation is mainly mediated by modulation in chromatin structure, histone modifications, and DNA methylation. It has been seen that the expression of HbF is regulated by the methylation status of cytosine, loss of histone modifications, and reduced chromatin accessibility in adult life (Ginder, 2015). The induction of gamma chain expression occurs when genetic or chemical inhibition of DNA methylation takes place. Inhibition of DNA methylation is mediated by methyl-cytosine binding protein, histone arginine methyltransferase PRMT5, the histone methyltransferases, and histone deacetylases (HDACs) etc. Chromatin regulators like DNMT3A, NuRD are also found to occupy the promoter site as repressive complexes. Lysine-specific demethylases (LSDs) are associated with BCL11A's activity of silencing the gamma globin gene. It has been seen that inhibitors of LSDs increase the expression of gamma globin gene (Ginder, 2015). MicroRNAs (miRNAs) have also been found to regulate the switching of globin genes. Lin28B is a protein that targets let-7 miRNA family (Lee et al. 2013). Let-7 miRNA family is known to be involved in the regulation of globin gene switching (Lee et al. 2013). Several miRNAs have been found to have significant impact on the process of regulating the gamma globin expression and Hb switching, although the mechanism is still not clearly understood (Saki et al. 2016; Cokic et al. 2012).

Therapeutic Induction of HbF

The major modalities for the treatment of beta thalassemia include blood transfusion and use of iron chelating agents. Allogenic bone marrow transplantation (BMT) and gene therapy provide the only curative option for beta thalassemia. However, there are limitations associated with BMT which requires a normal or carrier sibling donor, and also demands expertise and good infrastructure. Moreover, BMT is an expensive process which makes its feasibility limited for most of the patients. Gene therapy till now is at experimental levels and has not been implemented on a mass scale as a treatment option. Thus, developing novel therapeutic strategies have become extremely important in the last two decades.

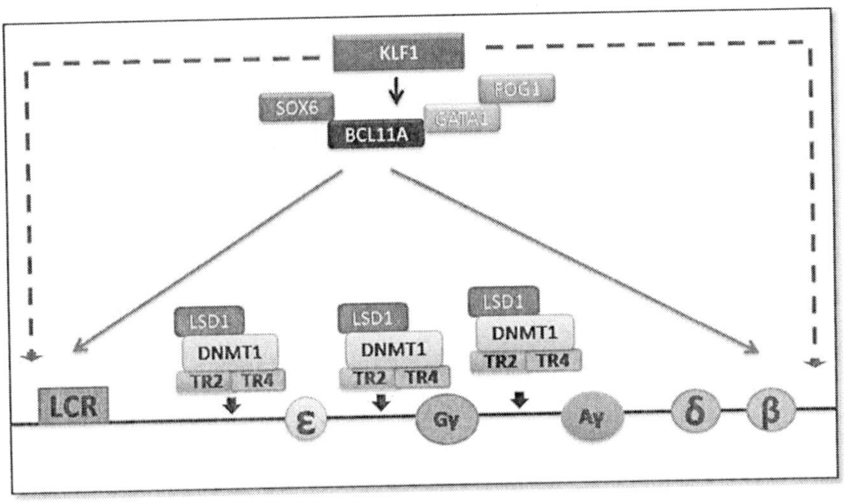

Figure 1. Key Transcription Factors Regulating HbF Expression.

The use of HbF inducing agents in recent years has been regarded as extremely important to treat patients with hemoglobinopathies. The fact that afflicted individuals remain asymptomatic at birth when their HbF level remains elevated led to the idea of using HbF inducing agents to treat thalassemia (Sankaran et al. 2010b). The phenotypic severity in thalassemia is seen to have a direct correlation with the level of HbF. The individuals having hereditary persistence of fetal hemoglobin (HPFH) often have high HbF level and milder phenotype (Musallam et al. 2012; Galanello et al. 2009). Thus treatment with HbF inducing agent serves as an important strategy in reducing the severe clinical phenotypes of beta thalassemia patients.

Hydroxyurea Therapy

Hydroxyurea has emerged as an important HbF inducing agent for the treatment of beta thalassemia. Hydroxyurea (HU) is known to inhibit ribonucleotide reductase (a key enzyme in the DNA repair and synthesis mechanism). Hydroxyurea has initially been used as a cytotoxic agent in the treatment of myeloproliferative neoplasms. HU is also one of the most common drugs used in the treatment of hemoglobinopathies like sickle cell disease as the drug is observed to increase the HbF levels (Galanello et al. 2009; Campbell et al. 2006; Fibach et al. 1993; Platt et al. 1984; Mabaera et al. 2008b). The exact mechanism by which HU increases the HbF level has not been fully explained. It has been observed that HU may increase the expression of gamma globin by up to 10-folds and results in balancing the globin chain ratio (Cokic et al. 2008; Ikuta et al. 2001). Further, it has also been observed that the increase in HbF level does not always correlate with increase of total hemoglobin level (Ikuta et al. 2001). Evidence suggests that the effect of HU may be beyond the increase the HbF levels. Effects like improvement of erythrocyte morphology, reduction of rate of hemolysis and increase in local NO release are also seen and all of these together may contribute to the better clinical outcomes in sickle cell patients treated with HU (Fucharoen et al. 1996). Furthermore, HU has been found to lower the phosphatidylserine externalization on the red cells when administered to splenectomized HbE-beta thalassemia patients (Zeng et al. 1995). Whether the beneficial effects of hydroxyurea can be attributed to HbF induction or an associated decrease in α-globin aggregation also remains to be elucidated.

There are numerous studies which examined the efficacy and safety of HU in different patient populations. A study reported that the increase in HbF levels varies extensively between 1% and 90% (with an average of 20%) and it is seen to be more in beta thalassemia patients than in sickle cell disease (Ware, 2010). Response to HU is usually considered positive if the level of hemoglobin increases from the baseline by a minimum of 0.9 g/dl. While some studies have reported a good correlation between the rise of HbF and hemoglobin (Hb) (Ware, 2010; Ataga et al. 2007); others did not (Ehsani et al. 2007). In one of our studies on HbE-beta thalassemia patients we observed a significant rise of HbF from the baseline (Biswas et al. 2019). However, there was no correlation between the rise of HbF & total hemoglobin in the study. In transfusion dependent thalassemia patients, the response to HU therapy is evaluated by a significant reduction of transfusion requirement from 15 to 50%. In our study some of the patients became transfusion independent even though the increase in hemoglobin was not significant. This could possibly be attributed to the fact that such patients could tolerate lower hemoglobin levels while on treatment with hydroxyurea. Hydroxyurea therapy is also known to decrease serum ferritin levels. Reduction in the level of ineffective erythropoiesis (in non transfusion dependent thalassemia) and decrease in transfusion requirement (in transfusion dependent thalassemia) could be the possible reasons for this. In non transfusion dependent patients, the main effect of hydroxyurea therapy is increase in hemoglobin level. An increase of hemoglobin level by >1g/dl is seen in around 40 to 70% patients; although level of increase is seen to vary widely (Mancuso et al. 2006; Dixit et al. 2005). Response to hydroxyurea therapy usually takes around 3 to 6 months. The maximum effect may be obtained in 12 months and the response is sustained in most of the cases (McGann et al. 2006). The initial dose of hydroxyurea is usually 10 mg/kg/day and may be increased to a maximum of 20 mg/kg/day, if there is no response (provided there is no toxicity). Patients with sickle cell disease are often seen to require higher doses (Italia et al. 2009; Karimi et al. 2005). Parameters like baseline HbF percentage, beta globin gene mutation, beta globin haplotypes and various polymorphisms have been studied as determinants of response to hydroxyurea therapy. Results from these studies are not very consistent. The presence of XmnI polymorphism has been found to be the most consistent determinant and this has been associated with a favorable response (Alebouyeh et al. 2004). Additionally, the presence of rs766432 polymorphism at intron 2 of BCL11A gene has also been found to be associated with a favorable response (Banan et al. 2012). In a study conducted by us, an associated alpha gene deletion was found to be the most important determinant for HU response in patients of HbE-beta thalassemia (Biswas et al. 2019). HU therapy is usually well tolerated by patients; however, myelosuppression is known to occur in some cases.

Myelosuppression thus observed is usually dose dependent and is known to revert back upon dose reduction (dePaula et al. 2003). To date only one case of leukemia has been reported in a beta-thalassemia patient under HU therapy for 3 years (Bradai et al. 2007). Hyperpigmentation, discoloration of nails, alopecia, and headache are the other side effects of this drug (Karimi et al. 2010).

Short Chain Fatty Acids

Perrini and co-workers found that infants of diabetic mothers have a delayed fetal-to-adult hemoglobin switch (Perrine et al. 1985). Further, researchers have also found that the level of β-hydroxybutryrate is elevated in mothers with diabetes. These findings motivated researchers to investigate the efficacy of butyrate as an inducer of HbF. Subsequently, Perrine et al. started investigating the effect of butyrate and other similar short-chain fatty acids on HbF induction. Studies observed an increase in gamma globin at mRNA level and this further led to the reduction of imbalance in globin chain ratio (Perrine et al. 1993; Perrine et al. 1994). Furthermore, therapy with oral sodium phenylbutyrate also showed increased levels of total hemoglobin, in addition to an increased production of HbF. Other derivatives of these agents, such as 2, 2-dimethylbutyrate have also been investigated.

Long term response to butyrate therapy is found to be lost and this may possibly be because of an antiproliferative effect of the drug on bone marrow. An intermittent therapy was thus found to be more effective in patients of sickle cell disease than continuous therapy (Atweh et al. 1999). An oral derivative of the drug - isobutyramide was also investigated later. An increase in HbF level by 3-6% from the baseline has been observed when isobutyramide is used (Domenica et al. 2000; Reich et al. 2000). This response is seen to be associated with high pre-therapy HbF and erythropoietin levels. Relatively poor response of the drug in beta thalassemia compared to SCD is attributed to the fact that it increases alpha globin gene expression in beta thalassemia (Fathallah et al. 2009).

DNA Methylation/Histone Deacetylation Inhibition

DNA methylation and Histone deacetylation play very significant roles in the epigenetic regulation of HbF synthesis. DNA methyl transferase (DNMT) inhibitors like azacitidine and decitabine and Histone Deacetylation complex (HDAC) inhibitors and are seen to result in increasing the HbF level and are thus are used in the treatment of hemoglobinopathies. Following the discovery of the inhibitory effect of 5-azacytidine on DNA methylation, it was investigated subsequently whether 5-azacytidine may have the efficacy to induce HbF. It was found that 5-azacytidine was able to induce gamma globin expression and reduce the globin chain imbalance resulting in an increment in total hemoglobin. The efficacy of 5-azacytidine as an inducer of HbF was found to be due to a localized demethylation in the promoter region of the gamma globin gene and is not due to result of global DNA demethylation (Mabaera et al. 2008a). However, questions have been raised about the cytotoxic effects and safety of this drug in conventional therapy (Ley et al. 1982; Ley et al. 1983).

Decitabine (5-aza-2'-deoxycytidine) is a potent HbF inducing agent which has been found to demethylate and reactivate the expression of gamma globin gene. The drug is not known to cause significant DNA damage at low concentrations and is thus considered favorable for usage (Saunthararajah et al. 2003; Koshy et al. 2000). Decitabine was found to increase the level of HbF and total Hb significantly in several studies in which the patients were treated with this drug and the treatment was well tolerated (Saunthararajah et al. 2003; Koshy et al. 2000). In a study conducted by Olivieri et al. in 2011, it was shown that Decitabine at a concentration of 0.2 mg/kg dose (twice weekly) elevated the level of HbF and total Hb among thalassemia intermedia patients (Olivieri et al. 2011). Recently in a study conducted by us, Decitabine was found to be safe and effective in increasing total Hb and HbF levels in HbE-beta thalassemia patients. The effect was sustained even after 6 months of withdrawal of the drug (Kalantri et al. 2018).

Immunomodulators and Anti-Angiogenic Agents

Thalidomide: Thalidomide is a drug known for its immunomodulating and antiangiogenic properties in multiple myeloma and has been found to be effective in HbF induction (Moutouh-de et al. 2008). The exact mechanism of action of this drug is yet to be fully understood. However, it may be due to the suppression of NF-KB induction. This results in an elevated level of reactive oxygen species (ROS) which in turn activates the signaling molecules like p38/MAPK and consequently increases the production of HbF (Aerbajinai et al. 2007; Fard et al. 2012). Three cases of beta thalassemia were reported to have significant increase in fetal hemoglobin with Thalidomide (Masera et al. 2010). However, Thalidomide is known to be associated with adverse side-effects like DVT, phocomelia etc. Since many patients with thalassemia are in the reproductive age group, they should be counselled properly to avoid conception while on therapy with Thalidomide.

Pomalidomide is another drug that is known to be an effective inducer of HbF production. It has been studied that the epigenetic mechanism by which this drug induces HbF synthesis is associated with acetylation of H3K9 and H3K14 at LCR region of gamma globin gene (Moutouh-de et al. 2008).

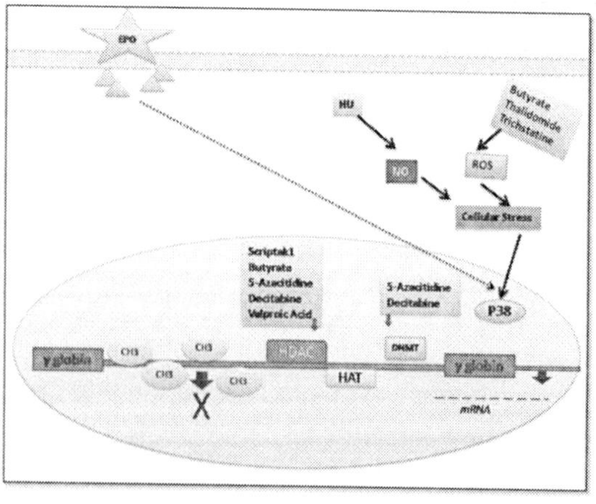

Figure 2. Key Pharmacological Agents Inducing HbF Expression.

Combination Therapy

Combination therapy by using more than one HbF inducing agent and targeting different pathways to induce HbF has been a very useful strategy for the treatment of thalassemia patients. In a recent study from Iran, Thalidomide was studied alone and in combination with Butyrate on umbilical cord blood derived erythroid differentiated cells. Thalidomide was found to have the highest effect on H3K27 methylation when compared with butyrate alone or in combination (Fard et al. 2012). Furthermore, erythropoietic stimulating agents have been found to be associated with increase in Hb level in beta thalassemia patients. Better effects have been seen when HU is used in combination with butyrate (McDonagh et al. 1992). Subsequently, various trials have investigated the combination of these 3 drugs as well. A combination of Erythropoietin with hydroxyurea was found to have a stronger effect than hydroxyurea alone (Loukopoulos et al. 1998). Further, Olivieri et al. demonstrated a remarkable improvement in level of hemoglobin in a beta thalassemia patient, when hydroxyurea is used in addition to Butyrate. However later studies have not shown consistency in such results.

Future Direction

Recent studies using microarray data have revealed that patients with hereditary persistence of fetal hemoglobin (HPFH) and beta thalassemia carriers carrying high HbF levels control the aberrant expression of numbers of mRNAs, miRNAs and long non coding RNAs (lncRNAs). Such aberrant expression was specifically observed among patients with high HbF in comparison with the control groups. It has been found that six long non coding RNAs could competitively bind to miR-486-3p and result in elevated HbF levels (Ketong et al. 2017). Further, it has been shown that the MAPK signaling pathway acts through SCF to induce HbF (Hagh et al. 2011). Furthermore, TGF-β is known to activate fetal Krüppel-like transcription factor (FKLF) and thereby increase gamma globin gene expression.

Additionally, it has also been observed that SCF, TGF-β and erythropoietin (EPO) act synergistically to favor the induction of HbF.

MicroRNAs (miRNAs) also play a significant role in the induction of gamma globin gene. Hojjati et al. studied expression of miRNAs in HU responders and non-responders. They found that HU responders showed higher expression of gamma globin mRNA, miR-210 and miR-486-3p levels (Hojjati et al. 2017). This may help us in selecting patients who may be put on HU treatment (Hojjati et al. 2017). Sodium butyrate is another known potent inducer of HbF which has been found to up-regulate several miRNAs and therefore can be used for inducing HbF with miRNAs as the promising surrogates (Tayebi et al. 2017).

SIRT1 is a protein deacetylase which has been found to up-regulate the expression of gamma globin mRNA by suppressing the expression of the suppressors of gamma globin genes. The gamma globin gene suppressors like BCL11A, KLF1, HDAC1 and HDAC2 are found to be suppressed by SIRT1 (Dai et al. 2017). SIRT1 acts via binding to the locus control region (LCR) of beta globin gene and to the promoter regions of gamma globin genes. By facilitating LCR looping to the gamma globin promoters, SIRT1 indirectly increases histone acetylation in the promoter regions of gamma globin genes and increases its expression (Dai et al. 2017). The ectopic expression of SIRT1 has been found to upregulate the gamma globin gene expression in primary human erythroid cells and in K562 cells (Dai et al. 2017). SIRT1 thus may well be a novel therapeutic target for HbF induction (Dai et al. 2017).

Conclusion

Induction of HbF has become a part of management of beta thalassemia patients for more than three decades now. Recent understanding of the molecular basis of reactivation of gamma globin gene by different pathways suggests that many of these pathways can be targeted in future for development of more effective HbF inducing agents.

REFERENCES

Aerbajinai W, Zhu J, Gao Z, Chin K, Rodgers GP. Thalidomide induces gamma-globin gene expression through increased reactive oxygen species-mediated p38 MAPK signaling and histone H4 acetylation in adult erythropoiesis. *Blood.* 2007 Oct 15;110(8):2864-71.

Alebouyeh M, Moussavi F, Haddad-Deylami H, et al. Hydroxyurea in the treatment of major betathalassemia and importance of genetic screening. *Ann Hematol.* 2004; 83(7):430-433.

Ataga KI, Cappellini MD, Rachmilewitz EA. Betathalassaemia and sickle cell anaemia as paradigms of hypercoagulability. *Br J Haematol.* 2007; 139(1):3-13.

Atweh GF, Sutton M, Nassif I, et al. Sustained induction of fetal hemoglobin by pulse butyrate therapy in sickle cell disease. *Blood.* 1999; 93(6): 1790-1797.

Banan M, Bayat H, Azarkeivan A, et al. The XmnI and BCL11A single nucleotide polymorphisms may help predict hydroxyurea response in Iranian b-thalassemia patients. *Hemoglobin.* 2012;36(4):371-380.

Biswas S, Nag A, Ghosh K, Ray R, Roy K, Bandyopadhyay A, Bhattacharyya M. Genetic determinants related to pharmacological induction of foetal haemoglobin in transfusion-dependent HbE-β thalassaemia. *Annals of hematology* 98, 2 (2019): 289-299.

Bradai M, Pissard S, Abad MT, et al. Decreased transfusion needs associated with hydroxyurea therapy in Algerian patients with thalassemia major or intermedia. *Transfusion.* 2007;47(10): 1830-1836.

Campbell PJ, Green AR. The myeloproliferative disorders. *N Engl J Med.* 2006;355(23): 2452-2466.

Cokic VP, Andric SA, Stojilkovic SS, et al. Hydroxyurea nitrosylates and activates soluble guanylyl cyclase in human erythroid cells. *Blood.* 2008; 111(3):1117-1123.

Cokic Vladan P, Bhattacharya Bhaskar, Beleslin-Cokic Bojana B, Noguchi Constance T, Puri Raj K, and Schechter Alan N. JAK-STAT and AKT pathway-coupled genes in erythroid progenitor cells through ontogeny. *Journal of translational medicine* 10, no. 1 (2012): 116.

Yan Dai, Chen Tyngwei, Ijaz Heba, Cho Elizabeth H, and Steinberg Martin H. SIRT1 activates the expression of fetal hemoglobin genes. *Am J Hematol.* 2017; 92(11): 1177–1186.

dePaula EV, Lima CS, Arruda VR, et al. Longterm hydroxyurea therapy in beta-thalassaemia patients. *Eur J Haematol.* 2003;70(3):151-155.

Dixit A, Chatterjee TC, Mishra P, et al. Hydroxyurea in thalassemia intermedia— a promising therapy. *Ann Hematol.* 2005; 84(7): 441-446.

Domenica CM, Graziadei G, Ciceri L, et al. Oral isobutyramide therapy in patients with thalassemia intermedia: results of a phase II open study. *Blood Cells Mol Dis.* 2000; 26(1): 105-111.

Ehsani MA, Hedayati-Asl AA, Bagheri A, et al. Hydroxyurea-induced hematological response in transfusion-independent beta-thalassemia intermedia: case series and review of literature [published correction appears in PediatrHematol Oncol. 2011;28(6):538]. *Pediatr Hematol Oncol.* 2009;26(8):560-565.

Fan AX, Mir AH, Jared S, Ekaterina G, Jörg B. Regulation of erythroid cell differentiation by transcription factors, chromatin structure alterations, and noncoding RNA. *In Epigenetic Gene Expression and Regulation,* pp. 237-264. Academic Press, 2015.

Fard AD, Kaviani S, Noruzinia M, Soleimani M, Abroun S, Hajifathali A, et al. Changing the pattern of histone H3 methylation following treatment of erythroid progenitors derived from cord blood CD133+ cells with sodium butyrate and thalidomide. *Iranian Journal of Blood and Cancer*. 2012. 4(2): 53-59.

Fard AD, Seyed AH, Mohammad S, Fatemeh S, Kaveh J. Evaluation of novel fetal hemoglobin inducer drugs in treatment of β-hemoglobinopathy disorders. *International journal of hematology-oncology and stem cell research* 7, no. 3 (2013): 47.

Fathallah H, Taher A, Bazarbachi A, et al. Differences in response to fetal hemoglobin induction therapy in beta-thalassemia and sickle cell disease. *Blood Cells Mol Dis.* 2009; 43(1): 58-62.

Eitan Fibach, Burke Lillian P, Schechter Alan N, Noguchi CT, and Rodgers GP. Hydroxyurea increases fetal hemoglobin in cultured erythroid cells derived from normal individuals and patients with sickle cell anemia or beta-thalassemia. *Blood.* 8(6) (1993): 1630-1635.

Fucharoen S, Siritanaratkul N, Winichagoon P, et al. Hydroxyurea increases hemoglobin F levels and improves the effectiveness of erythropoiesis in beta-thalassemia/hemoglobin E disease. *Blood.* 1996;87(3):887-892.

Galanello R, Sanna S, Perseu L, et al. Amelioration of Sardinian beta0 thalassemia by genetic modifiers. *Blood.* 2009; 114(18):3935-3937.

Ginder Gordon D. Epigenetic regulation of fetal globin gene expression in adult erythroid cells. *Translational Research* 165, 1 (2015): 115-125.

Hagh MF, Fard AD, Saki N, Shahjahani M, Kaviani S. Molecular Mechanisms of hemoglobin F induction. *International Journal of Hematology-Oncology and Stem Cell Research.* 2011; 5(4).

Hojjati MT, Azarkeivan A, Pourfathollah AA, Amirizadeh N. Comparison of MicroRNAs Mediated in Reactivation of the γ-Globin in β-Thalassemia Patients, Responders and Non-Responders to Hydroxyurea. *Hemoglobin.* 2017; 41(2):110-115.

Ikuta T, Ausenda S, Cappellini MD. Mechanism for fetal globin gene expression: role of the soluble guanylate cyclase-cGMP-dependent protein kinase pathway. *Proc Natl Acad Sci USA.* 2001; 98(4):1847-1852.

Italia KY, Jijina FJ, Merchant R, et al. Response to hydroxyurea in beta thalassemia major and intermedia: experience in western India. *Clin Chim Acta.* 2009; 407(1-2):10-15.

Jackson DA, Jennifer CM, Dean A. β-Globin locus control region HS2 and HS3 interact structurally and functionally. *Nucleic acids research* 31, no. 4 (2003): 1180-1190.

Kalantri Arun, Siddhesh& Ray, Rudra Bhattacharyya, Maitreyee. (2018). Efficacy of decitabine as hemoglobin F inducer in HbE/β-thalassemia. *Annals of Hematology.* 97. 10.1007/s00277-018-3357-y.

Karimi M, Cohan N, Mousavizadeh K, et al. Adverse effects of hydroxyurea in betathalassemia intermedia patients: 10 years' experience. *PediatrHematol Oncol.* 2010;27(3): 205-211.

Karimi M, Darzi H, Yavarian M. Hematologic and clinical responses of thalassemia intermedia patients to hydroxyurea during 6 years of therapy in Iran. *J PediatrHematol Oncol.* 2005; 27(7): 380-385.

Ketong L, Siyuan J, Shanjuan Y, Jianming L and Yunyan H. Genome-wide analysis of aberrantly expressed lncRNAs and miRNAs with associated co-expression and ceRNA networks in β-thalassemia and hereditary persistence of fetal hemoglobin. *Oncotarget,* 2017, Vol. 8, (No. 30), pp: 49931-49943.

Koshy M, Dorn L, Bressler L, et al. 2-deoxy 5azacytidine and fetal hemoglobin induction in sickle cell anemia. *Blood.* 2000;96(7): 2379-2384.

Lee, Y. Terry, Jaira F. de Vasconcellos, Joan Yuan, Colleen Byrnes, Seung-Jae Noh, Emily R. Meier, Ki Soon Kim et al. LIN28B-mediated expression of fetal hemoglobin and production of fetal-like erythrocytes from adult human erythroblasts ex vivo. *Blood* 122, 6 (2013): 1034-1041.

Ley TJ, Anagnou NP, Young NS, et al. 5-azacytidine for beta thalassaemia? *Lancet.* 1983; 1(8322):467.

Ley TJ, DeSimone J, Anagnou NP, et al. 5-azacytidine selectively increases gammaglobin synthesis in a patient with beta1 thalassemia. *N Engl J Med.* 1982; 307(24): 1469-1475.

Loukopoulos D, Voskaridou E, Stamoulakatou A, et al. Hydroxyurea therapy in thalassemia. *Ann N Y Acad Sci.* 1998; 850:120-128.

Mabaera R, Greene MR, Richardson CA, et al. Neither DNA hypomethylation nor changes in the kinetics of erythroid differentiation explain 5azacytidine's ability to induce human fetal hemoglobin. *Blood.* 2008a; 111(1):411-420.

Mabaera R, West RJ, Conine SJ, et al. A cell stress signaling model of fetal hemoglobin induction: what doesn't kill red blood cells may make them stronger. *Exp Hematol.* 2008b;36(9): 1057-1072.

Mancuso A, Maggio A, Renda D, et al. Treatment with hydroxycarbamide for intermedia thalassaemia: decrease of efficacy in some patients during long-term follow up. *Br J Haematol.* 2006; 133(1):105-106.

Masera N, Tavecchia L, Capra M et al. Optimal response to thalidomide in a patient with thalassaemia major resistant to conventional therapy. *Blood Transfus,* 2010; 8: 63-5.

McDonagh KT, Dover GJ, Donahue RE, et al. Hydroxyurea-induced HbF production in anemic primates: augmentation by erythropoietin, hematopoietic growth factors, and sodium butyrate. *Exp Hematol.* 1992;20(10):1156-1164.

McGann PT, Ware RE. Hydroxyurea for sickle cell anemia: what have we learned and what questions still remain? *Curr Opin Hematol.* 2011; 18(3):158-165.

Miri-Moghaddam E, Bahrami S, Naderi M, Bazi A, Karimipoor M. Xmn1-158 γGvariant in B-thalassemia intermediate patients in South-East of Iran. *International journal of hematology-oncology and stem cell research.* 11. 2 (2017): 165.

Moutouh-de Parseval LA, Verhelle D, Glezer E, et al. Pomalidomide and lenalidomide regulate erythropoiesis and fetal hemoglobin production in human CD34+ cells. *J Clin Invest* 2008; 118: 248-58.

Musallam KM, Sankaran VG, Cappellini MD, et al. Fetal hemoglobin levels and morbidity in untransfused patients with b-thalassemia intermedia. *Blood.* 2012; 119(2):364-367.

Olivieri NF. The β-thalassemias. *New England journal of medicine* 341, no. 2 (1999): 99-109.

Olivieri NF, Saunthararajah Y, Thayalasuthan V, et al; Thalassemia Clinical Research Network. A pilot study of subcutaneous decitabine in b-thalassemia intermedia. *Blood.* 2011;118(10): 2708-2711.

Paikari A and Vivien AS. Fetal haemoglobin induction in sickle cell disease. *British journal of haematology* 180, no. 2 (2018): 189-200.

Perrine SP, Greene MF, Faller DV. Delay in the fetal globin switch in infants of diabetic mothers. *N Engl J Med.* 1985;312(6):334-338.

Perrine SP, Ginder GD, Faller DV, et al. A shortterm trial of butyrate to stimulate fetal-globingene expression in the beta-globin disorders. *N Engl J Med.* 1993; 328(2):81-86.

Perrine SP, Olivieri NF, Faller DV. Butyrate derivatives. New agents for stimulating fetal globin production in the beta-globin disorders. *Am J Pediatr Hematol Oncol.* 1994; 16(1):67-71.

Platt OS, Orkin SH, Dover G, et al. Hydroxyurea enhances fetal hemoglobin production in sickle cell anemia. *J Clin Invest.* 1984;74(2):652-656.

Reich S, Bührer C, Henze G, et al. Oral isobutyramide reduces transfusion requirements in some patients with homozygous betathalassemia. *Blood.* 2000; 96(10):3357-3363.

Saki Najmaldin, Saeid Abroun, Masoud Soleimani, Maria Kavianpour, Mohammad Shahjahani, Javad Mohammadi-Asl, and Saeideh Hajizamani. MicroRNA expression in β-thalassemia and sickle cell disease: a role in the induction of fetal hemoglobin. *Cell Journal (Yakhteh)* 17, 4 (2016): 583.

Sankaran VG and Orkin SH. The switch from fetal to adult hemoglobin. *Cold Spring Harbor perspectives in medicine* 3, no. 1 (2013): a011643.

Sankaran VG, Jian X, Orkin SH. Advances in the understanding of haemoglobin switching. *British journal of haematology* 149, no. 2 (2010a): 181-194.

Sankaran VG, Nathan DG. Thalassemia: an overview of 50 years of clinical research. *Hematol Oncol Clin North Am.* 2010b;24(6):1005-1020.

Saunthararajah Y, Hillery CA, Lavelle D, et al. Effects of 5-aza-29-deoxycytidine on fetal hemoglobin levels, red cell adhesion, and hematopoietic differentiation in patients with sickle cell disease. *Blood.* 2003; 102(12): 3865-3870.

Tayebi Behnoosh, Fatemeh Abrishami, Shaban Alizadeh, Neda Minayi, Mozhdeh Mohammadian et al. Modulation of microRNAs expression in hematopoietic stem cells treated with sodium butyrate in inducing fetal hemoglobin expression. *Artificial Cells, Nanomedicine, and Biotechnology.* 2017; 45:1, 146-156.

Thein SL. Pathophysiology of β thalassemia—A guide to molecular therapies. *ASH Education Program Book* 2005, no. 1 (2005): 31-37.

Thein SL. The molecular basis of β-thalassemia. *Cold Spring Harbor perspectives in medicine* 3, no. 5 (2013): a011700.

vanDijk Thamar Bryn, Gillemans N, Stein C, Fanis P, Demmers J, van deCorput M, Essers J, Grosveld F, Bauer UM, Philipsen S. Friend of Prmt, a novel chromatin target of protein arginine methyltransferases. *Molecular and cellular biology*. 30, 1 (2010): 260-272.

Ware RE. How I use hydroxyurea to treat young patients with sickle cell anemia. *Blood*. 2010; 115(26):5300-5311.

Wilber A, Arthur WN, Derek AP. Transcriptional regulation of fetal to adult hemoglobin switching: new therapeutic opportunities. *Blood* 11715 (2011): 3945-3953.

Jian Xu, Sankaran Vijay G, Ni Min, Menne Tobias F, Puram Rishi V, Kim Woojin and Orkin Stuart H. Transcriptional silencing of γ-globin by BCL11A involves long-range interactions and cooperation with SOX6. *Genes & development* 24, (2010): 783-798.

Yien Yvette Y, James JB. Functional interactions between erythroid Kruppel-like factor (EKLF/KLF1) and protein phosphatase PPM1B/PP2Cβ. *Journal of Biological Chemistry* 287, no. 19 (2012): 15193-15204.

Zeng YT, Huang SZ, Ren ZR, et al. Hydroxyurea therapy in beta-thalassaemia intermedia: improvement in haematological parameters due to enhanced beta-globin synthesis. *Br J Haematol.* 1995; 90(3):557-563.

In: Fetal Hemoglobin
Editor: Nishant Chakravorty

ISBN: 978-1-53617-955-2

Chapter 9

GENETIC MODIFIERS OF β-THALASSEMIA AND THEIR THERAPEUTIC IMPLICATIONS

J. W. Nirmani Yasara and Sachith Mettananda*
Department of Paediatrics, University of Kelaniya, Ragama, Sri Lanka

ABSTRACT

β-thalassemia is one of the most common single gene disorders in the world. It occurs due to mutations in and around the β-globin gene that affect gene expression at multiple stages. More than 300 mutations responsible for β-thalassemia have been discovered so far. The remarkable diversity of this disease's phenotype is due to numerous genetic modifiers which act either primarily or secondarily, ameliorating or worsening the severity of β-thalassemia. Primary modifiers impact the clinical characteristics by modifying the extent of globin chain imbalance; whereas, secondary modifiers modulate complications associated with the disease without interfering with the globin balance directly. The α- and β-globin genotypes play major roles in clinical heterogeneity. Inheritance of milder β-thalassemia mutations (β^{+} or β^{++}) and α-thalassemia mutations are known to reduce the severity of the disease. Conversely, dominantly

* Corresponding Author's Email: sachith.mettananda@kln.ac.lk.

inherited β-thalassemia mutations and co-inheritance of excess (triplicated - ααα or quadruplicated - αααα) α-globin genes, worsen the condition by exacerbating the globin chain imbalance. Additionally, through recent advances in genome analysis, numerous modifying genetic variants have also been unraveled. These include polymorphisms in the *Xmn1-HBG2* region, the *BCL11A* enhancer, the intergenic region of the *HBS1L-MYB* and *KLF1* which give an innate ability to upregulate γ-globin and thus fetal hemoglobin. The major secondary genetic modifiers that were delineated recently are those that affect bilirubin metabolism, iron homeostasis and cause bone and cardiac diseases. Even though, hematopoietic stem cell transplantation remains the only cure for β-thalassemia at present, current understanding of the molecular pathology governing the heterogeneity of the clinical phenotype has uncovered several novel therapeutic pathways which have shown promise in preclinical models and clinical trials.

Keywords: thalassemia, genetic modifiers, fetal hemoglobin, α-globin

INTRODUCTION

β-thalassemia is one of the commonest single gene disorders in the world (Taher, Weatherall, and Cappellini 2018). It's regional prevalence varies considerably across the globe, with higher prevalence seen in the traditional thalassemia belt which includes Mediterranean, Sub-Saharan Africa, Middle-East, Indian subcontinent and Southeast Asia (Flint et al. 1998, Mettananda and de Silva 2017). However, due to population migration, β-thalassemia is becoming a major health burden to many developed countries including United States, United Kingdom and Canada (Fucharoen and Weatherall 2016). According to global statistics, 70,000 children with different types of thalassemia are born each year; over half of them suffer from severe forms of β-thalassemia (Weatherall 2010).

The human adult hemoglobin (HbA) molecule is a tetramer of two α- and two β-globin polypeptide chains that are combined with heme groups. This heme moiety is an iron containing porphyrin derivative which plays an essential role in oxygen transportation (Weatherall and Clegg 2001). A reduction or a complete cessation of the synthesis of structurally normal β-globin, commonly due to molecular defects of genes encoding β-globin, gives rise to β-thalassemia (Mettananda and Higgs 2018).

Even though β-thalassemia is an autosomal recessive prototypical Mendelian monogenic disorder, it shows a remarkable heterogeneity in the disease phenotype due to the presence of numerous genetic modifiers (Thein 2013a). Depending on the clinical and hematological presentation, three distinguishable subcategories, with increasing severity of β-thalassemia, have been identified: β-thalassemia trait, non-transfusion dependent β-thalassemia (β-thalassemia intermedia) and transfusion dependent β-thalassemia (β-thalassemia major) (Cao and Galanello 2010, Mettananda 2018a). The clinical management of β-thalassemia mainly depends on red blood cell (RBC) transfusion to replace the depleted hemoglobin, and iron chelation therapy (Mettananda et al. 2019, Suriapperuma et al. 2018). Unfortunately, current standard treatment only provides supportive care and allogenic hematopoietic stem cell transplantation (HSCT) remains the only definite cure, despite revolutionary advances in the field of molecular medicine (Mettananda 2017).

Human β-Globin Locus

Among thousands of loci in human genome, the globin genes are possibly the most immensely studied and delineated genes (Higgs, Engel, and Stamatoyannopoulos 2012). A single gene, located in the short arm of chromosome 11 (11p 15.15), which clusters along with other β-like globin genes, encodes for β-globin (Stamatoyannopoulos 2005). The five functional genes in this cluster – ε (*HBE*), Gγ (*HBG2*), Aγ (*HBG1*), δ (*HBD*) and β (*HBB*) – are located according to the order of expression during development (telomere-ε-$^{G}\gamma$-$^{A}\gamma$-δ- β -centromere) (Smith and Orkin 2016).

The cumulative efforts of numerous research groups have provided a clear insight into how the genes in the β-globin locus are switched on and off during development and erythropoiesis (Sankaran and Orkin 2013). This knowledge on the regulation of expression of β-globin genes is essential to understand the molecular basis of β-thalassemia. The β-globin gene (1.6kb) which contains 5′ and 3′ untranslated regions (UTR) and three exons, is regulated by a local promoter and upstream enhancers known as locus control region (LCR) (Thein 2018). This *cis*- acting LCR is a cluster of five DNaseI hypersensitive enhancers (designated HS-1 to HS-5) which span 6-20kb proximal to the β-globin gene (Stamatoyannopoulos 2005). The importance in the regulation of LCR in β-globin gene has been studied using natural deletions that remove the HS sites partially or completely (Forrester et al. 1990, Bender et al. 2000).

The core sequence of each HS site (~250 nucleotides) contains several motifs which bind to erythroid specific transcriptional factors (TF) (Stamatoyannopoulos 2005) – GATA binding factor 1 (GATA-1), nuclear factor erythroid-2 (NF-E2), Krüppel-like factor (KLF-1), friend of GATA-1 (FOG1), stem cell leukemia (SCL) factor – and other factors with ubiquitous tissue distributions (Lee et al. 2017, Mettananda and Higgs 2018). Among several models tested experimentally, the looping model has been identified to delineate how LCR controls the expression of genes in the β-globin locus (Grosveld 1999). Formation of three-dimensional chromosomal interaction occurs as a result of folding of LCR to bring it to close proximity of the gene promoters to facilitate binding of TFs and other co-activators (Deng et al. 2014).

Developmental-Stage-Specific Expression of Genes in the β-Globin Locus

During early embryonic period, erythroid lineage cell formation occurs in the blood islands of yolk sac, which functions as a transient extramedullary tissue (Higgs and Wood 2011). Definitive hematopoietic stem cells which originate from the ventral wall of the dorsal aorta, migrate towards the fetal liver during the middle of the first trimester, which then at the time of birth, migrate and reside in the bone marrow, the principle erythropoietic destination for the rest of the life (Dzierzak and Philipsen

2013). Similarly, the predominant hemoglobin subtype also varies throughout the embryonic and fetal development due to the well-coordinated developmental-stage-specific expression of genes in the α- and β-globin loci. Among the five genes located in the β-globin cluster, ε-globin gene is the first to express, and its expression is limited to early embryonic stage. This is followed by the expression of two γ-globin genes that are located 3′ to the ε-globin gene, which predominates during fetal life. After birth, the expression is switched from γ-globin to β-globin (and to a lesser extent to δ-globin) - the globin chains found in hemoglobins of adults (Bauer, Kamran, and Orkin 2012) (Figure 1).

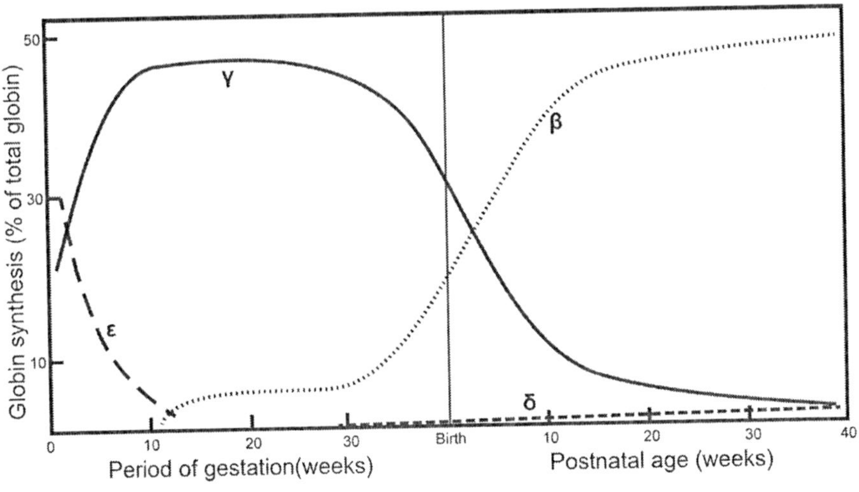

Figure 1. Developmental specific expression of genes in the human β-globin locus. ε-globin is the predominant type of globin during the embryonic period, which is then replaced by γ-globin during the fetal stage. Within the first few months after birth, the second globin switch occurs with the domination of β-globin over γ-globin (Mettananda 2015).

Hemoglobin Gower-I ($\zeta_2\varepsilon_2$), Gower-II ($\alpha_2\varepsilon_2$) and Portland ($\zeta_2\beta_2$) are predominantly produced during the first few weeks of gestation, followed by fetal hemoglobin (HbF, $\alpha_2\gamma_2$) - the major hemoglobin subtype produced during rest of the fetal life. Adult hemoglobin, HbA ($\alpha_2\beta_2$), is a tetramer of two α- and two β-globin chains, whereas, the combination of α- and δ-globin chains form hemoglobin A_2 ($\alpha_2\delta_2$), which is synthesized at lower level in adults (Weatherall and Clegg 2001, Mettananda, Gibbons, and Higgs 2016). The hemoglobin switch from HbF to HbA occurs approximately at around six-months of age. It is at this stage that the diseases affecting β-globin gene start to manifest (Galanello and Origa 2010, Nuinoon et al. 2010). By two years of age, HbF roughly comprises only about 0.1-1% of the total hemoglobin, as HbA becomes the predominant hemoglobin subtype.

Regulation of Expression of Genes in the β-Globin Locus

Two mechanisms are responsible for the developmental-stage-specific hemoglobin switch; gene competition and autonomic silencing. Two independent research teams discovered in early 1990s that when LCR is linked either to β- or γ-globin gene alone, it fails in the developmental control, suggesting that γ- and β-globin genes compete for the interactions with LCR (Enver et al. 1990, Behringer et al. 1990). As LCR preferentially interacts with the γ-globin gene during fetal life, the expression of β-globin gene is completely turned off. Likewise, a preferential interaction has been discovered between LCR and the β-globin gene during adult stage. In addition, during embryonic erythropoiesis, competition between the upstream ε- and γ-globin genes silence the β-globin gene (Stamatoyannopoulos 2005).

In autonomous silencing, the gene silencing sequences located proximal or distal to the genes disrupt the interactions between the genes and the LCR by forming a repressor complex in association with TFs. The silencing element of γ-globin gene maps from −378 to −730 region upstream of the promoter, while ε-globin gene is switched off by sequences that are located both 5′ and 3′ to the promoter (Stamatoyannopoulos 2005, Stamatoyannopoulos et al. 1993). The sequential activation of five functional genes in the β-globin locus reflects their arrangement along the chromosome 11. *In-vivo* studies done on transgenic mice have discovered that the physical proximity to the LCR increases the probability of interactions with regulatory elements of the gene (Stamatoyannopoulos 2005).

Furthermore, recent research has identified a number of TFs that govern the hemoglobin switch in human β-globin locus. B-cell lymphoma/leukemia 11A (BCL11A) acts as a potent repressor of γ-globin transcription and regulates the hemoglobin switch with the aid of multiprotein complexes recruited to the β-globin gene cluster (Xu et al. 2010, Sankaran et al. 2008). Direct repeat erythroid specific definitive (DRED) complex, DNA methyltransferase 1 (DNMT1) and lysine-specific histone demethylase 1 (LSD1) also suppress γ-globin by binding to the direct repeat element of the γ-globin promoter (Shi et al. 2013). Moreover, Krüppel-like factor 1 (KLF1) directly activates the β-globin and upregulates the expression of BCL11A, whereas MYB transcription factor (MYB) indirectly represses γ-globin transcription via activation of testicular nuclear receptor (TR) 2 and TR4 in the DRED complex and KLF (Zhou et al. 2010b, Bauer, Kamran, and Orkin 2012, Suzuki, Yamamoto, and Engel 2014).

Molecular Basis of β-Thalassemia

Even though more than 300 β-thalassemia mutations have been identified, only about 40 are responsible for over 90% of the cases worldwide (Kountouris et al. 2014). These mutations could significantly vary in their severity depending on the type and location. They range from point mutations to large deletions involving the entire β-globin cluster. Unlike α-thalassemia, non-deletional mutations are the main cause for β-thalassemia. These include point mutations - insertions, substitutions or deletions of a single nucleotide, - either in the β-globin gene or the flanking regions, which downregulate the β-globin production by interfering gene expression at different stages (Thein 2018, Mettananda and Higgs 2018) (Table 1).

Table 1. Mutations causing β-thalassemia
(Mettananda and Higgs 2018, Thein 2018)

Mechanism	Severity	Common mutation/s	Geographical distribution
Point mutations			
Transcriptional mutations			
Promoter regulatory elements	β^{+} or β^{++}	-101(C>T)	Mediterranean
5′ UTR	β^{+} or β^{++}	CAP +1 (A>C)	Asian Indian
Mutations involving RNA processing			
Splice junction	β^{0}	IVS1-1 (G>T)	Asian Indian, South-east Asian, Chinese
Consensus splice sites	β^{0} or β^{+}	IVS1-5 (G>C)	Asian Indian, South-east Asian
Cryptic splice sites map in introns	β^{0} or β^{+}	IVS1-110 (G>A)	Mediterranean
Cryptic splice sites map in exons	β^{0} or β^{+}	CD26 (GAG>AAG)	South-east Asian, European
RNA cleavage-Poly A signal	β^{0} or β^{++}	AATAAA>AATGAA	Mediterranean
Others in 3′ UTR	β^{++}	Term CD +6, C>G	Greek
Mutations involving RNA translation			
Initiation codon	β^{0}	ATG>ATA	Indian, Swedish
Nonsense codon	β^{0}	CD39 (CAG>TAG) CD17 (AAG>TAG)	Mediterranean Chinese, Japanese, Reunion Island
Frameshift	β^{0}	CD41/42 (-TTCT)	Chinese, South-east Asian, Indian
Deletions			
Deletions confined to β-globin gene	β^{0}	-619 bp deletions	
Deletions extending to other genes	β^{0}	$(\varepsilon\gamma\delta\beta)^{0}$ or $(\delta\beta)^{0}$	
Deletions of the LCR	β^{0}		

Even though, nearly 50% of non-deletional mutations completely inactivate the β-globin gene, some transcriptional mutations only cause mild to moderate decline in β-globin production. Non-deletional mutations affecting exon-intron splice junction and poly-adenylation signal, interfere with primary mRNA transcript processing. Mutations that occur at cryptic splice sites and GT or AG sequences at exon-intron splice sites result in abnormal splicing, which could exhibit mild to severe hematological phenotype relying on the quantity of abnormal mRNA transcripts produced (Thein 2004). β-Globin gene output could be completely abolished if a premature stop codon is introduced either by a single base substitution or a frame shift mutation. The resultant abnormal mRNA transcript goes through nonsense mediated RNA decay (He and Jacobson 2015, Thein 2004). Introduction of a stop codon near the 3' end of the gene gives the transcript a power of escaping this mechanism, resulting in a highly unstable β-globin chain (Hall and Thein 1994). Mutations associated with translation affecting initiation codon (ATG) are mainly single base substitutions (Thein and Wood 2009). In addition to common mutations within the β-globin gene, mutations in the conserved DNA sequences of the promoter or 5' UTR also affect the transcription of β-globin (Maragoudaki et al. 1999, Thein 2013b).

A very small proportion of β-thalassemia is caused by deletional mutations. Altogether, eighteen deletional mutations limited to the β-globin gene have been discovered (Thein 2013b). Deletion of β-globin promoter enhances the interaction of the LCR with the γ- and δ-globin genes, removing the competition for the 5' LCR (Huisman 1997).

Even though β-thalassemia follows the Mendelian recessive inheritance pattern, more than 30 mutations causing dominant inheritance have been discovered in which the presence of a single mutated β-globin gene exhibit the disease phenotype. In contrast to recessively inherited disease, these mutations are commonly found in regions where gene frequency of β-thalassemia is relatively low and occurs mainly as *de-novo* mutations (Thein 2018).

Apart from these typical β-thalassemia genotypes, some rare forms of unusual causes have been discovered. They include *trans*-acting mutations, insertion of retrotransposons within the globin gene and somatic deletions of β-globin gene (Viprakasit et al. 2001, Yu et al. 2002, Kimberland et al.

1999, Galanello et al. 2004). Rarely, a heterozygous individual also can develop β-thalassemia intermedia due to mosaic somatic deletions of the chromosome 11p15.5 which could inactivate the β-locus in *trans* in a subpopulation of hematopoietic cells (Galanello et al. 2004).

PATHOPHYSIOLOGY OF β-THALASSEMIA

The direct link between the α- and β-globin chain imbalance and the pathophysiology of β-thalassemia was discovered in early 1960s when improved techniques for separation and analysis of different hemoglobin variants were available (Clegg, Naughton, and Weatherall 1968, Weatherall and Clegg 2001). From late 1970s to 1980s, a blooming of molecular era occurred with isolation of human mRNA and globin genes. By 1990, a majority of the mutations causing β-thalassemia were identified with the help of polymerase chain reaction techniques (Ho et al. 1998). With this, the molecular pathology of β-thalassemia started to unravel (Weatherall and Clegg 2001).

In β-thalassemia, despite the reduction in the synthesis of β-globin, the production of α-globin continues normally. Consequently, cytoplasms of RBCs are loaded with free α-globin chains that have deleterious effects. Precipitation of free α-globin chains in RBCs and their precursors lead to formation of toxic α-hemichromes and reactive oxygen species resulting in oxidative destruction of these cells (Mettananda, Gibbons, and Higgs 2016). Destruction of mature RBCs in the circulation leads to hemolysis while that in erythroid precursors within bone marrow leads to ineffective erythropoiesis. The clinically significant anemia in β-thalassemia is a cumulative effect of the ineffective erythropoiesis, premature hemolysis and the direct effect of the reduction of hemoglobin synthesis (Schrier 2002, Thein 2005).

Genetic Modifiers of β-Thalassemia

The clinical heterogeneity observed in β-thalassemia is largely explained by the pathophysiology which revolves around the destructive effects of the unbalanced synthesis of α- and β-like globin chains. A number of studies on the disease pathophysiology discovered two main modifiers of β-thalassemia: co-inheritance of α-thalassemia and upregulation of γ-globin expression (Mettananda, Gibbons, and Higgs 2016). However, with time, the possibility of other genetic factors that could alter disease severity was suggested by studying the unexplained heterogeneity of the disease phenotype observed among patients with identical β- and α-globin genotypes. With dramatic advancement of high throughput genome analysis techniques, including genome wide association studies (GWAS); numerous other genetic modifiers were delineated. GWAS had a major advantage over traditional study methods that included linkage analyses and association studies of a specific genetic variant, due to its ability to reveal unexpected genetic interactions (Higgs, Engel, and Stamatoyannopoulos 2012, Thein 2013a). Using these methods, a large number of genetic modifiers which could ameliorate or worsen the severity of β- thalassemia have been discovered (Table 2).

Genetic modifiers of β-thalassemia act either primarily or secondarily, modifying disease pathophysiology. Primary modifiers affect the degree of β- and α-globin imbalance directly. Among those, most delineated ones are the mutations associated with β- and α-globin genotype, and genetic variants which have an innate ability to upregulate γ-globin and HbF. In contrast, secondary modifiers are located outside the globin gene loci and modify the disease phenotype without altering the globin chain imbalance. They moderate complications associated with the disease (Thein 2018, Nienhuis and Nathan 2012) (Figure 2).

Table 2. Genetic modifiers of β-thalassemia
(Mettananda and Higgs 2018, Thein 2008)

Genetic Modifier	Mechanism
Primary modifiers	
1. β-globin genotype	Directly responsible for the amount of β-globin synthesized
2. α-globin genotype	Directly affect the globin chain imbalance
a. Co-inheritance of α-thalassemia	Reduce free α-globin level
b. Co-inheritance of excess α-chains (ααα/αα, ααα/ααα, αααα/αα)	Increase free α-globin level
3. Innate ability to upregulate γ-globin	Forms HbF which reduces excess α-globin pool and chain imbalance
a. $\delta\beta^0$ thalassemia	
b. *Xmn1-HBG2* polymorphism at $^G\gamma$ (-158 C>T)	
c. *BCL11A* polymorphism (chromosome 2p16)	
d. *HBS1L-MYB* polymorphism (intergenic region or *MYB* on chromosome 6q23)	
e. Mutations of *KLF1*	
4. Other potential modifiers	
a. Variants associated with α-chain proteolysis	Proteolysis of extra α chains
b. α-hemoglobin stabilizing protein	Chaperons extra α-chains
Secondary modifiers	
1. Serum bilirubin level and the risk of gall stones	*UGT1A1* promoter polymorphism- (TA)n
2. Iron metabolism	*HFE-H63D* variant
3. Bone diseases (osteopenia and osteoporosis)	Variant associated with bone mass -*VDR, COL1A1, COL2A2, TGFB1*
4. Cardiovascular diseases	Apolipoprotein -(APOE) ε4 allele
5. Cardiomyopathy (cardiac iron deposition)	Glutathione-S-transferase M1
6. Susceptibility to infections	*HLA* alleles

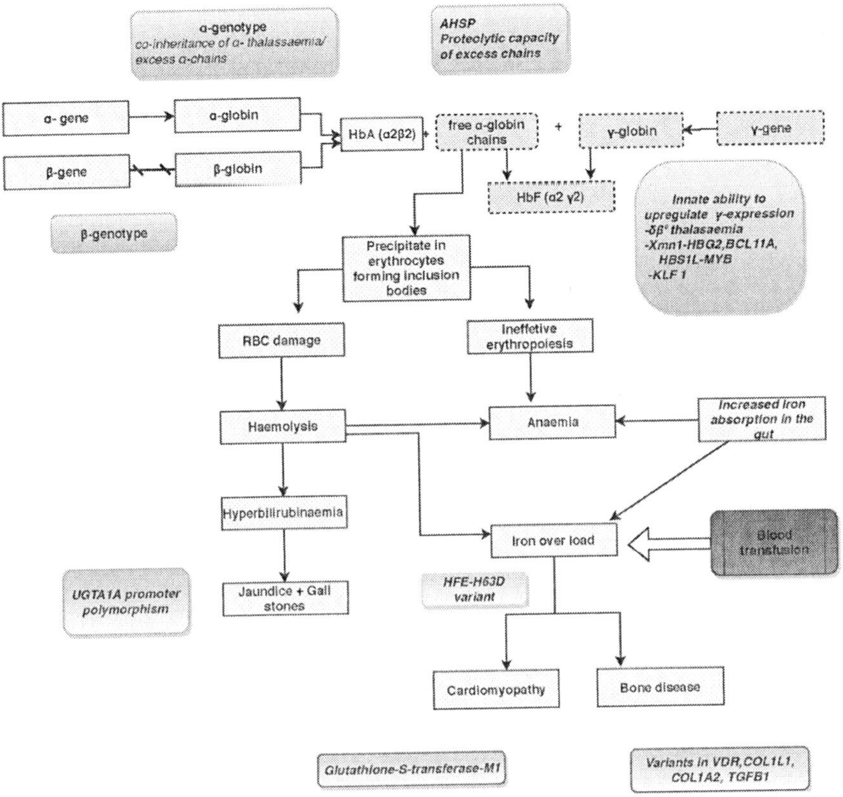

Figure 2. Effects of genetic modifiers of β-thalassemia (Mettananda and Higgs 2018, Thein 2018, Nienhuis and Nathan 2012).

β-Globin Genotype

As β-thalassemia is associated with a quantitative reduction in β-globin, the severity of the disease relies on the extent of imbalance between α- and β-globin chains. The amount of β-globin production depends on the β-globin genotype, hence, β-globin genotype plays a major role in the disease severity. Depending on the expression levels of the affected β-globin genes and the extent to which the β-globin production is altered, mutations are categorized in to three types: mutations that completely abolish the β-globin production (β^{o}), mutations that moderately reduce β-globin production (β^{+})

and mutations which cause minimal reduction in the β-globin output (β^{++}) (Mettananda and Higgs 2018, Weatherall 2001). Normally, heterozygous mutations are associated with β-thalassemia minor (carrier/ trait) and the individuals are completely asymptomatic, while, co-inheritance of homozygosity or compound heterozygosity leads to transfusion dependent β-thalassemia (Musallam et al. 2013). However, some individuals with heterozygous β-thalassemia mutations, suffer from severe anemia and become transfusion dependent due to dominantly inherited β-thalassemia mutations (Weatherall et al. 1973). These dominant mutations could produce shortened or elongated β-globin chains with abnormal carboxyl ends, due to single base substitution (missense), nonsense or frameshift mutations. Their unstable nature results in very rapid post synthetic degradation of β-globin chains, which then precipitate in erythroid precursors and result in hemolysis and ineffective erythropoiesis, causing hematological manifestation of severe β-thalassemia (Thein 2008).

α-Globin Genotype

Human α-globin genes are located in a multi-gene cluster on the short arm of chromosome 16 (16p13.3) and are found to be associated with numerous mutations which interfere the α-globin gene expression. These mutations could either reduce or enhance the α-globin production (Harteveld and Higgs 2010). Even though the central mechanism of disease pathophysiology underlying β-thalassemia is associated with the quantitative reduction of β-globin production, α-globin genotype is crucial for the balance between α/β-globin chains and the severity of β-thalassemia. According to numerous studies, co-inheritance of α-thalassemia ameliorates the disease severity by reducing the free α-globin pool, which precipitate in erythrocyte and their precursors, resulting in hemolysis and ineffective erythropoiesis (Mettananda, Gibbons, and Higgs 2015). Co-inheritance of α- and β-thalassemia is common among populations where prevalence of both types of thalassemia are high and *Plasmodium falciparum* malaria is endemic. This is due to the protective advantage among individuals with heterozygous thalassemia against the malaria infection (Mettananda, Suranjan, et al. 2018, Premawardhena et al. 2017). Several cohort, family and case control studies done on various ethnic groups have shown that deletion of a single α-globin gene (-α/αα) ameliorates the severity of β-thalassemia except in individuals who are homozygous for β^{o} mutations (β^{o}/β^{o}). Likewise, deletion of two α-globin genes improves the hematological parameters in a majority of patients with β-thalassemia (Mettananda 2015, Kan and Nathan 1970, Wainscoat et al. 1983). According to the findings of GWAS done on Sardinian population, even a deletion of a single gene enhances the probability of a less severe β-thalassemia phenotype (Galanello et al. 2009). However, the effect of co-inheritance of α-thalassemia is most significant among patients with hemoglobin E β-thalassemia, the commonest subtype of β-thalassemia (Sripichai et al. 2008, Vichinsky 2007, Fucharoen and Weatherall 2012).

In contrast, excess α-globin genes in the form of triplicated or quadruplicated α-globin genes, worsen β-thalassemia by shifting globin chain imbalance more towards excessive α-globin chains (Premawardhena, Fisher, Olivieri, de Silva, Sloane-Stanley, et al. 2005). A study done on a Sardinian family with triplicated α-globin genes demonstrated that co-inheritance of two additional genes (ααα/ααα) with heterozygous β-thalassemia genotype, mimicked thalassemia intermedia phenotype with a significant α/β-globin chain synthesis imbalance (Galanello et al. 1983). However, the effect of an additional single α-globin gene is highly variable (Traeger-Synodinos et al. 1996, Camaschella et al. 1997).

Innate Ability to Upregulate γ-Globin

HbA is the principle hemoglobin subtype synthesized throughout adult life. However, HbF production is not completely abolished at birth. In fact, a residual amount of HbF persists in adult life as well. HbF in adults is mainly contained in F cells, a subgroup of erythrocytes. Genetic factors that control γ-globin production in adults could modify the clinical severity of β-thalassemia because γ-globin chains could combine with free α-globin chains to prevent their deposition in erythrocytes as insoluble hemichromes (Mettananda, Gibbons, and Higgs 2016). Therefore, cells with high HbF escape from harmful effects of free α-globin chains and have prolonged life spans in patients with β-thalassemia compared to other erythrocytes. The elevated HbF levels in patients with β-thalassemia are believed to be due to the survival of F cells (Thein 2018). The role of genetics in the significant heterogeneity of HbF levels in β-thalassemia patients has been extensively studied during past few years.

Hereditary persistence of fetal hemoglobin (HPFH) includes conditions which give rise to persistence of HbF during adult life without major hematological manifestations (Menzel and Thein 2018). This could be due to non-deletional point mutations or deletions in the promoter regions of γ-globin genes. Long deletions of the β-globin gene cluster that remove β- and δ-globin genes without affecting either one or both of the γ-globin genes result in δβ-thalassemia. This type of thalassemia was first discovered in 1952 and homozygous individuals have 100% HbF with complete absence of HbA and HbA_2 (Musallam et al. 2013).

Non-deletional forms of HPFH include '+' allele of *XmnI-HBG2* polymorphism (rs7482144) in the *HBG2* promoter. This is a sequence variant (C→T) at position -158, located proximal to the γ-globin gene transcription start site (Garner et al. 2004). This polymorphism has a gene frequency of 0·32–0·35 and could be seen in all populations (Thein 2018). Although the influence of *Xmn1-HBG2* on HbF levels is minimal in healthy adults, it accounts for the increment of HbF in patients with homozygous β-thalassemia (Thein 2004).

Attention towards genetic variants that could give rise to high HbF levels that is independent of the β-globin gene cluster has gained through various studies, including family studies. As a result, quantitative trait loci of *HBS1L-MYB* intergenic polymorphism (HMIP) locus on chromosome 6q23 and *BCL11A* on chromosome 2p16 which interfere with HbF levels in adults were discovered. These variants, along with *Xmn1-HBG2,* are known to account for 50% of total variants in HbF levels among northern Europeans (Thein 2018).

MYB, which encodes MYB transcription factor that regulates hematopoiesis and erythroid differentiation, has been characterized as the genetic variant in HMIP (Thein et al. 2007). Apart from these essential roles, two mechanisms through which MYB could regulate HbF have been identified: firstly, when MYB levels are low, it stimulates erythroid differentiation, resulting in release of erythroid progenitor cells synthesizing principally HbF to the circulation and secondly, it could act as a direct activator of *KLF1* and other repressors of γ-globin gene (Thein 2018).

BCL11A exerts significant effect on HbF level through single nucleotide polymorphism (SNP) (rs11886868) in the intron 2 of the *BCL11A* gene. The C allele has been especially identified to be associated with relatively high HbF levels in several different ethnic groups (Uda et al. 2008, Menzel et al. 2007). *BCL11A* encodes for a zinc finger protein transcription factor which is responsible for globin switch during the ontogeny. By interacting with GATA-1 and sex determining region Y-box (SOX6), that bind to upstream of the γ-globin promoter, BCL11A is able to reconfigure β-globin locus to downregulate the expression of γ-globin (Xu et al. 2010, Thein 2018).

The remaining variants that modulate HbF levels escape from even advanced genetic analysis studies. The most significant genetic variant from this category identified so far is *KLF1,* which was first described in 1993 (Miller and Bieker 1993). Genome-wide SNP scan and linkage analysis studies conducted on a Maltese family with HPFH delineated the locus of *KLF1* gene on chromosome 19p13. The expression profile on erythroid progenitors confirmed the capacity of KLF1 in modifying the γ-gene in this family (Borg et al. 2010). KLF1 plays an essential role in hemoglobin switching as it could indirectly silence γ-globin gene through activating *BCL11A* and directly activate β-globin gene (Esteghamat et al. 2013, Zhou et al. 2010a).

In addition to the above, some other genetic factors that could have roles in HbF regulation have been uncovered. These include chromatin modeling factors, nucleosome remodeling deacetylase (NURD) complex, leukemia/lymphoma-related factor (LRF), SOX6, orphan nuclear receptor TR2 and TR4, epigenetic enzymes histone deacetylase (HDAC) 1 and HDAC2 (Suzuki, Yamamoto, and Engel 2014, Thein 2018). More recently, several microRNAs that interfere with the regulation of γ-globin were described. These include miR-15a, miR-16-1, miR-186, miR-301b and miR-125a (Suzuki, Yamamoto, and Engel 2014, Das, Mitra, and Chakravorty 2019).

Other Primary Modifiers

Apart from these main genetic modifiers, some other genetic factors which affect the disease severity have been discovered through numerous studies done in β-thalassemia patients as well as through *in-vitro* and animal studies. Alpha hemoglobin stabilizing protein (AHSP) is an essentially required erythroid specific protein in erythropoiesis, which could prevent free α-globin chain precipitation inside erythrocytes, by binding to free α-globin chains inside the cell (Lai et al. 2006). Therefore, it acts as a molecular chaperone. Reduced erythrocyte life span was discovered in AHSP knockout mice, and apoptosis rate of erythrocyte precursors was higher despite the increment in the number exhibiting hematological manifestations similar to β-thalassemia. Loss of AHSP in heterozygous β-thalassemia mice exacerbated the disease severity (Kong et al. 2004). A quantitative polymerase chain reaction assay (qPCR) conducted using mRNA levels in circulating reticulocytes of 103 healthy individuals, discovered that AHSP expression is independent of age and sex. Among eight sequence variants of AHSP locus, six commonly occurring variants have been identified. The gene frequency of short homopolymer T15, which is associated with low AHSP expression, is higher (on the basis of the allele frequency in the healthy population) in heterozygotes for β-thalassemia and also in heterozygotes for the triplicated α-globin genes (Galanello et al. 2006, Lai et al. 2006). Yet, the correlation between expression levels of AHSP and clinical manifestations of β-thalassemia are inconclusive.

Furthermore, the proteolytic capacity to catabolize excess α-globin chains, preventing their precipitation in erythroid precursors, has also been characterized as a genetic modifier of β-thalassemia. Adenosine triphosphate (ATP) and ubiquitin dependent proteolytic pathways are the primary mechanisms responsible for proteolysis of free α-chains (Shaeffer and Cohen 1997, Shaeffer 1988). Therefore, defects in these pathways could worsen severity by increasing free γ-globin pool. However, the exact effects of these are still not uncovered (Thein 2008).

Secondary Modifiers

The secondary modifiers, which are also known as modifiers of complications of β-thalassemia, moderate complications that are directly associated with anemia or transfusion iron overload. These genetic variants are located outside the β-globin locus. Most characterized secondary modifiers so far are the ones that affect bilirubin metabolism, iron homeostasis, cardiac complications and bone diseases (Weatherall and Clegg 2001).

Precipitation of free α-globin chains results in destruction of erythrocytes and its precursors in patients with β-thalassemia. This leads to increase breakdown of heme and thereby production of bilirubin. Hence, jaundice and gall stone formation are common complications among patients with β-thalassemia. Mutations in UDP-glucuronosyltransferase (*UGT1A1)* gene – the gene that controls the hepatic glucuronidation of bilirubin, is an important secondary modifier of β-thalassemia. Homozygotes for these mutations develop Gilbert syndrome in whom high levels of bilirubin and predisposition to gall stone formation is common (Farheen et al. 2006). Several studies conducted on β-thalassemia patients have discovered that these mutations are common among individuals who develop hyperbilirubinemia, jaundice and gall stones (Galanello et al. 2001, Galanello et al. 1997).

Iron overload is an inevitable consequence of blood transfusions in β-thalassemia. This leads to numerous complications that include cirrhosis, cardiomyopathy, pituitary failure, hypothyroidism and diabetes mellitus (Mettananda 2018a). Among the genetic factors that are involved in iron metabolism, genes which encode for transferrin receptor 2, ferroportin, human hemochromatosis protein (HFE), hepcidin and hemojuvelin, are important modifiers of β-thalassemia (Hentze, Muckenthaler, and Andrews 2004). HFE plays a major role in the iron metabolism in β-thalassemia and, the H63D variant, a common polymorphism in *HFE* gene has been identified as a modifier of iron absorption. Inappropriately high ferritin levels were observed in β-thalassemia carriers who were homozygous for H63D mutation, proposing a modulatory effect on iron absorption by H63D mutation (Melis et al. 2002). In addition, osteoporosis, which develops secondary to endocrine abnormalities in β-thalassemia, is believed to be related to the polymorphic variant in genes which are involved in bone metabolism, vitamin D receptor and *COL1A1* (Cao and Galanello 2010).

Novel Therapeutic Approaches

Even though β-thalassemia is one of the most characterized and well documented genetic disorder in the world, allogenic HSCT remains the only existing curative treatment option (Angelucci et al. 2014). Several thousand patients with β-thalassemia have undergone HSCT since the first HSCT in 1982 (Baronciani et al. 2016). However, allogenic HSCT is not practical for all patients, not only because of serious complications, including graft versus host disease (GVHD) and graft rejection, but also due to the difficulty of finding human leukocyte antigen (HLA) matched donors (Mettananda 2018b). In HLA matched donor transplants, disease free survival rates are between 80-90% (Taher, Weatherall, and Cappellini 2018). Although HLA matched sibling donors provide most desirable outcomes, due to improvement in the management of complications, HSCT from matched-unrelated donors, haplo-matched related donors and cord blood, are dragging attention in many countries (Baronciani et al. 2016, Gaziev et al. 2013).

All curative options of β-thalassemia attempt to address following issues: globin chain imbalance, ineffective erythropoiesis and complications of disease that include iron overload. Gene therapy has been rising as a promising cure for thalassemia in which hematopoietic stem cells (HSC) are transduced with a lentiviral vector to introduce exogenous genetic material (Thompson et al. 2018). The ultimate goal of gene therapy is to express the transduced β-globin gene in HSC of the patient. Although gene therapy has made reasonable progress into clinical trials, the results have been equivocal (Persons 2010, Thompson et al. 2018, Cavazzana, Antoniani, and Miccio 2017). Unfortunately, there are many challenges that need to be addressed before gene therapy becomes standard of care for all patients with β-thalassemia. Therefore, the hunt for a better cure has been a priority in the scientific community during the past few years (Mettananda 2018b).

Induction of Fetal Hemoglobin

In the search for novel therapeutic approaches, induction of HbF has been a promising pathway due to the presence of numerous validated molecular targets including *BCL11A, KLF1* and *MYB,* which could be manipulated to modify gene expression. BCL11A is a well characterized γ-globin repressor and is an exciting molecular target to induce HbF. Most work has been on down regulating *BCL11A,* through genome editing of its enhancers to induce γ-globin by eliminating repression by BCL11A (Canver et al. 2015, Smith and Orkin 2016). However, manipulation of *BCL11A* has been a challenge as it also plays essential roles in neuronal and lymphoid development (Liu et al. 2003, John et al. 2012). With the recent identification of erythroid specific part of the enhancer sequence located in an intronic region, it has been possible to target only the erythroid function of the *BCL11A* (Bauer et al. 2013). Therefore, this developmental-stage-specific lineage restricted *BCL11A* enhancer, has been delineated as a promising molecular target for therapeutic genome editing.

Apart from the three main TFs – MYB, KLF and BCL11A – which are involved in γ-globin repression, several other factors have also been identified. As stated before, these include a number of epigenetic enzymes for example, DNMT, LSD1, HDAC1 and HDAC2, which have been proposed as molecular targets to upregulate γ-globin. Inhibition of LSD1 using tranylcypromine and HDAC1 and 2, using HDAC inhibitors have been shown to upregulate γ-globin in pre-clinical studies (Bradner et al. 2010, Shi et al. 2013, Okam et al. 2015). LIM-domain-binding protein (LDB) 1 is another TF that aids LCR to interact with the genes in the β-globin locus in a developmental-stage-specific way through chromatin looping. Modification of LDB1 through artificial zinc finger protease to bind to γ-globin promoter instead of β-globin has been shown to reactivate the γ-globin gene (Deng et al. 2014). Yet, the effectiveness of these as molecular targets to cure β-thalassemia, relies on our ability to modify these variants without affecting their essential roles in erythropoiesis or other tissues.

RNA interference (RNAi) is another method that has been tested to enhance HbF. It is mediated either by miRNA or short interfering RNAs (siRNA). This advanced genome editing method has proven successful in knocking down *BCL11A* in murine models, exhibiting the possibility of using this technique to upregulate γ-globin expression (Guda et al. 2015).

Apart from genetic modifications, few novel pharmacological agents that can enhance HbF level are also under investigation. Among them, hydroxyurea, 5-azacytidine and sodium butyrate have stolen the spotlight. 5-Azacystidine is the first pharmacological agent to be tested in clinical trials. Since then, efficacy and safety of a number of pharmacological agents have been tested; however, none has been efficacious enough for routine use (Fard et al. 2013, Atweh and Schechter 2001). The only exception is hydroxyurea, which has demonstrated variable results in clinical studies worldwide (Foong et al. 2016, Saxon, Rees, and Olivieri 1998, Algiraigri and Kassam 2017, Zamani et al. 2009). In a recent meta-analysis of eleven observational studies involving 859 transfusion dependent β-thalassemia patients, hydroxyurea had a complete response rate of 26% and an overall response rate of 60%, showing its promising future as a novel therapeutic agent for β-thalassemia (Algiraigri et al. 2017). Furthermore, some of the compounds like, 5-azacytidine and sodium butyrate act as epigenetic modifiers. Butyrate inhibits the activity of HDAC enzymes, while azacytidine induces HbF by inhibiting the effect of DNMT enzymes in the β-globin cluster (Thein 2018).

Down Regulation of α-Globin

Numerous studies done on β-thalassemia have described the beneficial effects of co-inheritance of α-thalassemia, as it could transform the phenotype from severe transfusion dependent to milder form of thalassemia (Mettananda, Gibbons, and Higgs 2015, Premawardhena, Fisher, Olivieri, de Silva, Arambepola, et al. 2005, de Silva et al. 2000). Although the attention has predominantly been gained by the upregulation of γ-globin expression, downregulation of α-globin gene is also an exciting therapeutic approach due to the well characterized role of excess α-globin in disease pathophysiology (Mettananda, Gibbons, and Higgs 2016, Voon and Vadolas 2008). Suppression of α-globin could be achieved through RNAi, genome editing or pharmaceutical methods.

RNAi has been tested *in vitro* as well as *in vivo,* to knock down α-globin expression (Voon, Wardan, and Vadolas 2008). In 2007, Xie et al. micro-injected lentiviral vectors containing short hairpin RNA (shRNA) targeting α-globin gene into heterozygous β-thalassemia mouse single cell embryos (Xie et al. 2007). This resulted in transgenic mice which had reduced levels of α-globin (20-30% less than the normal) and exhibited improved hematological manifestations (Xie et al. 2011).

Genome editing using clustered, regularly interspaced, short palindromic repeat (CRISPR)-CRISPR associated system (CRISPR-Cas9), has revolutionized the field of genome engineering in the recent years (Ledford 2015). This method has been utilized successfully to down regulate the α-globin by disrupting the erythroid specific MCS-R2 enhancer of the α-globin gene (Mettananda, Fisher, Hay, et al. 2017). Using this method a natural deletion of MCS-R2 was emulated in human HSCs and the differentiated erythroid cells produced lower α-globin levels than normal (Mettananda, Fisher, Hay, et al. 2017, Coelho et al. 2010). These genome edited cells were present in long-term repopulating HSCs in mouse models following xeno-transplantation, supporting the hope of advancing this as a realistic therapeutic approach in the future.

Pharmacological compounds, which act through specific epigenetic pathways to modify expression of genes via modifying the chromosomal environment, is drawing significant attention over the last few years, due to their ability to overcome numerous complications associated with genome editing and RNAi. A recent *in vitro* study revealed that a novel histone demethylase inhibitor compound IOX1, has the potential to down regulate α-globin in human erythroid cells, by altering the chromatin environment of associated genes (Mettananda, Fisher, Sloane-Stanley, et al. 2017, Mettananda, Clark, et al. 2018). Although these are not be amenable to clinical practice at present, development of newer and more specific inhibitors will invariably have a potential to cure β-thalassemia.

Other Therapeutic Approaches

Apart from the above mentioned therapeutic approaches, compounds that could address dyserythropoiesis have also been identified as potential targets. This category of pharmacological compounds includes activin receptor-II trap ligands, JAK2 inhibitors and the agents that could restrict the availability of iron for erythropoiesis (Dussiot et al. 2014, El-Beshlawy and El-Ghamrawy 2019).

Hepcidin is an essential hormone that regulates iron absorption. By manipulation of hepcidin, it has been possible to control iron absorption, thus, it has been found to be beneficial in the management of iron overload in β-thalassemia. Stimulation of hepcidin could be achieved either through hepcidin agonists (minihepcidin-MH), or *in-vivo* inhibition of TMPRSS6, an essential protease enzyme required for the activation of hepcidin. The effectiveness of inhibiting TMPRSS6, in order to enhance hepcidin level, has been evaluated using both siRNA and antisense oligonucleotide in murine models (Sebastiani, Wilkinson, and Pantopoulos 2016, Guo et al. 2013, Schmidt et al. 2013).

CONCLUSION

β-Thalassemia is a monogenic disorder that occurs mainly due to non-deletional mutations of the β-globin gene or *cis* acting enhancers, declining the β-globin chain production. The wide heterogeneity of the disease phenotype is attributed to various genetic modifiers that could ameliorate or worsen the clinical and hematological phenotype. Some of the genetic variants directly modify the globin chain imbalance, whereas, others manage the complications associated with the disease. Therefore, it is important to consider these variants in making an accurate molecular diagnosis and to correlate with the disease phenotype in patients with β-thalassemia. The identification of genetic factors which could modify the severity helps to improve the disease management as these variants could be used as novel therapeutic targets. Numerous novel therapeutic approaches including, genome editing, RNAi and pharmacological agents, are being tested in *in vivo* and *in vitro* models. Some pharmaceutical agents are already in clinical trials. Upregulation of γ-globin production in β-thalassemia patients is turning out to be a promising and exciting therapeutic approach. However, most of these methods are still in their infancy and require several optimizations before they can be routinely used in clinical practice.

REFERENCES

Algiraigri, A. H., and A. Kassam. 2017. "Hydroxyurea for hemoglobin E/beta-thalassemia: a systematic review and meta-analysis." *Int J Hematol* 106 (6):748-756. doi: 10.1007/s12185-017-2307-0.

Algiraigri, A. H., N. A. M. Wright, E. O. Paolucci, and A. Kassam. 2017. "Hydroxyurea for lifelong transfusion-dependent beta-thalassemia: A meta-analysis." *Pediatr Hematol Oncol* 34 (8):435-448. doi: 10.1080/ 08880018.2017.1354948.

Angelucci, E., S. Matthes-Martin, D. Baronciani, F. Bernaudin, S. Bonanomi, M. D. Cappellini, J. H. Dalle, P. Di Bartolomeo, C. D. de Heredia, R. Dickerhoff, C. Giardini, E. Gluckman, A. A. Hussein, N. Kamani, M. Minkov, F. Locatelli, V. Rocha, P. Sedlacek, F. Smiers, I. Thuret, I. Yaniv, M. Cavazzana, C. Peters, Ebmt Inborn Error, and Ebmt Paediatric Working Parties. 2014. "Hematopoietic stem cell transplantation in thalassemia major and sickle cell disease: indications and management recommendations from an international expert panel." *Haematologica* 99 (5):811-20. doi: 10.3324/haematol.2013.099747.

Atweh, G. F., and A. N. Schechter. 2001. "Pharmacologic induction of fetal hemoglobin: raising the therapeutic bar in sickle cell disease." *Curr Opin Hematol* 8 (2):123-30.

Baronciani, D., E. Angelucci, U. Potschger, J. Gaziev, A. Yesilipek, M. Zecca, M. G. Orofino, C. Giardini, A. Al-Ahmari, S. Marktel, J. de la Fuente, A. Ghavamzadeh, A. A. Hussein, C. Targhetta, F. Pilo, F. Locatelli, G. Dini, P. Bader, and C. Peters. 2016. "Hemopoietic stem cell transplantation in thalassemia: a report from the European Society for Blood and Bone Marrow Transplantation Hemoglobinopathy Registry, 2000-2010." *Bone Marrow Transplant* 51 (4):536-41. doi: 10.1038/bmt.2015.293.

Bauer, D. E., S. C. Kamran, and S. H. Orkin. 2012. "Reawakening fetal hemoglobin: prospects for new therapies for the beta-globin disorders." *Blood* 120 (15):2945-53. doi: 10.1182/blood-2012-06-292078.

Bauer, Daniel E., Sophia C. Kamran, Samuel Lessard, Jian Xu, Yuko Fujiwara, Carrie Lin, Zhen Shao, Matthew C Canver, Elenoe C Smith, and Luca Pinello. 2013. "An erythroid enhancer of BCL11A subject to genetic variation determines fetal hemoglobin level." *Science* 342 (6155):253-257.

Behringer, Richard R., Thomas M. Ryan, Richard D. Palmiter, Ralph L. Brinster, and Tim M. Townes. 1990. "Human gamma-to beta-globin gene switching in transgenic mice." *Genes & Development* 4 (3):380-389.

Bender, M. A., Michael Bulger, Jennie Close, and Mark Groudine. 2000. "β-globin gene switching and DNase I sensitivity of the endogenous β-globin locus in mice do not require the locus control region." *Molecular cell* 5 (2):387-393.

Borg, J., P. Papadopoulos, M. Georgitsi, L. Gutierrez, G. Grech, P. Fanis, M. Phylactides, A. J. Verkerk, P. J. van der Spek, C. A. Scerri, W. Cassar, R. Galdies, W. van Ijcken, Z. Ozgur, N. Gillemans, J. Hou, M. Bugeja, F. G. Grosveld, M. von Lindern, A. E. Felice, G. P. Patrinos, and S. Philipsen. 2010. "Haploinsufficiency for the erythroid transcription factor KLF1 causes hereditary persistence of fetal hemoglobin." *Nat Genet* 42 (9):801-5. doi: 10.1038/ng.630.

Bradner, J. E., R. Mak, S. K. Tanguturi, R. Mazitschek, S. J. Haggarty, K. Ross, C. Y. Chang, J. Bosco, N. West, E. Morse, K. Lin, J. P. Shen, N. P. Kwiatkowski, N. Gheldof, J. Dekker, D. J. DeAngelo, S. A. Carr, S. L. Schreiber, T. R. Golub, and B. L. Ebert. 2010. "Chemical genetic strategy identifies histone deacetylase 1 (HDAC1) and HDAC2 as therapeutic targets in sickle cell disease." *Proc Natl Acad Sci U S A* 107 (28):12617-22. doi: 10.1073/pnas.1006774107.

Camaschella, C., A. C. Kattamis, D. Petroni, A. Roetto, P. Sivera, L. Sbaiz, A. Cohen, K. Ohene-Frempong, P. Trifillis, S. Surrey, and P. Fortina. 1997. "Different hematological phenotypes caused by the interaction of triplicated α-globin genes and heterozygous β-thalassemia." *American Journal of Hematology* 55 (2):83-88. doi: 10.1002/(sici)1096-8652(199706)55:2<83::Aid-ajh6>3.0.Co;2-z.

Canver, M. C., E. C. Smith, F. Sher, L. Pinello, N. E. Sanjana, O. Shalem, D. D. Chen, P. G. Schupp, D. S. Vinjamur, S. P. Garcia, S. Luc, R. Kurita, Y. Nakamura, Y. Fujiwara, T. Maeda, G. C. Yuan, F. Zhang, S. H. Orkin, and D. E. Bauer. 2015. "BCL11A enhancer dissection by Cas9-mediated in situ saturating mutagenesis." *Nature* 527 (7577):192-7. doi: 10.1038/nature15521.

Cao, A., and R. Galanello. 2010. "Beta-thalassemia." *Genet Med* 12 (2):61-76. doi: 10.1097/GIM.0b013e3181cd68ed.

Cavazzana, M., C. Antoniani, and A. Miccio. 2017. "Gene Therapy for beta-Hemoglobinopathies." *Mol Ther* 25 (5):1142-1154. doi: 10.1016/j.ymthe.2017.03.024.

Clegg, J. B., M. A. Naughton, and D. J. Weatherall. 1968. "Separation of the α and β-chains of human haemoglobin." *Nature* 219 (5149):69.

Coelho, A., I. Picanco, F. Seuanes, M. T. Seixas, and P. Faustino. 2010. "Novel large deletions in the human alpha-globin gene cluster: Clarifying the HS-40 long-range regulatory role in the native chromosome environment." *Blood Cells Mol Dis* 45 (2):147-53. doi: 10.1016/j.bcmd.2010.05.010.

Das, S. S., A. Mitra, and N. Chakravorty. 2019. "Diseases and their clinical heterogeneity - Are we ignoring the SNiPers and micRomaNAgers? An illustration using Beta-thalassemia clinical spectrum and fetal hemoglobin levels." *Genomics* 111 (1):67-75. doi: 10.1016/j.ygeno.2018.01.002.

de Silva, S., C. A. Fisher, A. Premawardhena, S. P. Lamabadusuriya, T. E. Peto, G. Perera, J. M. Old, J. B. Clegg, N. F. Olivieri, and D. J. Weatherall. 2000. "Thalassaemia in Sri Lanka: implications for the future health burden of Asian populations. Sri Lanka Thalassaemia Study Group." *Lancet* 355 (9206):786-91.

Deng, W., J. W. Rupon, I. Krivega, L. Breda, I. Motta, K. S. Jahn, A. Reik, P. D. Gregory, S. Rivella, A. Dean, and G. A. Blobel. 2014. "Reactivation of developmentally silenced globin genes by forced chromatin looping." *Cell* 158 (4):849-60. doi: 10.1016/j.cell.2014.05.050.

Dussiot, M., T. T. Maciel, A. Fricot, C. Chartier, O. Negre, J. Veiga, D. Grapton, E. Paubelle, E. Payen, Y. Beuzard, P. Leboulch, J. A. Ribeil, J. B. Arlet, F. Cote, G. Courtois, Y. Z. Ginzburg, T. O. Daniel, R. Chopra, V. Sung, O. Hermine, and I. C. Moura. 2014. "An activin receptor IIA ligand trap corrects ineffective erythropoiesis in beta-thalassemia." *Nat Med* 20 (4):398-407. doi: 10.1038/nm.3468.

Dzierzak, Elaine, and Sjaak Philipsen. 2013. "Erythropoiesis: development and differentiation." *Cold Spring Harbor perspectives in medicine* 3 (4):a011601.

El-Beshlawy, A., and M. El-Ghamrawy. 2019. "Recent trends in treatment of thalassemia." *Blood Cells Mol Dis* 76:53-58. doi: 10.1016/j.bcmd. 2019.01.006.

Enver, Tariq, Natacha Raich, Allen J. Ebens, Thalia Papayannopoulou, Frank Costantini, and George Stamatoyannopoulos. 1990. "Developmental regulation of human fetal-to-adult globin gene switching in transgenic mice." *Nature* 344 (6264):309.

Esteghamat, Fatemehsadat, Nynke Gillemans, Ivan Bilic, Emile van den Akker, Ileana Cantù, Teus van Gent, Ursula Klingmüller, Kirsten van Lom, Marieke von Lindern, and Frank Grosveld. 2013. "Erythropoiesis and globin switching in compound Klf1:: Bcl11a mutant mice." *Blood* 121 (13):2553-2562.

Fard, Ali Dehghani, Seyed Ahmad Hosseini, Mohammad Shahjahani, Fatemeh Salari, and Kaveh Jaseb. 2013. "Evaluation of Novel Fetal Hemoglobin Inducer Drugs in Treatment of β-Hemoglobinopathy Disorders." *International journal of hematology-oncology and stem cell research* 7 (3):47-54.

Farheen, Shabana, Sanghamitra Sengupta, Amal Santra, Suparna Pal, Gopal Krishna Dhali, Meenakshi Chakravorty, Partha P Majumder, and Abhijit Chowdhury. 2006. "Gilbert' s syndrome: High frequency of the (TA)~ 7 TAA allele in India and its interaction with a novel CAT insertion in promoter of the gene for bilirubin UDP-glucuronosyl-transferase 1 gene." *World journal of gastroenterology* 12 (14):2269.

Flint, J., R. M. Harding, A. J. Boyce, and J. B. Clegg. 1998. "The population genetics of the haemoglobinopathies." *Baillieres Clin Haematol* 11 (1):1-51.

Foong, W. C., J. J. Ho, C. K. Loh, and V. Viprakasit. 2016. "Hydroxyurea for reducing blood transfusion in non-transfusion dependent beta thalassaemias." *Cochrane Database Syst Rev* 10:CD011579. doi: 10. 1002/14651858.CD011579.pub2.

Forrester, William C., Elliot Epner, M. Catherine Driscoll, Tariq Enver, Martha Brice, Thalia Papayannopoulou, and Mark Groudine. 1990. "A deletion of the human beta-globin locus activation region causes a major alteration in chromatin structure and replication across the entire beta-globin locus." *Genes & development* 4 (10):1637-1649.

Fucharoen, S., and D. J. Weatherall. 2012. "The hemoglobin E thalassemias." *Cold Spring Harb Perspect Med* 2 (8). doi: 10.1101/cshperspect.a011734.

Fucharoen, S., and D. J. Weatherall. 2016. "Progress Toward the Control and Management of the Thalassemias." *Hematol Oncol Clin North Am* 30 (2):359-71. doi: 10.1016/j.hoc.2015.12.001.

Galanello, R., L. Perseu, M. A. Melis, L. Cipollina, S. Barella, N. Giagu, M. P. Turco, O. Maccioni, and A. Cao. 1997. "Hyperbilirubinaemia in heterozygous β-thalassaemia is related to co-inherited Gilbert' s syndrome." *British journal of haematology* 99 (2):433-436.

Galanello, R., S. Piras, S. Barella, G. B. Leoni, M. D. Cipollina, L. Perseu, and A. Cao. 2001. "Cholelithiasis and Gilbert' s syndrome in homozygous β-thalassaemia." *British journal of haematology* 115 (4):926-928.

Galanello, R., R. Ruggeri, E. Paglietti, M. Addis, M. A. Melis, and A. Cao. 1983. "A family with segregating triplicated alpha globin loci and beta thalassemia." *Blood* 62 (5):1035-1040.

Galanello, R., and R. Origa. 2010. "Beta-thalassemia." *Orphanet J Rare Dis* 5:11. doi: 10.1186/1750-1172-5-11.

Galanello, R., S. Sanna, L. Perseu, M. C. Sollaino, S. Satta, M. E. Lai, S. Barella, M. Uda, G. Usala, G. R. Abecasis, and A. Cao. 2009. "Amelioration of Sardinian beta0 thalassemia by genetic modifiers." *Blood* 114 (18):3935-7. doi: 10.1182/blood-2009-04-217901.

Galanello, Renzo, Lucia Perseu, Chiara Perra, Liliana Maccioni, Susanna Barella, Maurizio Longinotti, Antonio Cao, and Mario Cazzola. 2004. "Somatic deletion of the normal β-globin gene leading to thalassaemia intermedia in heterozygous β-thalassaemic patients." *British journal of haematology* 127 (5):604-606.

Galanello, Renzo, Stefania Satta, Maria Dolores Cipollina, Chiara Perra, and Luciana Perseu. 2006. *AHSP Promoter Homopolymer in Beta-Thalassemia Syndromes*. Am Soc Hematology.

Garner, C., N. Silver, S. Best, S. Menzel, C. Martin, T. D. Spector, and S. L. Thein. 2004. "Quantitative trait locus on chromosome 8q influences the switch from fetal to adult hemoglobin." *Blood* 104 (7):2184-6. doi: 10.1182/blood-2004-02-0527.

Gaziev, Javid, Marco Marziali, Antonella Isgrò, Pietro Sodani, Katia Paciaroni, Cristiano Gallucci, Marco Andreani, Manuela Testi, Gioia De Angelis, and Cecilia Alfieri. 2013. "Bone marrow transplantation for thalassemia from alternative related donors: improved outcomes with a new approach." *Blood* 122 (15):2751-2756.

Grosveld, Frank. 1999. "Activation by locus control regions?" *Current opinion in genetics & development* 9 (2):152-157.

Guda, Swaroopa, Christian Brendel, Raffaele Renella, Peng Du, Daniel E. Bauer, Matthew C. Canver, Jennifer K. Grenier, Andrew W. Grimson, Sophia C. Kamran, and James Thornton. 2015. "miRNA-embedded shRNAs for lineage-specific BCL11A knockdown and hemoglobin F induction." *Molecular Therapy* 23 (9):1465-1474.

Guo, Shuling, Carla Casu, Sara Gardenghi, Sheri Booten, Mariam Aghajan, Raechel Peralta, Andy Watt, Sue Freier, Brett P. Monia, and Stefano Rivella. 2013. "Reducing TMPRSS6 ameliorates hemochromatosis and β-thalassemia in mice." *The Journal of Clinical Investigation* 123 (4):1531-1541. doi: 10.1172/JCI66969.

Hall, Georgina W., and Sweelay Thein. 1994. "Nonsense codon mutations in the terminal exon of the beta-globin gene are not associated with a reduction in beta-mRNA accumulation: a mechanism for the phenotype of dominant beta-thalassemia." *Blood* 83 (8):2031-2037.

Harteveld, C. L., and D. R. Higgs. 2010. "Alpha-thalassaemia." *Orphanet J Rare Dis* 5:13. doi: 10.1186/1750-1172-5-13.

He, Feng, and Allan Jacobson. 2015. "Nonsense-mediated mRNA decay: degradation of defective transcripts is only part of the story." *Annual review of genetics* 49:339-366.

Hentze, Matthias W., Martina U. Muckenthaler, and Nancy C Andrews. 2004. "Balancing acts: molecular control of mammalian iron metabolism." *cell* 117 (3):285-297.

Higgs, D. R., J. D. Engel, and G. Stamatoyannopoulos. 2012. "Thalassaemia." *Lancet* 379 (9813):373-83. doi: 10.1016/S0140-6736(11)60283-3.

Higgs, D. R., and W. G. Wood. 2011. "Erythropoiesis." In *Postgraduate Haematology*, edited by AV Hoffbrand, D Catovsky, EGD Tuddenham and AR Green, 12-25. NJ: Wiley-Blackwell.

Ho, P. J., G. W. Hall, L. Y. Luo, D. J. Weatherall, and S. L. Thein. 1998. "Beta-thalassaemia intermedia: is it possible consistently to predict phenotype from genotype?" *Br J Haematol* 100 (1):70-8.

Huisman, T. H. J. 1997. "Levels of Hb A2 in heterozygotes and homozygotes for beta-thalassemia mutations: influence of mutations in the CACCC and ATAAA motifs of the beta-globin gene promoter." *Acta haematologica* 98 (4):187-194.

John, Anita, Heike Brylka, Christoph Wiegreffe, Ruth Simon, Pentao Liu, René Jüttner, E. Bryan Crenshaw, Frank P. Luyten, Nancy A. Jenkins, and Neal G. Copeland. 2012. "Bcl11a is required for neuronal morphogenesis and sensory circuit formation in dorsal spinal cord development." *Development* 139 (10):1831-1841.

Kan, Y. W., and D. G. Nathan. 1970. "Mild thalassemia: the result of interactions of alpha and beta thalassemia genes." *J Clin Invest* 49 (4):635-42. doi: 10.1172/JCI106274.

Kimberland, Michelle L., Vladimir Divoky, Josef Prchal, Uwe Schwahn, Wolfgang Berger, and Haig H Kazazian Jr. 1999. "Full-length human L1 insertions retain the capacity for high frequency retrotransposition in cultured cells." *Human molecular genetics* 8 (8):1557-1560.

Kong, Y. I., Suiping Zhou, Anthony J. Kihm, Anne M Katein, Xiang Yu, David A. Gell, Joel P. Mackay, Kazuhiko Adachi, Linda Foster-Brown, and Calvert S. Louden. 2004. "Loss of α-hemoglobin–stabilizing protein impairs erythropoiesis and exacerbates β-thalassemia." *The Journal of clinical investigation* 114 (10):1457-1466.

Kountouris, Petros, Carsten W. Lederer, Pavlos Fanis, Xenia Feleki, John Old, and Marina Kleanthous. 2014. "IthaGenes: an interactive database for haemoglobin variations and epidemiology." *PloS one* 9 (7):e103020.

Lai, Mei I., Jie Jiang, Nicholas Silver, Steve Best, Stephan Menzel, Aleksandar Mijovic, Stefano Colella, Jiannis Ragoussis, Chad Garner, and Mitchell J. Weiss. 2006. "α-Haemoglobin stabilising protein is a quantitative trait gene that modifies the phenotype of β-thalassaemia." *British journal of haematology* 133 (6):675-682.

Ledford, H. 2015. "CRISPR, the disruptor." *Nature* 522 (7554):20-4. doi: 10.1038/522020a.

Lee, W. S., B. McColl, J. Maksimovic, and J. Vadolas. 2017. "Epigenetic interplay at the beta-globin locus." *Biochim Biophys Acta* 1860 (4):393-404. doi: 10.1016/j.bbagrm.2017.01.014.

Liu, Pentao, Jonathan R. Keller, Mariaestela Ortiz, Lino Tessarollo, Rivka A. Rachel, Takuro Nakamura, Nancy A. Jenkins, and Neal G. Copeland. 2003. "Bcl11a is essential for normal lymphoid development." *Nature immunology* 4 (6):525.

Maragoudaki, E, E. Kanavakis, J. Traeger-Synodinos, C. Vrettou, M. Tzetis, A. Metaxotou-Mavrommati, and C. Kattamis. 1999. "Molecular, haematological and clinical studies of the– 101 C→ T substitution of the β-globin gene promoter in 25 β-thalassaemia intermedia patients and 45 heterozygotes." *British journal of haematology* 107 (4):699-706.

Melis, Maria Antonietta, Milena Cau, Federica Deidda, Susanna Barella, Antonio Cao, and Renzo Galanello. 2002. "H63D mutation in the HFE gene increases iron overload in beta-thalassemia carriers." *haematologica* 87 (3):242-245.

Menzel, S., C. Garner, I. Gut, F. Matsuda, M. Yamaguchi, S. Heath, M. Foglio, D. Zelenika, A. Boland, H. Rooks, S. Best, T. D. Spector, M. Farrall, M. Lathrop, and S. L. Thein. 2007. "A QTL influencing F cell production maps to a gene encoding a zinc-finger protein on chromosome 2p15." *Nat Genet* 39 (10):1197-9. doi: 10.1038/ng2108.

Menzel, S., and S. L. Thein. 2018. "Genetic Modifiers of Fetal Haemoglobin in Sickle Cell Disease." *Mol Diagn Ther*. doi: 10.1007/s40291-018-0370-8.

Mettananda, Doluweera Sachith Gopeetha. 2015. *Ameliorating β-thalassaemia by manipulating expression of the α-globin gene* Oxford: University of Oxford.

Mettananda, S. 2017. "Thalassaemia: In a quest towards an ultimate cure." *Sri Lanka Journal of Child Health* 46 (3):203-210.

Mettananda, S. 2018a. "Management of Thalassaemia." *Sri Lanka Journal of Child Health* 47 (2):159-165.

Mettananda, S. 2018b. "Thalassaemia: Current research may provide a cure to it in the future." *Sri Lanka Journal of Child Health* 47 (4):372. doi: http://doi.org/10.4038/sljch.v47i4.8606.

Mettananda, S., K. Clark, C. A. Fisher, J. A. Sloane-Stanley, R. J. Gibbons, and D. R. Higgs. 2018. "Phenotypic and molecular characterization of a serum-free miniature erythroid differentiation system suitable for high-throughput screening and single-cell assays." *Exp Hematol* 60:10-20. doi: 10.1016/j.exphem.2018.01.001.

Mettananda, S., and D. G. de Silva. 2017. "Anaemia in children: are we using the correct prevention strategies?" *Ceylon Med J* 62 (2):73-76. doi: 10.4038/cmj.v62i2.8469.

Mettananda, S., C. A. Fisher, D. Hay, M. Badat, L. Quek, K. Clark, P. Hublitz, D. Downes, J. Kerry, M. Gosden, J. Telenius, J. A. Sloane-Stanley, P. Faustino, A. Coelho, J. Doondeea, B. Usukhbayar, P. Sopp, J. A. Sharpe, J. R. Hughes, P. Vyas, R. J. Gibbons, and D. R. Higgs. 2017. "Editing an alpha-globin enhancer in primary human hematopoietic stem cells as a treatment for beta-thalassemia." *Nat Commun* 8 (1):424. doi: 10.1038/s41467-017-00479-7.

Mettananda, S., C. A. Fisher, J. A. Sloane-Stanley, S. Taylor, U. Oppermann, R. J. Gibbons, and D. R. Higgs. 2017. "Selective silencing of alpha-globin by the histone demethylase inhibitor IOX1: a potentially new pathway for treatment of beta-thalassemia." *Haematologica* 102 (3):e80-e84. doi: 10.3324/haematol.2016.155655.

Mettananda, S., R. J. Gibbons, and D. R. Higgs. 2015. "alpha-Globin as a molecular target in the treatment of beta-thalassemia." *Blood* 125 (24):3694-701. doi: 10.1182/blood-2015-03-633594.

Mettananda, S., R. J. Gibbons, and D. R. Higgs. 2016. "Understanding alpha-globin gene regulation and implications for the treatment of beta-thalassemia." *Ann N Y Acad Sci* 1368 (1):16-24. doi: 10.1111/nyas.12988.

Mettananda, S., and D. R. Higgs. 2018. "Molecular Basis and Genetic Modifiers of Thalassemia." *Hematol Oncol Clin North Am* 32 (2):177-191. doi: 10.1016/j.hoc.2017.11.003.

Mettananda, S., H. Pathiraja, R. Peiris, N. Wickramarathne, D. Bandara, U. de Silva, C. Mettananda, and A. Premawardhena. 2019. "Blood transfusion therapy for beta-thalassemia major and hemoglobin E beta-thalassemia: Adequacy, trends, and determinants in Sri Lanka." *Pediatr Blood Cancer* 66 (5):e27643. doi: 10.1002/pbc.27643.

Mettananda, S., M. Suranjan, R. Fernando, T. Dias, C. Mettananda, R. Rodrigo, L. Perera, R. Gibbons, A. Premawardhena, and D. Higgs. 2018. "Anaemia among females in child-bearing age: Relative contributions, effects and interactions of alpha- and beta-thalassaemia." *PLoS One* 13 (11):e0206928. doi: 10.1371/journal.pone.0206928.

Miller, I. J., and James J. Bieker. 1993. "A novel, erythroid cell-specific murine transcription factor that binds to the CACCC element and is related to the Krüppel family of nuclear proteins." *Molecular and cellular biology* 13 (5):2776-2786.

Musallam, K. M., S. Rivella, E. Vichinsky, and E. A. Rachmilewitz. 2013. "Non-transfusion-dependent thalassemias." *Haematologica* 98 (6):833-44. doi: 10.3324/haematol.2012.066845.

Nienhuis, A. W., and D. G. Nathan. 2012. "Pathophysiology and Clinical Manifestations of the beta-Thalassemias." *Cold Spring Harb Perspect Med* 2 (12):a011726. doi: 10.1101/cshperspect.a011726.

Nuinoon, Manit, Wattanan Makarasara, Taisei Mushiroda, Iswari Setianingsih, Pustika Amalia Wahidiyat, Orapan Sripichai, Natsuhiko Kumasaka, Atsushi Takahashi, Saovaros Svasti, and Thongperm Munkongdee. 2010. "A genome-wide association identified the common genetic variants influence disease severity in β 0-thalassemia/hemoglobin E." *Human genetics* 127 (3):303-314.

Okam, M. M., E. B. Esrick, E. Mandell, F. Campigotto, D. S. Neuberg, and B. L. Ebert. 2015. "Phase 1/2 trial of vorinostat in patients with sickle cell disease who have not benefitted from hydroxyurea." *Blood* 125 (23):3668-9. doi: 10.1182/blood-2015-03-635391.

Persons, D. A. 2010. "Gene therapy: Targeting beta-thalassaemia." *Nature* 467 (7313):277-8. doi: 10.1038/467277a.

Premawardhena, A., A. Allen, F. Piel, C. Fisher, L. Perera, R. Rodrigo, G. Goonathilaka, L. Ramees, T. Peto, N. Olivieri, and D. Weatherall. 2017. "The evolutionary and clinical implications of the uneven distribution of the frequency of the inherited haemoglobin variants over short geographical distances." *Br J Haematol* 176 (3):475-484. doi: 10.1111/bjh.14437.

Premawardhena, A., C. A. Fisher, N. F. Olivieri, S. de Silva, M. Arambepola, W. Perera, A. O' Donnell, T. E. Peto, V. Viprakasit, L. Merson, G. Muraca, and D. J. Weatherall. 2005. "Haemoglobin E beta thalassaemia in Sri Lanka." *Lancet* 366 (9495):1467-70. doi: 10.1016/S0140-6736(05)67396-5.

Premawardhena, A., C. A. Fisher, N. F. Olivieri, S. de Silva, J. Sloane-Stanley, W. G. Wood, and D. J. Weatherall. 2005. "A novel molecular basis for beta thalassemia intermedia poses new questions about its pathophysiology." *Blood* 106 (9):3251-5. doi: 10.1182/blood-2005-02-0593.

Sankaran, V. G., and S. H. Orkin. 2013. "The switch from fetal to adult hemoglobin." *Cold Spring Harb Perspect Med* 3 (1):a011643. doi: 10.1101/cshperspect.a011643.

Sankaran, Vijay G., Tobias F Menne, Jian Xu, Thomas E. Akie, Guillaume Lettre, Ben Van Handel, Hanna K. A. Mikkola, Joel N. Hirschhorn, Alan B. Cantor, and Stuart H. Orkin. 2008. "Human fetal hemoglobin expression is regulated by the developmental stage-specific repressor BCL11A." *Science* 322 (5909):1839-1842.

Saxon, Ben R., David Rees, and Nancy F. Olivieri. 1998. "Regression of extramedullary haemopoiesis and augmentation of fetal haemoglobin concentration during hydroxyurea therapy in β thalassaemia." *British journal of haematology* 101 (3):416-419.

Schmidt, Paul J., Iva Toudjarska, Anoop K. Sendamarai, Tim Racie, Stuart Milstein, Brian R. Bettencourt, Julia Hettinger, David Bumcrot, and Mark D. Fleming. 2013. "An RNAi therapeutic targeting Tmprss6 decreases iron overload in <em>Hfe</em>^{–/–} mice and ameliorates anemia and iron overload in murine β-thalassemia intermedia." *Blood* 121 (7):1200-1208. doi: 10.1182/blood-2012-09-453977.

Schrier, S. L. 2002. "Pathophysiology of thalassemia." *Curr Opin Hematol* 9 (2):123-6.

Sebastiani, Giada, Nicole Wilkinson, and Kostas Pantopoulos. 2016. "Pharmacological Targeting of the Hepcidin/Ferroportin Axis." *Frontiers in pharmacology* 7:160-160. doi: 10.3389/fphar.2016.00160.

Shaeffer, J. R. 1988. "ATP-dependent proteolysis of hemoglobin alpha chains in beta-thalassemic hemolysates is ubiquitin-dependent." *J Biol Chem* 263 (27):13663-9.

Shaeffer, Joseph R., and Robert E. Cohen. 1997. "Ubiquitin Aldehyde Increases Adenosine Triphosphate–Dependent Proteolysis of Hemoglobin α-Subunits in β-Thalassemic Hemolysates." *Blood* 90 (3):1300-1308.

Shi, L., S. Cui, J. D. Engel, and O. Tanabe. 2013. "Lysine-specific demethylase 1 is a therapeutic target for fetal hemoglobin induction." *Nat Med* 19 (3):291-4. doi: 10.1038/nm.3101.

Smith, E. C., and S. H. Orkin. 2016. "Hemoglobin genetics: recent contributions of GWAS and gene editing." *Hum Mol Genet* 25 (R2):R99-R105. doi: 10.1093/hmg/ddw170.

Sripichai, O., T. Munkongdee, C. Kumkhaek, S. Svasti, P. Winichagoon, and S. Fucharoen. 2008. "Coinheritance of the different copy numbers of alpha-globin gene modifies severity of beta-thalassemia/Hb E disease." *Ann Hematol* 87 (5):375-9. doi: 10.1007/s00277-007-0407-2.

Stamatoyannopoulos, G., B. Josephson, J. W. Zhang, and Qiliang Li. 1993. "Developmental regulation of human gamma-globin genes in transgenic mice." *Molecular and Cellular Biology* 13 (12):7636-7644.

Stamatoyannopoulos, G. 2005. "Control of globin gene expression during development and erythroid differentiation." *Exp Hematol* 33 (3):259-71. doi: 10.1016/j.exphem.2004.11.007.

Suriapperuma, T., R. Peiris, C. Mettananda, A. Premawardhena, and S. Mettananda. 2018. "Body iron status of children and adolescents with transfusion dependent beta-thalassaemia: trends of serum ferritin and associations of optimal body iron control." *BMC Res Notes* 11 (1):547. doi: 10.1186/s13104-018-3634-9.

Suzuki, M., M. Yamamoto, and J. D. Engel. 2014. "Fetal globin gene repressors as drug targets for molecular therapies to treat the beta-globinopathies." *Mol Cell Biol* 34 (19):3560-9. doi: 10.1128/MCB. 00714-14.

Taher, A. T., D. J. Weatherall, and M. D. Cappellini. 2018. "Thalassaemia." *Lancet* 391 (10116):155-167. doi: 10.1016/S0140-6736(17)31822-6.

Thein, S. L. 2008. "Genetic modifiers of the beta-haemoglobinopathies." *British Journal of Haematology* 141 (3):357-366. doi: DOI 10.1111/j. 1365-2141.2008.07084.x.

Thein, S. L. 2013a. "Genetic association studies in beta-hemoglobinopathies." *Hematology Am Soc Hematol Educ Program* 2013:354-61. doi: 10.1182/asheducation-2013.1.354.

Thein, S. L. 2013b. "The molecular basis of beta-thalassemia." *Cold Spring Harb Perspect Med* 3 (5):a011700. doi: 10.1101/cshperspect.a011700.

Thein, S. L., S. Menzel, X. Peng, S. Best, J. Jiang, J. Close, N. Silver, A. Gerovasilli, C. Ping, M. Yamaguchi, K. Wahlberg, P. Ulug, T. D. Spector, C. Garner, F. Matsuda, M. Farrall, and M. Lathrop. 2007. "Intergenic variants of HBS1L-MYB are responsible for a major quantitative trait locus on chromosome 6q23 influencing fetal hemoglobin levels in adults." *Proc Natl Acad Sci U S A* 104 (27):11346-51. doi: 10.1073/pnas.0611393104.

Thein, Swee Lay. 2004. "Genetic insights into the clinical diversity of β thalassaemia." *British journal of haematology* 124 (3):264-274.

Thein, Swee Lay. 2005. "Pathophysiology of β Thalassemia—A Guide to Molecular Therapies." *ASH Education Program Book* 2005 (1):31-37. doi: 10.1182/asheducation-2005.1.31.

Thein, Swee Lay. 2018. "Molecular basis of β thalassemia and potential therapeutic targets." *Blood Cells, Molecules, and Diseases* 70:54-65.

Thein, Swee Lay, and William G Wood. 2009. "The Molecular Basis of 8 Thalassemia, 88 Thalassemia, and Hereditary Persistence of Fetal Hemoglobin." *Disorders of hemoglobin: genetics, pathophysiology, and clinical management*:323.

Thompson, A. A., M. C. Walters, J. Kwiatkowski, J. E. J. Rasko, J. A. Ribeil, S. Hongeng, E. Magrin, G. J. Schiller, E. Payen, M. Semeraro, D. Moshous, F. Lefrere, H. Puy, P. Bourget, A. Magnani, L. Caccavelli, J. S. Diana, F. Suarez, F. Monpoux, V. Brousse, C. Poirot, C. Brouzes, J. F. Meritet, C. Pondarre, Y. Beuzard, S. Chretien, T. Lefebvre, D. T. Teachey, U. Anurathapan, P. J. Ho, C. von Kalle, M. Kletzel, E. Vichinsky, S. Soni, G. Veres, O. Negre, R. W. Ross, D. Davidson, A. Petrusich, L. Sandler, M. Asmal, O. Hermine, M. De Montalembert, S. Hacein-Bey-Abina, S. Blanche, P. Leboulch, and M. Cavazzana. 2018. "Gene Therapy in Patients with Transfusion-Dependent beta-Thalassemia." *N Engl J Med* 378 (16):1479-1493. doi: 10.1056/NEJMoa1705342.

Traeger-Synodinos, J., E. Kanavakis, C. Vrettou, E. Maragoudaki, T. H. Michael, A. Metaxotou-Mavromati, and C. Kattamis. 1996. "The triplicated α-globin gene locus in β-thalassaemia heterozygotes: clinical, haematological, biosynthetic and molecular studies." *British journal of haematology* 95 (3):467-471.

Uda, M., R. Galanello, S. Sanna, G. Lettre, V. G. Sankaran, W. Chen, G. Usala, F. Busonero, A. Maschio, G. Albai, M. G. Piras, N. Sestu, S. Lai, M. Dei, A. Mulas, L. Crisponi, S. Naitza, I. Asunis, M. Deiana, R. Nagaraja, L. Perseu, S. Satta, M. D. Cipollina, C. Sollaino, P. Moi, J. N. Hirschhorn, S. H. Orkin, G. R. Abecasis, D. Schlessinger, and A. Cao. 2008. "Genome-wide association study shows BCL11A associated with persistent fetal hemoglobin and amelioration of the phenotype of beta-thalassemia." *Proc Natl Acad Sci U S A* 105 (5):1620-5. doi: 10.1073/pnas.0711566105.

Vichinsky, Elliott. 2007. "Hemoglobin E syndromes." *ASH Education Program Book* 2007 (1):79-83.

Viprakasit, V., R. J. Gibbons, B. C. Broughton, J. L. Tolmie, D. Brown, P. Lunt, R. M. Winter, S. Marinoni, M. Stefanini, L. Brueton, A. R. Lehmann, and D. R. Higgs. 2001. "Mutations in the general transcription factor TFIIH result in beta-thalassaemia in individuals with trichothiodystrophy." *Hum Mol Genet* 10 (24):2797-802.

Voon, H. P., and J. Vadolas. 2008. "Controlling alpha-globin: a review of alpha-globin expression and its impact on beta-thalassemia." *Haematologica* 93 (12):1868-76. doi: 10.3324/haematol.13490.

Voon, Hsiao Phin Joanna, Hady Wardan, and Jim Vadolas. 2008. "siRNA-mediated reduction of α-globin results in phenotypic improvements in β-thalassemic cells." *Haematologica* 93 (8):1238-1242.

Wainscoat, J. S., J. M. Old, D. J. Weatherall, and S. H. Orkin. 1983. "The molecular basis for the clinical diversity of beta thalassaemia in Cypriots." *Lancet* 1 (8336):1235-7.

Weatherall, D. J. 2010. "The inherited diseases of hemoglobin are an emerging global health burden." *Blood* 115 (22):4331-6. doi: 10.1182/blood-2010-01-251348.

Weatherall, D. J., and J. B. Clegg. 2001. *The thalassaemia syndromes*. 4 ed. Oxford: Blackwell Science.

Weatherall, D. J. 2001. "Phenotype—genotype relationships in monogenic disease: lessons from the thalassaemias." *Nature reviews genetics* 2 (4):245.

Weatherall, D. J., J. B. Clegg, H. H. M. Knox-Macaulay, C. Bunch, C. R. Hopkins, and I. J. Temperley. 1973. "A genetically determined disorder with features both of thalassaemia and congenital dyserythropoietic anaemia." *British journal of haematology* 24 (6):681-702.

Xie, S. Y., W. Li, Z. R. Ren, S. Z. Huang, F. Zeng, and Y. T. Zeng. 2011. "Correction of beta654-thalassaemia mice using direct intravenous injection of siRNA and antisense RNA vectors." *Int J Hematol* 93 (3):301-10. doi: 10.1007/s12185-010-0727-1.

Xie, Shu-Yang, Zhao-Rui Ren, Jing-Zhi Zhang, Xin-Bin Guo, Qing-Xue Wang, Shu Wang, Dan Lin, Xiu-Li Gong, Wei Li, and Shu-Zhen Huang. 2007. "Restoration of the balanced α/β-globin gene expression in β 654-thalassemia mice using combined RNAi and antisense RNA approach." *Human molecular genetics* 16 (21):2616-2625.

Xu, Jian, Vijay G. Sankaran, Min Ni, Tobias F. Menne, Rishi V. Puram, Woojin Kim, and Stuart H. Orkin. 2010. "Transcriptional silencing of γ-globin by BCL11A involves long-range interactions and cooperation with SOX6." *Genes & development* 24 (8):783-798.

Yu, Channing, Kathy K. Niakan, Mark Matsushita, George Stamatoyannopoulos, Stuart H. Orkin, and Wendy H. Raskind. 2002. "X-linked thrombocytopenia with thalassemia from a mutation in the amino finger of GATA-1 affecting DNA binding rather than FOG-1 interaction." *Blood* 100 (6):2040-2045.

Zamani, F., R. Shakeri, S. M. Eslami, S. M. Razavi, and A. Basi. 2009. "Hydroxyurea therapy in 49 patients with major beta-thalassemia." *Arch Iran Med* 12 (3):295-7.

Zhou, D., K. Liu, C. W. Sun, K. M. Pawlik, and T. M. Townes. 2010a. "KLF1 regulates BCL11A expression and gamma- to beta-globin gene switching." *Nat Genet* 42 (9):742-4. doi: 10.1038/ng.637.

Zhou, Dewang, Kaimao Liu, Chiao-Wang Sun, Kevin M. Pawlik, and Tim M. Townes. 2010b. "KLF1 regulates BCL11A expression and γ-to β-globin gene switching." *Nature genetics* 42 (9):742.

In: Fetal Hemoglobin
Editor: Nishant Chakravorty

ISBN: 978-1-53617-955-2

Chapter 10

CLINICAL MANAGEMENT OF β-THALASSEMIA

Sachith Mettananda[1,*] and Hashan Pathiraja[2]
[1]Department of Paediatrics, University of Kelaniya, Ragama, Sri Lanka
[2]National Hospital of Sri Lanka, Colombo, Sri Lanka

ABSTRACT

β-thalassemia is one of the most common genetic disorders in the world. Annually, approximately 45,000 new patients are diagnosed worldwide for β-thalassemia. Patients with severe disease present during infancy, coupled with profound anemia, and the disease is invariably fatal without treatment. The currently available standard treatments include regular red blood cell transfusions and management of complications associated with the disease and therapy. Blood transfusions are administered every 2 to 5 week, with an aim to maintain pre-transfusion hemoglobin levels between 9.0 and 10.5 g/dl. Iron overload is an invariable complication of regular transfusions of β-thalassemia. It is estimated that with each unit of packed red blood cells, 200 mg of iron is infused into the body. Unfortunately, the body does not possess an effective mechanism to excrete this excess iron. Iron overload in β-thalassemia is treated with iron

* Corresponding Author's Email: sachith.mettananda@kln.ac.lk.

chelator medications, namely desferrioxamine, deferasirox and deferiprone which are either used as monotherapy or in combination. In order to be able to manage iron overload related complications like, cardiac dysfunction, chronic liver disease, diabetes mellitus and multiple endocrinopathies appropriately, regular follow ups are necessary. The only cure for β-thalassemia is allogenic hematopoietic stem cell transplantation, which is considered as the standard treatment in a scenario of availability of a matched sibling donor. However, the clinical usefulness of this treatment is limited by the lack of availability of suitable donors, its complications and its cost. Results of gene therapy for β-thalassemia are promising, with several patients achieving transfusion independency in clinical trials. Additionally, novel therapeutic pathways that include de-repression of fetal hemoglobin and down regulation of α-globin through pharmacological mechanisms and genome editing, have shown considerable promise in pre-clinical and early clinical trials.

Keywords: β-thalassemia, transfusion dependent thalassemia, iron overload, iron chelation, β-globin, α-globin, γ-globin, gene therapy, genome editing

INTRODUCTION

β-thalassemia is one of the most common inheritable disorders of hemoglobin in the world (Higgs, Engel, and Stamatoyannopoulos 2012). It is extremely prevalent in the tropical regions which extend from the Mediterranean through Middle East and sub Saharan Africa to South and South East Asia (Weatherall 2012). The annual global incidence of severe forms of β-thalassemia is reported as 45,000, of which, majority of the births take place in the developing countries (Weatherall 2011). Therefore, β-thalassemia presents a significant health burden to many resource deficit countries in the world (Weatherall 2010).

The original description of β-thalassemia was provided by an American pediatrician by the name Thomas B. Cooley, who reported five children with anemia, splenomegaly and bone changes in 1925 (Cooley and Lee 1925). The etiopathology and molecular basis of β-thalassemia was discovered in the mid part of the 20th century. In fact, it was one of the very first genetic diseases to be characterized at the molecular level (Weatherall and Clegg 2001).

β-Thalassemia is an autosomal recessive disease, in which over 300 disease causing mutations have been characterized so far [a complete updated list is available at http://globin.cse.psu.edu/hbvar/] (Mettananda and Higgs 2018, Thein 2013). A majority of these mutations are point mutations within the human β-globin gene or its immediate flanking regions in chromosome 11 which causes either reduced or no synthesis of β-globin chains {β-globin chains combine with α-globin chains to produce adult hemoglobin (Hb), HbA ($\alpha_2\beta_2$)}. β-thalassemia mutations are categorized into β^0-, β^+- and β^{++}-thalassemia alleles depending on the extent to which the β-globin chain output is reduced. β^0-Thalassemia mutations, which is of the severe type, cause complete cessation of the production of β-globin chains, while β^+-thalassemia mutations result in milder reductions in β-globin chain synthesis. Mutations which are categorized as β^{++}-thalassemia will only result in subtle decrement in the production of β-globin chains (Thein, 2017). In addition, a point mutation at the 26th codon of the β-globin gene that substitutes glutamic acid with lysine, results in structurally abnormal hemoglobin, hemoglobin E ($\alpha_2\beta^E{}_2$), which is synthesized at reduced rates and gives rise to the phenotype of β-thalassemia. The co-inheritance of a β-thalassemia allele and an HbE allele from either of the parents results in HbE β-thalassemia which accounts for up to 50% of patients with β-thalassemia worldwide (Fucharoen and Weatherall 2012).

The basic pathophysiology of β-thalassemia is well understood. Irrespective of the reduced production of β-globin, the synthesis of α-globin continues normally, resulting in an excess of α-globin chains within the red blood cells (RBCs) and the erythroid precursors. This unbalanced synthesis of α-globin chain leads to the production of toxic α-hemichromes and reactive oxygen species resulting in the oxidative destruction of RBCs, causing hemolysis (destruction of RBCs in the circulation) and ineffective erythropoiesis (destruction of erythroid precursors within the bone marrow).

Despite being one of the first diseases to be described at molecular level, the progress in the management of β-thalassemia has been painfully slow. The current standard treatments provide only supportive care to majority of the patients (Mettananda 2017). Nonetheless, during the last few years, significant improvements have been made in many of the treatment options and genetics based therapies in particular. In this chapter, we aim to discuss the current standard management of β-thalassemia and to provide an insight into the emerging therapies.

Clinical Presentation

Children with β-thalassemia are normal at birth. This is because, at the time of birth, the predominant type of hemoglobin is HbF and the synthesis of HbA occurs only at low levels. Physiological switch of hemoglobin from HbF to HbA happens around the age of 6 months, during which the symptoms and signs of β-thalassemia start to manifest (Mettananda, Gibbons, and Higgs 2016). Due to lack of synthesis of HbA, hemolysis and ineffective erythropoiesis, patients with β-thalassemia become progressively anemic towards the latter half of the first year. Most symptoms are due to profound anemia (Cao and Galanello 2010). Common symptoms of β-thalassemia include lethargy, poor feeding, irritability, fatigue, faltering growth and shortness of breath. The characteristic physical signs are pallor, mild jaundice, signs of extramedullary hematopoiesis, which include hepatomegaly and splenomegaly, and signs of bone marrow expansion, for example, frontal bossing, maxillary hyperplasia and dental malocclusion.

Most children with severe β-thalassemia develop profound anemia, which is invariably fatal, if not treated with regular RBC transfusions (Origa 2017). These patients are known to have transfusion-dependent β-thalassemia (TDBT), and include ones with β-thalassemia major and severe HbE β-thalassemia. Some patients with β-thalassemia have an intermediate clinical phenotype and present mild symptoms and signs only after two years of age. They require no or only occasional RBC transfusions and are known to have non-transfusion dependent β-thalassemia (NTDBT) or β-thalassemia intermedia (Mettananda 2018a). This clinical heterogeneity can be explained, at least partly, by the β-globin genotype and other genetic modifiers which are described in another chapter of this book (see *'Genetic modifiers of β-thalassemia and their therapeutic implications'*).

DIAGNOSIS

Most patients with β-thalassemia are diagnosed within the first two years of life. Once β-thalassemia is clinically suspected, the initial evaluation includes a full blood count with RBC indices, peripheral blood smear and qualitative and quantitative hemoglobin analysis (either by high-performance liquid chromatography or capillary electrophoresis). The full blood count will reveal low hemoglobin, mean corpuscular volume and mean corpuscular hemoglobin values. Mean corpuscular hemoglobin concentration will be low or low-normal. Peripheral blood smear shows microcytosis, hypochromia, anisopoikilocytosis, and nucleated RBCs (Taher, Weatherall, and Cappellini 2018). The diagnosis of β-thalassemia is confirmed by qualitative and quantitative hemoglobin analysis. In β^0/β^0 thalassemia, HbA is absent, HbF is 95–98% and HbA_2 is 2–5%. In β^0/β^+- or β^+/β^+-thalassemia, the hemoglobin pattern shows HbA of 10–30%, HbF of 70–90% and HbA_2 of 2–5% (Cao and Galanello 2010). In patients with hemoglobin E β-thalassemia, HbE constitutes 30%-70%, while the remainder is composed of HbF. Depending on the type of the associated β-thalassemia mutation, variable levels of HbA can be found (Fucharoen and Weatherall 2012).

PRINCIPALS OF MANAGEMENT

All patients with β-thalassemia should be ideally managed at a dedicated thalassemia centre by a specialist pediatrician or a pediatric hematologist who has expertise and experience in managing patients with thalassemia. The current supportive treatment for β-thalassemia includes RBC transfusion, iron chelation and screening and management of complications. In a subset of patients, curative therapy is available in the form of allogenic hematopoietic stem cell transplantation (HSCT) (Cappellini et al. 2014).

Red blood Cell Transfusion Therapy

The mainstay of management in β-thalassemia is RBC transfusion, which aims to correct anemia and related complications. Additionally, RBC transfusions have beneficial effects on suppressing bone marrow activity and extramedullary erythropoiesis, promoting normal growth and physical activity, and reducing intestinal absorption of iron whilst minimizing transfusion related iron accumulation. The transfusion protocol and frequency depend on whether the patient has TDBT or NTDBT, and this should be decided by an experienced physician during the first few years of treatment. However, this categorization will require frequent revisits throughout life, as the patients can transform from TDBT to NTDBT categories and *vice versa,* due to variations in clinical management and changes in disease modifiers (Taher, Weatherall, and Cappellini 2018).

Before commencing the first RBC transfusion, it is mandatory to establish the proper diagnosis and to do extended phenotyping of RBC antigens (minor blood groups) to minimize future transfusion related complications. Transfusion guidelines for TDBT recommend regular RBC transfusions, usually every 2 to 5 weeks, in order to maintain a target pre-transfusion hemoglobin level of 9.0 – 10.5 g/dl (Cappellini et al. 2018). The current accepted protocol is to achieve a post-transfusion hemoglobin level of 14.0 g/dl by transfusing leuco-depleted (leucocytes reduced to $< 1 \times 10^6$ per unit) packed RBC. All patients with thalassemia should receive fresh blood packs (as far as possible) which are less than 2 weeks old. The volume of blood to be transfused is calculated using a standard formula (Equation 1) (Cappellini et al. 2014).

$$\frac{(14.0\ g/dL - \text{Pretransfusion haemoglobin}) \times \text{Weight (kg)} \times 3}{\text{Haematocrit of transfused blood}} = \begin{array}{c}\text{Volume of blood (mL)}\\ \text{to be transfused}\end{array}$$

Equation 1. Formula to calculate transfusion volume in β-thalassemia (Cappellini et al. 2014).

Patients with NTDBT do not require regular RBC transfusions. Lower base line hemoglobin levels (as low as 6-7g/dl) are acceptable in these patients and the frequency and need for transfusions should be decided individually (Mettananda et al. 2019). The indications for transfusion of NTDBT patients are not clearly defined; however, they include growth failure, reduced exercise tolerance, poor quality of life, and rapidly enlarging spleen (>3 cm per year). Transfusions may also be required during situations of acute stress or low hemoglobin such as infection and surgery (Sleiman et al. 2018). Low hemoglobin level solely is not an indication for transfusion, unless it is as low as Hb <5 g/dl. (Taher et al. 2013).

Iron Chelation

Iron overload is an inevitable complication of β-thalassemia (Figure 1), owing to their regular requirement for RBC transfusions. A single unit of transfused RBC contains approximately 200 mg of iron. During blood transfusion, this amount enters into the body parenterally, bypassing the intestinal regulation and leads to transfusion iron overload. This iron cannot be eliminated from the body due to the lack of any effective mechanism (Bou-Fakhredin et al. 2017). Furthermore, in spite of having high iron stores, the intestinal absorption of iron is increased in patients with β-thalassemia (Pasricha et al. 2013). The mechanism of this paradox is partly understood, and is believed to be due to inappropriate suppression of hepatic synthesis of hepcidin, which is the master regulator of iron absorption in humans (Nemeth 2013). In β-thalassemia, the increased ineffective erythropoiesis and RBC turn over results in raised erythroferrone levels which suppress the hepatic synthesis of hepcidin (Jones et al. 2015). The resultant reduction in hepcidin leads to increased intestinal absorption of iron despite having higher body iron stores (Kautz et al. 2014).

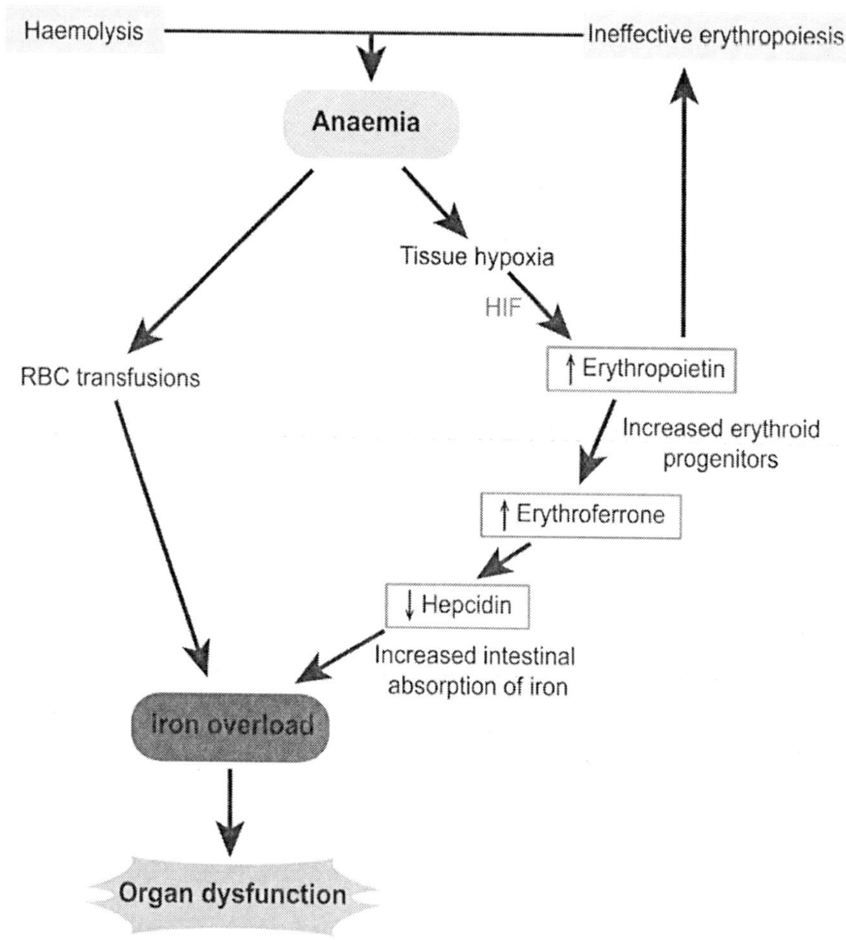

Figure 1. Mechanism of iron overload and organ dysfunction in β-thalassemia (Mettananda 2015).

Screening for iron overload should commence when patients receive 8 to 10 RBC transfusions (Cappellini et al. 2014). Of the several methods available, serum ferritin is the most commonly used marker of iron overload, which is generally measured every 3 months in all patients with β-thalassemia. In patients with TDBT, serum ferritin above 1000 μg/l is an indication to commence iron chelation, whilst serum ferritin above 2500 μg/l is associated with increased risk of cardiac and endocrine morbidity. In patients with NTDBT, iron chelation is commenced when serum ferritin is > 800 μg/l (Cappellini et al. 2018).

In addition to serial measurements of serum ferritin, several other non-invasive techniques are available to measure organ specific iron loading. Introduction of Magnetic Resonance Imaging (MRI) based techniques for assessment of liver and myocardial iron concentrations is considered as one of the most important advances in thalassemia care. Liver iron concentration is measured using two different MRI techniques, T2* and R2 (also known as FerriScan®). Assessment of cardiac iron overload is by T2* MRI. Cardiac T2* value less than 10 ms is associated with cardiac failure and extremely high mortality, while, a value below 20 ms is indicative of declining left ventricular function (Taher, Weatherall, and Cappellini 2018). However, due to limitations in the availability and the cost of MRI based techniques, serum ferritin still remains the most widely used investigation to assess iron overload in most low- and middle- income countries where β-thalassemia is highly prevalent (Suriapperuma et al. 2018).

Iron deposition is the primary reason for organ damage and morbidity in patients with β-thalassemia. Therefore, meticulous control of iron overload by regular and effective chelation is of paramount importance in the management. The three iron chelators which are currently available to treat iron overload in these patients are desferrioxamine, deferiprone and deferasirox (Table 1). Desferrioxamine is the oldest and the most effective iron chelator which has been in use for many decades. However, the need for intravenous or subcutaneous infusion for long hours and the unavailability of an oral dosage form are the most important limitations of this drug. Deferasirox, an oral chelator, with a similar efficacy to desferrioxamine, was introduced to the market in 2005. Due to its efficacy and ease of administration, deferasirox has substituted desferrioxamine as the first line iron chelator for iron overload in patients with β-thalassemia in many countries worldwide. While deferasirox is used as first line iron chelator in children above 2 years, desferrioxamine remains as the first line chelator in children less than 2 years. Desferrioxamine is also preferred over deferasirox in the presence of renal or hepatic dysfunction and as an emergency treatment for acute cardiac failure due to cardiac iron loading. Deferiprone, on the other hand, has the ability to effectively chelate myocardial iron compared to other drugs, however, is not recommended as monotherapy. Combination of deferasirox and desferrioxamine can be used when a single chelator is unable to control iron overload; and a combination of desferrioxamine and deferiprone is used especially when cardiac iron chelation becomes a priority. However, combination of two oral iron chelators, deferasirox and deferiprone has not been well studied (Cappellini et al. 2014).

Table 1. Pharmacological and adverse effect profile of iron chelator medications (Mettananda 2018a)

	Desferrioxamine (DFO)	**Deferasirox (DFX)**	**Deferiprone (DFP)**
Route of administration	subcutaneous or intravenous	oral	oral
Dose (mg/kg/d)	30-60	20-40	75-100
Frequency	8 to12-hour infusion for 5 to 7 days per week	once daily	three times daily
Iron excretory route	fecal and urinary	fecal	urinary
Adverse/ side effects	visual and hearing defects, bone growth retardation, local reactions and allergic reactions	gastrointestinal side effects, raised creatinine and elevated hepatic enzymes	gastrointestinal side effects, arthralgia, agranulocytosis and neutropenia
Main advantages	has been in use for several decades, low cost, higher availability	better compliance	effective in chelating cardiac iron
Main disadvantages	poor compliance	limited long-term safety data	not effective as monotherapy

SPLENECTOMY

Most patients with β-thalassemia underwent routine splenectomy as an adjunct to transfusion therapy in the past. However, splenectomy is rarely performed at present due to reports of post-splenectomy adverse events, that include, increased risk of sepsis, leg ulcers, pulmonary hypertension, venous thrombosis and strokes (Taher, Weatherall, and Cappellini 2018). The current recommendations for splenectomy are restricted to the patients with hypersplenism and symptomatic splenomegaly (Cappellini et al. 2014). Also, splenectomy is considered in patients who require very high transfusion volumes, especially when the annual transfusion requirement is over 200-250 ml/kg/year (Musallam, Angastiniotis, et al. 2013).

All patients who are undergoing splenectomy require pre-operative preparation and post-operative prophylaxis to minimize infections by encapsulated bacteria. Patients who are awaiting splenectomy should be vaccinated against *Streptococcus pneumoniae, Neisseria meningitidis* and *Haemophilus influenzae b* at least 2 weeks prior to the procedure. After splenectomy, patients should be commenced on prophylaxis with oral penicillin twice daily for at least 2 years or may be for life.

SCREENING FOR COMPLICATIONS

β-Thalassemia is a chronic progressive disease, with multiple long-term complications related to chronic anemia, blood transfusions, iron overload and iron chelator medications. Current standard treatment provides only supportive care and is not curative. Therefore, continuous surveillance of complications is pivotal to provide optimal care for patients with β-thalassemia (Table 2).

Table 2. An approach for screening for complications
(Mettananda 2018a)

Complication	**Parameter**	**Frequency (-monthly)**				
		1	3	6	12	24
Growth retardation	Weight		●			
Short stature	Height velocity			●		
Pubertal delay	Tanner staging			●		
Malocclusion of teeth	Dental assessment				▲	
Hypersplenism	Complete blood count	●				
Iron overload	Serum ferritin		●			
Hepatic dysfunction	AST/ALT/bilirubin/albumin		●			
	Abdominal ultrasonography				▲	
	Liver MRI (from 8yrs)				▲	
Transfusion related infections	Hepatitis A, B and C screen				▲	
Renal dysfunction	Serum creatinine		●			
	Urine full report			●		
Cardiac dysfunction	ECG				▲	
	Echocardiography				▲	
	Cardiac T2* MRI (from 8yrs)				▲	
Hypothyroidism	TSH/T4/T3 (from 5yrs)				▲	
Hypoparathyroidism	Serum calcium (from 5yrs)		●			
	PTH level (from 5yrs)				▲	
Diabetes mellitus	Fasting blood sugar (from 5yrs)		●			
	Glucose tolerance test (from 10yrs)				▲	
Osteoporosis	Vitamin D level			●		
	DEXA scan					▲
Adverse effects of drugs	Serum creatinine (DFX)	●				
	ALT (DFX/DFP)	●				
	White blood cell count (DFP)	●				
	Evaluation by an ophthalmologist (DFO/DFX/DFP)				▲	
	Evaluation of hearing (DFO/DFP/DFX)				▲	

Management of Complications

Transfusion Reactions

Different types of reactions occur as adverse events of RBC transfusions. The incidence of non-hemolytic febrile reactions and allergic reactions can be minimized by transfusing leucodepleted and washed RBCs respectively. Acute hemolytic reactions are commonly due to mis-matched incompatible transfusions, which can be avoided by ensuring meticulous attention to sample labeling and compatibility testing (Cappellini et al. 2014). Delayed hemolytic reactions occur after few days or weeks after transfusions and are characterized by unexpected low post-transfusion hemoglobin levels, with evidence of hemolysis. These reactions could be due to alloantibodies or autoantibodies. Hyperhemolysis syndrome, which occurs more frequently in patients who begin transfusion therapy later in life, is a very serious complication due to destruction of both donor and recipient RBCs resulting in extremely low hemoglobin levels (Hannema et al. 2010). Treatment options available include steroids, intravenous immunoglobulin and immunosuppressive drugs, which however, may not be effective in many patients.

Transfusion Related Infections

Several infections that include human immunodeficiency virus (HIV), hepatitis B, hepatitis C, human T-cell lymphotropic viruses (HTLV) and syphilis are transmitted by infected blood or blood products. Due to effective screening of donors and blood products, the incidence of these infections have markedly reduced in many countries in the world (Busch, Bloch, and Kleinman 2019).

Hepatitis C viral infection is the most common transfusion related infection among patients with β-thalassemia at present. Although hepatitis B is becoming less common due to the availability of effective vaccines, it may still be a problem in many parts of the world where the vaccine is not freely available. Patients with hepatitis C should be treated with direct-acting antivirals, whereas, interferon and nucleoside analogues are used in the treatment of hepatitis B infection (Marcon et al. 2018).

Cardiac Dysfunction

Despite advances in iron chelation, cardiac dysfunction and failure due to accumulation of iron in cardiac myocytes remain as the primary cause of mortality in patients with TDBT. Cardiac T2* MRI is the gold standard for assessment of cardiac iron burden, and echocardiography is useful to monitor ejection fraction and cardiac chamber dilatation. Patients with myocardial dysfunction are treated with angiotensin converting enzyme inhibitors and diuretics; however, they will require intensive chelation with continuous intravenous infusion of desferrioxamine and oral deferiprone when cardiac failure is present (Cappellini et al. 2014).

Chronic Liver Disease

Chronic liver disease in patients with TDBT can range from hepatic fibrosis to cirrhosis and hepatocellular carcinoma. Ferriscan and T2* MRI are the currently available methods to assess liver iron load (Viprakasit et al. 2018).

Short Stature

Short stature in patients with β-thalassemia is multifactorial and could be contributed by chronic anemia, growth hormone deficiency, other endocrinopathies, chelation toxicity and malnutrition. Treatment with subcutaneous recombinant growth hormone is indicated if deficiency is confirmed (Cappellini et al. 2014).

Delayed Puberty

Delay in pubertal development, defined as lack of sexual maturation by 13 years in females and 14 years in male, is due to hypogonadotropic hypogonadism that in turn occurs due to iron deposition in the pituitary gland. Low serum concentrations of sex hormones (testosterone or estrogen) and gonadotropins confirm the diagnosis. Hormone replacement therapy is given to induce sexual development (Marcon et al. 2018).

Hypothyroidism

Hypothyroidism in children with β-thalassemia is becoming rare due to improved chelation regimens. However children should be screened using thyroid function tests before overt symptoms are developed. All diagnosed patients should be treated with levothyroxine.

Hypoparathyroidism

Hypoparathyroidism is a clinically silent endocrinopathy in β-thalassemia, which is diagnosed by laboratory investigations that show hypocalcaemia, hyperphosphatemia and low parathyroid hormones levels. Treatment includes 25-hydroxyvitamin D, calcium supplements and phosphate binders.

Diabetes Mellitus

Diabetes is strongly associated with iron overload. Hemoglobin A_{IC}, which is commonly used to assess the glycemic control in non-thalassemia patients, cannot be used in patients with hemoglobinopathies because of alterations in hemoglobin balance and hence fructosamine is used instead.

Treatment and complications of diabetes are essentially same as for the general population.

Bone Disease

Osteopenia and osteoporosis are important complications of β-thalassemia. Pathophysiology includes iron toxicity on osteoblasts, sex hormone deficiency, chronic anemia and desferrioxamine treatment. Diagnosis is by bone mineral density measurement using bone densitometry and treatment includes vitamin D, calcium and bisphosphonates (Noureldine et al. 2018).

Quality of Life

With improvement in supportive therapies, the life expectancy of patients with β-thalassemia has risen remarkably and thus many patients survive beyond 40 years of age. This has transformed the outlook of β-thalassemia from an invariably fatal pediatric disorder to a chronic disease with morbidity and disability (Cappellini et al. 2018). Consequently, health related quality of life is an important parameter in the management (Amid et al. 2015). Several studies conducted in many countries have unequivocally shown that patients with β-thalassemia experience lower quality of life compared to age and sex matched controls in all physical, emotional, social and school functioning domains (Arian et al. 2019, Floris et al. 2018). This has been demonstrated in high-, middle- and low-income countries. Several factors including splenectomy, short stature, malnourishment and longer hospital stays have been shown to be associated with lower quality of life scores. Additionally, children with thalassemia have a higher risk of developing psychological problems, for example, conduct disorders, anxiety and depression, compared to normal counterparts (Mettananda, Pathiraja, et al. 2018). All these suggest that psychological counseling should be an integral part in the management of patients with β-thalassemia in the future.

Family Screening and Genetic Counseling

Parents of children with β-thalassemia should be screened to confirm their β-thalassemia carrier status. Once confirmed, they should be counseled regarding the recurrence risk, which is 25% for all future pregnancies (Mettananda and de Silva 2017). If facilities and legal provisions are available they should be offered antenatal diagnosis during subsequent pregnancies and termination in the case of an affected offspring.

Similarly, all siblings of patients with β-thalassemia should be screened to identify β-thalassemia carriers. Diagnosed β-thalassemia carriers are advised to check the β-thalassemia status of their partners before marriage. Marriages between β-thalassemia carriers should be highly discouraged as they carry 25% probability of producing a child with TDBT (Premawardhena et al. 2004).

Curative Therapies

Cure for β-thalassemia has been a major challenge of research during last few years. Any curative treatment method should eliminate globin chain imbalance in erythroid cells and premature RBC destruction, restore normal hemoglobin level without extramedullary hematopoiesis, render patients asymptomatic and allow them to live a normal life without unwanted effects (Mettananda 2018b). Allogenic HSCT is the only established cure for β-thalassemia at present. However, gene therapy has made a remarkable comeback with dozens of patients who have been cured in a number of clinical trials (Ghiaccio et al. 2019). Additionally, numerous attempts are underway to establish a cure that would be available to all patients with β-thalassemia (El-Beshlawy and El-Ghamrawy 2019).

Hematopoietic Stem Cell Transplantation

The first HSCT for thalassemia was performed in early 1980's and several thousand HSCT have been performed for β-thalassemia thus far. The outcome of HSCT primarily depends on the medical complication of the patient at the time of transplant and the nature of the hematopoietic stem cells (HSC) donor. The best outcome is reported when transplanting HSC from HLA-matched sibling donors which report overall survival of 90% and disease-free survival of over 80% (Baronciani et al. 2016). Unfortunately, only 10% of the β-thalassemia patients have HLA-matched sibling donors and this largely limits the usefulness of HSCT. Many centers around the world perform HSCT by using HLA matched-unrelated donors and umbilical cord blood as the HSC source with overall survival rates ranging between 60% and 65%. Additionally, several serious complications related to HSCT (like transplant related mortality, graft-versus-host disease and graft failure), limits the acceptability and usefulness of the procedure. Currently, HSCT is routinely indicated only in patients who are transfusion dependent and have well-matched sibling donors. For optimal results, HSCTs should be performed before patients develop iron-related complications and are younger than 14-years (Angelucci et al. 2014).

Gene Therapy

Gene therapy has been researched upon for many years as a cure for β-thalassemia in a number of clinical trials. Contrary to the initial promise, the progress in the field of gene therapy has been extremely slow. The steps of gene therapy involves harvesting HSCs from patients with β-thalassemia, insertion of the normal β-globin gene using viral vectors *in vitro*, and transplanting HSC with normal β-globin gene back to the patient as an autologous HSCT (Cavazzana, Antoniani, and Miccio 2017). The first gene therapy trial for thalassemia was started in 2007 and thus far dozens of patients have been treated (Cavazzana-Calvo et al. 2010). The most recent phase 1-2 trial demonstrated successful gene therapy in 22 patients with severe β-thalassemia, resulting in reduced or eliminated need for RBC transfusions, without serious adverse events (Thompson et al. 2018). However, further research is required to improve the efficiency of bone marrow harvesting, transfer of genetic material and level of gene expression, and to minimize oncogenic potential of the procedure (Persons 2010).

EMERGING THERAPIES

Several novel pharmacological and genetic therapies are being investigated as treatment options for β-thalassemia at present. Most of these treatment strategies aim to restore α/β-like globin chain imbalance in β-thalassemia RBCs. Induction of γ-globin with an intention to increase HbF using pharmacological agents has been attempted for several years (Musallam, Taher, et al. 2013). Although several drugs have reached different phases of clinical trials, none have been efficacious enough to be recommended for routine clinical use (Makis et al. 2016). Hydroxyurea, which has produced best results thus far, is widely used to increase HbF production to prevent sickling crisis in patients with sickle cell disease; however, it has not been shown to reduce the need for blood transfusion in patients with β-thalassemia (Foong et al. 2016, Algiraigri et al. 2017). Additionally, several other γ-globin de-repressors, that include histone deacetylase inhibitors, lysine-specific histone demethylase 1 inhibitors and DNA methyltransferase inhibitors are being tested in pre-clinical studies (Shi et al. 2013, Bradner et al. 2010).

An alternative approach that has reached clinical trials is improving ineffective erythropoiesis by inhibiting its mediators using JAK2 inhibitor and activin type IIB receptor ligand trap. A recent phase 2 clinical trial on 30 patients with TDBT concluded that treatment with ruxolitinib – a JAK 2 inhibitor – was associated with a significant reduction in splenic volumes, but not with a decline in transfusion requirement (Taher et al. 2018). Another trial which assessed the efficacy of luspatercept – an activin type IIB receptor ligand trap – in 31 patients with NTDBT and 32 patients with TDBT revealed that the treatment is associated with improvements in hemoglobin levels, transfusion requirements and quality of life scores (Piga et al. 2019).

Another approach which is still in pre-clinical stages is pharmacological down regulation of α-globin with an aim to improve globin chain balance (Mettananda, Gibbons, and Higgs 2016). This is backed by a large body of clinical evidence that demonstrated amelioration of the disease in patients who have natural reduction in α-globin output in the form of co-inherited α-thalassemia (Mettananda, Gibbons, and Higgs 2015, Mettananda, Suranjan, et al. 2018). Promising pre-clinical studies show that this may be achieved by inhibiting histone demethylase using epigenetic drugs (Mettananda, Fisher, Sloane-Stanley, et al. 2017).

Recently, genome editing, a new form of gene therapy has shown promise in pre-clinical studies (Ledford 2015). Contrary to traditional gene therapy, genome editing utilizes programmable nucleases to create an edit in a pre-determined target site in the human genome (Kim and Kim 2014). For β-thalassemia, one approach is editing genes that suppress γ-globin synthesis, such as BCL11A transcription factor, a strong silencer of the γ-globin gene, without affecting other hematopoietic lineages (Canver et al. 2015). Similarly, genome editing of the major α-globin enhancer has been successfully performed in human HSC to downregulate α-globin, thereby improving the globin chain imbalance in β-thalassemia erythroid cells (Mettananda, Clark, et al. 2018, Mettananda, Fisher, Hay, et al. 2017). Both these approaches have shown promise in *in vitro* studies; however, they require further validations before progressing in to clinical trials.

Conclusion

β-thalassemia is a life-long disease without an effective cure for majority of the patients affected. However, supportive care for patients with β-thalassemia has shown considerable improvements over the past decade due to availability of safe blood products for transfusion and invention of effective iron chelators. Average life expectancy of patients with β-thalassemia has risen considerably, with most living beyond the forth decade. Development of newer non-invasive methods to identify organ specific iron overload, better chelation and availability of oral iron chelators have dramatically improved the quality of life of these patients and many live near normal lives. Despite having several limitations, cure for β-thalassemia is currently possible through HSCT in patients who have well-matched donors. There are many exciting new developments in the field of β-thalassemia at therapeutic level. In particular, advances in gene therapy and genome editing will alleviate the need for HLA-matched donors for HSCT and may provide cure for all patients with β-thalassemia in the future.

REFERENCES

Algiraigri, A. H., N. A. M. Wright, E. O. Paolucci, and A. Kassam. 2017. "Hydroxyurea for lifelong transfusion-dependent beta-thalassemia: A meta-analysis." *Pediatr Hematol Oncol* 34 (8):435-448. doi: 10.1080/08880018.2017.1354948.

Amid, A., A. N. Saliba, A. T. Taher, and R. J. Klaassen. 2015. "Thalassemia in children: from quality of care to quality of life." *Arch Dis Child* 100 (11):1051-7. doi: 10.1136/archdischild-2014-308112.

Angelucci, E., S. Matthes-Martin, D. Baronciani, F. Bernaudin, S. Bonanomi, M. D. Cappellini, J. H. Dalle, P. Di Bartolomeo, C. D. de Heredia, R. Dickerhoff, C. Giardini, E. Gluckman, A. A. Hussein, N. Kamani, M. Minkov, F. Locatelli, V. Rocha, P. Sedlacek, F. Smiers, I. Thuret, I. Yaniv, M. Cavazzana, C. Peters, Ebmt Inborn Error, and Ebmt Paediatric Working Parties. 2014. "Hematopoietic stem cell transplantation in thalassemia major and sickle cell disease: indications and management recommendations from an international expert panel." *Haematologica* 99 (5):811-20. doi: 10.3324/haematol. 2013.099747.

Arian, M., M. Mirmohammadkhani, R. Ghorbani, and M. Soleimani. 2019. "Health-related quality of life (HRQoL) in beta-thalassemia major (beta-TM) patients assessed by 36-item short form health survey (SF-36): a meta-analysis." *Qual Life Res* 28 (2):321-334. doi: 10.1007/s11136-018-1986-1.

Baronciani, D., E. Angelucci, U. Potschger, J. Gaziev, A. Yesilipek, M. Zecca, M. G. Orofino, C. Giardini, A. Al-Ahmari, S. Marktcl, J. de la Fuente, A. Ghavamzadeh, A. A. Hussein, C. Targhetta, F. Pilo, F. Locatelli, G. Dini, P. Bader, and C. Peters. 2016. "Hemopoietic stem cell transplantation in thalassemia: a report from the European Society for Blood and Bone Marrow Transplantation Hemoglobinopathy Registry, 2000-2010." *Bone Marrow Transplant* 51 (4):536-41. doi: 10.1038/bmt.2015.293.

Bou-Fakhredin, R., A. H. Bazarbachi, B. Chaya, J. Sleiman, M. D. Cappellini, and A. T. Taher. 2017. "Iron Overload and Chelation Therapy in Non-Transfusion Dependent Thalassemia." *Int J Mol Sci* 18 (12). doi: 10.3390/ijms18122778.

Bradner, J. E., R. Mak, S. K. Tanguturi, R. Mazitschek, S. J. Haggarty, K. Ross, C. Y. Chang, J. Bosco, N. West, E. Morse, K. Lin, J. P. Shen, N. P. Kwiatkowski, N. Gheldof, J. Dekker, D. J. DeAngelo, S. A. Carr, S. L. Schreiber, T. R. Golub, and B. L. Ebert. 2010. "Chemical genetic strategy identifies histone deacetylase 1 (HDAC1) and HDAC2 as therapeutic targets in sickle cell disease." *Proc Natl Acad Sci USA* 107 (28):12617-22. doi: 10.1073/pnas.1006774107.

Busch, Michael P., Evan M. Bloch, and Steven H. Kleinman. 2019. "Prevention of transfusion transmitted infections." *Blood*: blood-2018-11-833996. doi: 10.1182/blood-2018-11-833996.

Canver, M. C., E. C. Smith, F. Sher, L. Pinello, N. E. Sanjana, O. Shalem, D. D. Chen, P. G. Schupp, D. S. Vinjamur, S. P. Garcia, S. Luc, R. Kurita, Y. Nakamura, Y. Fujiwara, T. Maeda, G. C. Yuan, F. Zhang, S. H. Orkin, and D. E. Bauer. 2015. "BCL11A enhancer dissection by Cas9-mediated in situ saturating mutagenesis." *Nature* 527 (7577):192-7. doi: 10.1038/nature15521.

Cao, A., and R. Galanello. 2010. "Beta-thalassemia." *Genet Med* 12 (2):61-76. doi: 10.1097/GIM.0b013e3181cd68ed.

Cappellini, M. D., J. B. Porter, V. Viprakasit, and A. T. Taher. 2018. "A paradigm shift on beta-thalassemia treatment: How will we manage this old disease with new therapies?" *Blood Rev* 32 (4):300-311. doi: 10.1016/j.blre.2018.02.001.

Cappellini, M. D., A. Cohen, J. Porter, A. Taher, and V. Viprakasit. 2014. *Guidelines for the management of transfusion dependent thalassemia (TDT)* Edited by M. D. Cappellini, A. Cohen, J. Porter, A. Taher and V. Viprakasit. 3rd Edition ed. Cyprus: Thalassemia International Federation.

Cavazzana, M., C. Antoniani, and A. Miccio. 2017. "Gene Therapy for beta-Hemoglobinopathies." *Mol Ther* 25 (5):1142-1154. doi: 10.1016/j.ymthe.2017.03.024.

Cavazzana-Calvo, Marina, Emmanuel Payen, Olivier Negre, Gary Wang, Kathleen Hehir, Floriane Fusil, Julian Down, Maria Denaro, Troy Brady, and Karen Westerman. 2010. "Transfusion independence and HMGA2 activation after gene therapy of human β-thalassemia." *Nature* 467 (7313):318.

Cooley, T. B., and P. Lee. 1925. "A Series of Cases of Splenomegaly in Children, with Anemia and Peculiar Bone Changes." *Transactions of the American Pediatric Society* 37:29-30.

El-Beshlawy, A., and M. El-Ghamrawy. 2019. "Recent trends in treatment of thalassemia." *Blood Cells Mol Dis* 76:53-58. doi: 10.1016/j.bcmd.2019.01.006.

Floris, F., F. Comitini, G. Leoni, P. Moi, M. Morittu, V. Orecchia, M. Perra, M. P. Pilia, A. Zappu, M. R. Casini, and R. Origa. 2018. "Quality of life in Sardinian patients with transfusion-dependent Thalassemia: a cross-sectional study." *Qual Life Res* 27 (10):2533-2539. doi: 10.1007/s11136-018-1911-7.

Foong, W. C., J. J. Ho, C. K. Loh, and V. Viprakasit. 2016. "Hydroxyurea for reducing blood transfusion in non-transfusion dependent beta thalassemias." *Cochrane Database Syst Rev* 10:CD011579. doi: 10.1002/14651858.CD011579.pub2.

Fucharoen, S., and D. J. Weatherall. 2012. "The hemoglobin E thalassemias." *Cold Spring Harb Perspect Med* 2 (8). doi: 10.1101/cshperspect.a011734.

Ghiaccio, V., M. Chappell, S. Rivella, and L. Breda. 2019. "Gene Therapy for Beta-Hemoglobinopathies: Milestones, New Therapies and Challenges." *Mol Diagn Ther* 23 (2):173-186. doi: 10.1007/s40291-019-00383-4.

Hannema, S. E., A. Brand, A. van Meurs, and F. J. Smiers. 2010. "Delayed hemolytic transfusion reaction with hyperhemolysis after first red blood cell transfusion in child with beta-thalassemia: challenges in treatment." *Transfusion* 50 (2):429-32. doi: 10.1111/j.1537-2995.2009.02399.x.

Higgs, D. R., J. D. Engel, and G. Stamatoyannopoulos. 2012. "Thalassemia." *Lancet* 379 (9813):373-83. doi: 10.1016/S0140-6736 (11)60283-3.

Jones, E., S. R. Pasricha, A. Allen, P. Evans, C. A. Fisher, K. Wray, A. Premawardhena, D. Bandara, A. Perera, C. Webster, P. Sturges, N. F. Olivieri, T. St Pierre, A. E. Armitage, J. B. Porter, D. J. Weatherall, and H. Drakesmith. 2015. "Hepcidin is suppressed by erythropoiesis in hemoglobin E beta-thalassemia and beta-thalassemia trait." *Blood* 125 (5):873-80. doi: 10.1182/blood-2014-10-606491.

Kautz, L., G. Jung, E. V. Valore, S. Rivella, E. Nemeth, and T. Ganz. 2014. "Identification of erythroferrone as an erythroid regulator of iron metabolism." *Nat Genet* 46 (7):678-84. doi: 10.1038/ng.2996.

Kim, H., and J. S. Kim. 2014. "A guide to genome engineering with programmable nucleases." *Nat Rev Genet* 15 (5):321-34. doi: 10.1038/nrg3686.

Ledford, H. 2015. "CRISPR, the disruptor." *Nature* 522 (7554):20-4. doi: 10.1038/522020a.

Makis, A., E. Hatzimichael, I. Papassotiriou, and E. Voskaridou. 2016. "2017 Clinical trials update in new treatments of beta-thalassemia." *Am J Hematol* 91 (11):1135-1145. doi: 10.1002/ajh.24530.

Marcon, A., I. Motta, A. T. Taher, and M. D. Cappellini. 2018. "Clinical Complications and Their Management." *Hematol Oncol Clin North Am* 32 (2):223-236. doi: 10.1016/j.hoc.2017.11.005.

Mettananda, Doluweera Sachith Gopeetha. 2015. *Ameliorating β-thalassemia by manipulating expression of the α-globin gene* Oxford: University of Oxford.

Mettananda, S. 2017. "Thalassemia: In a quest towards an ultimate cure." *Sri Lanka Journal of Child Health* 46 (3):203-210.

Mettananda, S. 2018a. "Management of Thalassemia." *Sri Lanka Journal of Child Health* 47 (2):159-165.

Mettananda, S. 2018b. "Thalassemia: Current research may provide a cure to it in the future." *Sri Lanka Journal of Child Health* 47 (4):372. doi: http://doi.org/10.4038/sljch.v47i4.8606.

Mettananda, S., K. Clark, C. A. Fisher, J. A. Sloane-Stanley, R. J. Gibbons, and D. R. Higgs. 2018. "Phenotypic and molecular characterization of a serum-free miniature erythroid differentiation system suitable for high-throughput screening and single-cell assays." *Exp Hematol* 60:10-20. doi: 10.1016/j.exphem.2018.01.001.

Mettananda, S., and D. G. de Silva. 2017. "Anemia in children: are we using the correct prevention strategies?" *Ceylon Med J* 62 (2):73-76. doi: 10.4038/cmj.v62i2.8469.

Mettananda, S., C. A. Fisher, D. Hay, M. Badat, L. Quek, K. Clark, P. Hublitz, D. Downes, J. Kerry, M. Gosden, J. Telenius, J. A. Sloane-Stanley, P. Faustino, A. Coelho, J. Doondeea, B. Usukhbayar, P. Sopp, J. A. Sharpe, J. R. Hughes, P. Vyas, R. J. Gibbons, and D. R. Higgs. 2017. "Editing an alpha-globin enhancer in primary human hematopoietic stem cells as a treatment for beta-thalassemia." *Nat Commun* 8 (1):424. doi: 10.1038/s41467-017-00479-7.

Mettananda, S., C. A. Fisher, J. A. Sloane-Stanley, S. Taylor, U. Oppermann, R. J. Gibbons, and D. R. Higgs. 2017. "Selective silencing of alpha-globin by the histone demethylase inhibitor IOX1: a potentially new pathway for treatment of beta-thalassemia." *Haematologica* 102 (3):e80-e84. doi: 10.3324/haematol.2016.155655.

Mettananda, S., R. J. Gibbons, and D. R. Higgs. 2015. "alpha-Globin as a molecular target in the treatment of beta-thalassemia." *Blood* 125 (24):3694-701. doi: 10.1182/blood-2015-03-633594.

Mettananda, S., R. J. Gibbons, and D. R. Higgs. 2016. "Understanding alpha-globin gene regulation and implications for the treatment of beta-thalassemia." *Ann NY Acad Sci* 1368 (1):16-24. doi: 10.1111/nyas.12988.

Mettananda, S., and D. R. Higgs. 2018. "Molecular Basis and Genetic Modifiers of Thalassemia." *Hematol Oncol Clin North Am* 32 (2):177-191. doi: 10.1016/j.hoc.2017.11.003.

Mettananda, S., H. Pathiraja, R. Peiris, D. Bandara, T. U. N. de Silva, C. Mettananda, and A. Premawardena. 2018. "Quality of life and psycological morbidity among children with transfusion dependent thalassemia and their parents." *Ceylon Med J* 63:28.

Mettananda, S., H. Pathiraja, R. Peiris, N. Wickramarathne, D. Bandara, U. de Silva, C. Mettananda, and A. Premawardhena. 2019. "Blood transfusion therapy for beta-thalassemia major and hemoglobin E beta-thalassemia: Adequacy, trends, and determinants in Sri Lanka." *Pediatr Blood Cancer* 66 (5):e27643. doi: 10.1002/pbc.27643.

Mettananda, S., M. Suranjan, R. Fernando, T. Dias, C. Mettananda, R. Rodrigo, L. Perera, R. Gibbons, A. Premawardhena, and D. Higgs. 2018. "Anemia among females in child-bearing age: Relative contributions, effects and interactions of alpha- and beta-thalassemia." *PLoS One* 13 (11):e0206928. doi: 10.1371/journal.pone.0206928.

Musallam, K. M., M. Angastiniotis, A. Eleftheriou, and J. B. Porter. 2013. "Cross-talk between available guidelines for the management of patients with beta-thalassemia major." *Acta Haematol* 130 (2):64-73. doi: 10. 1159/000345734.

Musallam, K. M., A. T. Taher, M. D. Cappellini, and V. G. Sankaran. 2013. "Clinical experience with fetal hemoglobin induction therapy in patients with beta-thalassemia." *Blood* 121 (12):2199-212; quiz 2372. doi: 10. 1182/blood-2012-10-408021.

Nemeth, Elizabeta. 2013. "Hepcidin and β-thalassemia major." *Blood* 122 (1):3-4. doi: 10.1182/blood-2013-05-502617.

Noureldine, M. H. A., A. T. Taher, A. A. Haydar, A. Berjawi, M. A. Khamashta, and I. Uthman. 2018. "Rheumatological complications of beta-thalassemia: an overview." *Rheumatology (Oxford)* 57 (1):19-27. doi: 10.1093/rheumatology/kex058.

Origa, R. 2017. "beta-Thalassemia." *Genet Med* 19 (6):609-619. doi: 10. 1038/gim.2016.173.

Pasricha, Sant-Rayn, David M. Frazer, Donald K. Bowden, and Gregory J. Anderson. 2013. "Transfusion suppresses erythropoiesis and increases hepcidin in adult patients with β-thalassemia major: a longitudinal study." *Blood* 122 (1):124-133. doi: 10.1182/blood-2012-12-471441.

Persons, D. A. 2010. "Gene therapy: Targeting beta-thalassemia." *Nature* 467 (7313):277-8. doi: 10.1038/467277a.

Piga, A., S. Perrotta, M. R. Gamberini, E. Voskaridou, A. Melpignano, A. Filosa, V. Caruso, A. Pietrangelo, F. Longo, I. Tartaglione, C. Borgna-Pignatti, X. Zhang, A. Laadem, M. L. Sherman, and K. M. Attie. 2019. "Luspatercept improves hemoglobin levels and blood transfusion requirements in a study of patients with beta-thalassemia." *Blood* 133 (12):1279-1289. doi: 10.1182/blood-2018-10-879247.

Premawardhena, Anuja, Shanthimala de Silva, Mahinda Arambepola, Nancy Olivieri, Laura Merson, Julia Muraco, Angela Allen, Christopher Fisher, Timothy Peto, Elliot Vichinsky, and David Weatherall. 2004. "Thalassemia in Sri Lanka: a progress report." *Human Molecular Genetics* 13 (suppl_2):R203-R206. doi: 10.1093/hmg/ddh250.

Shi, L., S. Cui, J. D. Engel, and O. Tanabe. 2013. "Lysine-specific demethylase 1 is a therapeutic target for fetal hemoglobin induction." *Nat Med* 19 (3):291-4. doi: 10.1038/nm.3101.

Sleiman, J., A. Tarhini, R. Bou-Fakhredin, A. N. Saliba, M. D. Cappellini, and A. T. Taher. 2018. "Non-Transfusion-Dependent Thalassemia: An Update on Complications and Management." *Int J Mol Sci* 19 (1). doi: 10.3390/ijms19010182.

Suriapperuma, T., R. Peiris, C. Mettananda, A. Premawardhena, and S. Mettananda. 2018. "Body iron status of children and adolescents with transfusion dependent beta-thalassemia: trends of serum ferritin and associations of optimal body iron control." *BMC Res Notes* 11 (1):547. doi: 10.1186/s13104-018-3634-9.

Taher, A. T., Z. Karakas, E. Cassinerio, N. Siritanaratkul, A. Kattamis, A. Maggio, S. Rivella, N. Hollaender, B. Mahuzier, B. Gadbaw, and Y. Aydinok. 2018. "Efficacy and safety of ruxolitinib in regularly transfused patients with thalassemia: results from a phase 2a study." *Blood* 131 (2):263-265. doi: 10.1182/blood-2017-06-790121.

Taher, A. T., D. J. Weatherall, and M. D. Cappellini. 2018. "Thalassemia." *Lancet* 391 (10116):155-167. doi: 10.1016/S0140-6736(17)31822-6.

Taher, A., E. Vichinsky, K. Musallam, M. D. Cappellini, and V. Viprakasit. 2013. *Guidelines for the management of non transfusion dependent thalassemia (NTDT)* Cyprus: Thalassemia International Federation.

Thein, S. L. 2013. "The molecular basis of beta-thalassemia." *Cold Spring Harb Perspect Med* 3 (5):a011700. doi: 10.1101/cshperspect.a011700.

Thein, S. L. 2017. "Molecular basis of beta thalassemia and potential therapeutic targets." *Blood Cells Mol Dis*. doi: 10.1016/j.bcmd.2017.06.001.

Thompson, A. A., M. C. Walters, J. Kwiatkowski, J. E. J. Rasko, J. A. Ribeil, S. Hongeng, E. Magrin, G. J. Schiller, E. Payen, M. Semeraro, D. Moshous, F. Lefrere, H. Puy, P. Bourget, A. Magnani, L. Caccavelli, J. S. Diana, F. Suarez, F. Monpoux, V. Brousse, C. Poirot, C. Brouzes, J. F. Meritet, C. Pondarre, Y. Beuzard, S. Chretien, T. Lefebvre, D. T. Teachey, U. Anurathapan, P. J. Ho, C. von Kalle, M. Kletzel, E. Vichinsky, S. Soni, G. Veres, O. Negre, R. W. Ross, D. Davidson, A. Petrusich, L. Sandler, M. Asmal, O. Hermine, M. De Montalembert, S. Hacein-Bey-Abina, S. Blanche, P. Leboulch, and M. Cavazzana. 2018. "Gene Therapy in Patients with Transfusion-Dependent beta-Thalassemia." *N Engl J Med* 378 (16):1479-1493. doi: 10.1056/NEJMoa1705342.

Viprakasit, V., A. Ajlan, Y. Aydinok, B. A. A. Al Ebadi, H. Dewedar, A. S. Ibrahim, L. Ragab, O. Trad, A. S. Wataify, L. L. L. Wong, and A. T. Taher. 2018. "MRI for the diagnosis of cardiac and liver iron overload in patients with transfusion-dependent thalassemia: An algorithm to guide clinical use when availability is limited." *Am J Hematol* 93 (6):E135-E137. doi: 10.1002/ajh.25075.

Weatherall, D. J. 2010. "The inherited diseases of hemoglobin are an emerging global health burden." *Blood* 115 (22):4331-6. doi: 10.1182/blood-2010-01-251348.

Weatherall, D. J. 2011. "The challenge of haemoglobinopathies in resource-poor countries." *Br J Haematol* 154 (6):736-44. doi: 10.1111/j.1365-2141.2011.08742.x.

Weatherall, D. J. 2012. "The definition and epidemiology of non-transfusion-dependent thalassemia." *Blood Rev* 26 Suppl 1:S3-6. doi: 10.1016/S0268-960X(12)70003-6.

Weatherall, D. J., and J. B. Clegg. 2001. *The thalassemia syndromes*. 4 ed. Oxford: Blackwell Science.

In: Fetal Hemoglobin
Editor: Nishant Chakravorty
ISBN: 978-1-53617-955-2

Chapter 11

CURRENT PRACTICES IN THE MANAGEMENT OF BETA-HEMOGLOBINOPATHIES

Manisha Jain, Ankit Jitani and Tuphan Kanti Dolai*
Department of Haematology,
Nil Ratan Sircar Medical College and Hospital, Kolkata, India

ABSTRACT

Thalassemia is an inherited hemoglobinopathy leading to quantitative defect in globin chain synthesis. On the other hand, qualitative defects in globin chain synthesis may result in structural variants of hemoglobinopathies e.g., sickle cell anemia. Thalassemia, as a disease, may manifest as a transfusion dependent phenotype (thalassemia major), a non-transfusion dependent phenotype (intermedia), or as a thalassemia trait. Patients with transfusion-dependent phenotype require regular red cell transfusion for survival. In non-transfusion dependent thalassemia, red cell transfusion may be required for vital functions like maintaining normal

* Corresponding Author's Email: tkdolai@gmail.com.

growth and development and during intercurrent illnesses, if not for survival. Blood transfusion is considered to be the most important component of thalassemia management, as it mitigates the symptoms and suppresses ineffective erythropoiesis, thus potentiating normal bone growth. However, blood transfusion comes at the cost of iron overload and is associated with the risk of transfusion-transmitted infections. Management of iron overload requires regular chelation therapy to remove excess iron. This excessive iron is the prime cause for most of the complications seen in thalassemia, like growth abnormality, endocrine dysfunction, delayed pubertal development and cardiac complications. Optimization of the chelation dosing schedule and regular monitoring of drug toxicity help in mitigating many of these complications. Non-transfusion dependent thalassemia cases may have some additional complications like, pulmonary hypertension, leg ulcers and increased risk of thrombosis, which need different therapeutic interventions. Such patients benefit from fetal hemoglobin (HbF) inducers. Individuals having thalassemia trait do not require any treatment as such; however, their identification helps in recognizing "at risk" pregnancies and thus in prevention of thalassemia by prenatal diagnosis. The definitive management of thalassemia is hematopoietic stem cell transplantation. Matched sibling and unrelated donor transplants have become the standard of care for eligible patients. Conditions like iron overload, age at the time of transplantation and liver size affect the transplantation outcome. Gene therapy and gene editing technology are potential treatment options which may change the management of thalassemia in the future.

Keywords: hemoglobinopathies, thalassemia, ineffective erythropoiesis, iron chelators

INTRODUCTION

Hemoglobinopathies are monogenic disorders that contribute to a significant global burden of disease (Colah et al. 2010). It includes disorders altering the expression of globin protein chains of hemoglobin. The types of defects in globin chain may include either structural variants (qualitative defects) in the globin protein chains (e.g., sickle cell anemia -HbS, Hemoglobin D - HbD^{Punjab}, Hemoglobin E - HbE) or absolute deficiency (quantitative defects) of one of the globin protein chains (e.g., α- or β-thalassemia). As a consequence of migration and mixing of populations leading to admixture of the above mentioned subtypes, several other hemoglobin (Hb) variants are seen. They may include compound heterozygous states like HbSβ, HbEβ, HbDβ thalassemia that have variable clinical phenotypes (Ghosh et al. 2014). Despite the near complete understanding of the genetic defects, definite treatment options are still limited. Allogenic stem cell transplant is considered as the sole curativeoption at present. Several researches are in the process of developing other feasible treatment options like the use of HbF inducers, erythropoiesis inducers, better iron chelators, as well as advanced treatment options like gene therapy. However, at present, life long regular blood transfusion, iron chelation therapy, and other supportive treatments are the mainstays of treatment available to most of the patients in the developing countries (Nancy et al. 2013).

Thalassemias (quantitative deficiency) are caused by genetic defects in the production of either α- or β- globin chains and lead to ineffective erythropoiesis due to the precipitation of excess (unmutated) globin chains in the erythroid precursors and RBCs. This results in a variable degree of anemia which is its typical clinical presentation. Management of the condition, therefore, includes treating the anemia and controlling the ineffective erythropoiesis by providing life longblood transfusion support at regular intervals.

Structural (qualitative) defects in globin chains have varied pathology and clinical manifestations. The most interesting among the structural defects is sickle cell disease, which has a totally different pathology and manifestations as compared to other structural defects. The basic defect in sickle cell disease is the defective stability and solubility of the globin chains under hypoxic conditions. This results in the polymerization of globin chains, and leads to an alteration in the shape (sickle-shaped) and function of red blood cells. Patients usually present with acute episodes of pain and other life threatening emergencies. Management of sickle cell anemia (depending on its presentation), relies on hydration, analgesics, blood transfusion support, HbF induction and other measures to prevent further such episodes.

Advancements in the field of medical care have led to prolonged survival of thalassemia patients. In present times, the disease is no longer considered to be limited to the pediatric age group. Consequently, the disease requires comprehensive care to address the varied needs and expectations of patients in the different age groups for a better quality of life (Chaplin et al. 2018). This chapter will focus on an overview of the management of the quantitative defects of hemoglobin disorder.

MANAGEMENT OF BETA-HEMOGLOBINOPATHIES

The spectrum of β hemoglobinopathies includes β-thalassemia homozygous and compound heterozygous conditions like, Hb-Sβ, Hb-Eβ, Hb-Dβ thalassemia. The clinical presentation depends on factors like the type of globin gene mutation, whether it is qualitative or quantitative, and on the presence of genetic modifiers. In β-thalassemia major, mutations in the two alleles may either be similiar or different. The requirement of life long regular blood transfusion for survival is an essential factor in distinguishing the severity of these hemoglobinopathies. The term transfusion dependent thalassemia (TDT), is conventionally used to describe severe phenotypes which need regular and life long blood transfusion support; while, non-transfusion dependent thalassemia (NTDT) is used for those requiring occasional or infrequent transfusion. Transfusion requirement in NTDT patients may be triggered due to certain clinical conditions, like infections, growth spurts, surgery, pregnancy etc. Blood transfusion requirement in NTDT may also arise due to certain complications like, extramedullary hematopoiesis, leg ulcers, etc. (Khaled et al. 2013). Based on the phenotypic severity and the characteristic differences in the pathology of the disease, management strategies are different for patients with TDT and NTDT.

Management of Transfusion-Dependent Thalassemia

TDT patients require regular red cell transfusion to survive. They suffer from several serious complications without adequate transfusion support, which may be fatal in early childhood if left untreated. Ideally, these patients should be managed in specialized centers with expertise available for management of thalassemia. Confirmation of the precise diagnosis of the disease, including the type of molecular defect is desirable prior to initiation of treatment. The pathophysiology of the disease and hence the phenotype depends on both - the deficiency/reduction of β-globin chain and the excessive accumulation of the unpaired α-globin chains (Colah et al. 2010). TDT patients usually present at the age of 6 to 8 months with symptoms of severe anemia, failure to thrive, and/ or fever and infections. Such patients require regular blood transfusion and its frequency depends on the level of hemoglobin and the presence of clinical characteristics that include the patterns of growth and growth velocity, performance status, progressive splenomegaly, skeletal deformity or presence of extramedullary hematopoiesis. The overall management of these patients (summarized in Figure 1) is based on the treatment of the disease per-se and the complications that arise due to long-term blood transfusion. These complications generally include iron overload and transfusion-transmitted infections (Ghosh et al. 2014).

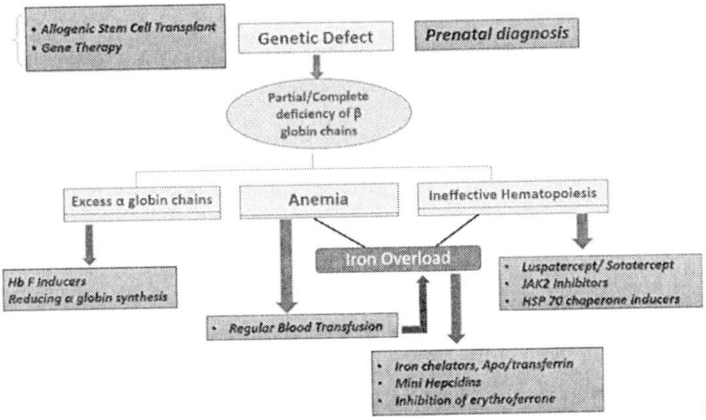

Figure 1. Overall management of transfusion-dependent Thalassemia patient.

Blood Transfusion

Regular transfusion therapy provides physiologically normal erythrocytes for normal tissue oxygenation. Further they also aid in suppressing the ineffective erythropoiesis taking place in the bone marrow. Effcctive regular blood transfusion to achieve appropriate hemoglobin level and optimal safety of transfused blood is the goal of blood transfusion therapy in TDT patients.

Transfusion Program: The frequency of blood transfusion is variable in TDT patients. Transfusion requirement is usually one unit every 2-5 weeks, but it may be even higher, especially in certain stress conditions. The target of regular blood transfusion is to maintain the pre-transfusion hemoglobin level of at least 9 to 10.5 g/dl, a strategy known as "hypertransfusion" (Colah et al. 2010). The goal is to promote normal growth, allow normal physical activities and to adequately suppress the ineffective bone marrow activity. However, this strategy often leads to an iron overload state, usually after 10-20 transfusion units. Maintaining a higher target pre-transfusion hemoglobin level of 11-12 g/dl, which is known as "supertransfusion" (Colah et al. 2010), may be required for patients with cardiac comorbidities, or those who develop extramedullary hematopoiesis or who do not achieve adequate suppression of the ineffective bone marrow activity by the hypertransfusion protocol.

Points to remember (Domenica et al. 2014)

- Blood donation should be encouraged from voluntary, non-remunerated donors.
- It should be collected, processed, stored and dispensed by quality assured blood banks.
- Before starting the transfusion program, extended alloantigen profiling including at least ABO, Rh (C, c, D, E, e) and Kell antigens should be done. Provide phenotype matched blood to the patient.
- Avoid blood from the first-degree relatives.
- Maximum period of blood storage should be 10-14 days, to achieve mean 24-hour post transfusion RBC survival ≥75%

- Red cell product with hematocrit (HCT) 75% and leucodepleted (preferably pre-storage) to a WBC count of < $1X10^6$ leucocytes/unit should be used.
- Pre-transfusion Hb should be ≥ 9-10.5 gm/dl, but post transfusion Hb should not exceed 14-15 gm/dl.
- For patients who develop severe allergic reactions, washed red cells may be used.
- Ideally, Hepatitis B vaccination should be completed prior to starting blood transfusion therapy.
- Viral serology should be done every 3 months to detect viral infection at the earliest.
- A detailed record of red cell transfusions should be kept and should include date, number of units, any reactions during transfusion.
- Alloantibody screening at regular intervals is necessary and ideally, it should be done during pre-transfusion testing at every visit. The risk of alloantibody formation reduces with increasing number of transfusions.

Iron Overload

Regular blood transfusion, gastrointestinal iron absorption and ineffective erythropoiesis, lead to gradual and continuous iron accumulation in TDT patients. The rate of iron accumulation ranges from 0.3-0.6 mg/kg/day. Humans lack an efficient mechanism to excrete excess iron from the body (Colah et al. 2010). The labile iron pool (non-transferrin bound plasma iron) is redox active and can generate free radicals (reactive oxygen species). These free radicals damage cellular membranes causing lipid peroxidation and thus lead to end-organ dysfunction. This phenomenon predominates in the liver, endocrine organs, and myocardium (suffers the most). Pharmacological agents having iron-chelating properties are the only option of removing this excess iron in TDT patients (Ghosh et al. 2014). Figure 2 illustrates the pathology of iron overload in TDT patients. Studies have shown that hypertransfusion programme may control ineffective erythropoiesis in the bone marrow and thus, reduce the gastrointestinal absorption of iron, contributing indirectly to reducing iron overload (Modell et al. 1974). Unfortunately, the iron content of the blood product itself is the major source of iron overload in these patients.

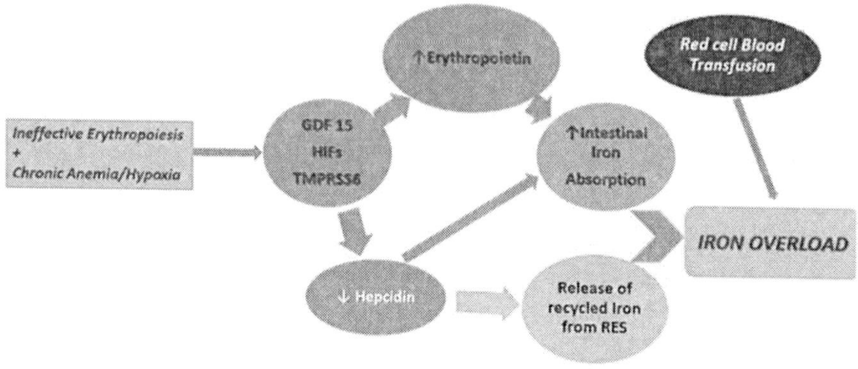

Figure 2. Pathophysiology of Iron Overload.

Iron overload is associated with increased morbidity. The mechanism of iron overload is different in various organs, affecting some at early stages and some at late stages. Liver and endocrine organs are among the first few organs to be affected. Endocrinopathies like hypothyroidism, hypoparathyroidism, growth retardation, hypogonadism and diabetes mellitus are all attributed to excessive iron accumulation in the organs producing these hormones. Other complications related to iron overload in TDT patients include heart failure secondary to cardiac siderosis, liver fibrosis, and osteoporosis (Taher et al. 2017). The stringent monitoring of excess iron in different organs, at regular interval, and effective iron chelator therapy, may prevent and delay these comorbidities. The identification and quantification of iron overload in different organs may be done with different modalities. These tools may be invasive or non-invasive (Musallam et. al 2013). The plasma or serum ferritin concentration is the most commonly used non-invasive test to estimate body iron and to monitor the TDT patient on iron chelator therapy. In TDT patients, serum ferritin levels ≥ 1000ng/ml has been considered as the threshold for initiating iron chelating therapy. In contrast to this value of serum ferritin ≥800ng/ml is considered as the threshold to start chelation therapy in NTDT patients. The quantification of liver iron concentration (LIC) using liver biopsy is the gold standard to diagnose liver iron overload. However, it is an invasive and painful procedure requiring expertise and is associated with inherent risks like bleeding. Non-invasive imaging tools, like magnetic resonance imaging (MRI) using R2 or T2* techniques have almost replaced the liver biopsy for quantification of liver iron overload in clinical practice. These methods are fast, safe and quite reliable (Galanello et al. 2010). Table 1 summarizes the various methods used to diagnose and quantify iron overload.

Table 1. Methods used for the diagnosis and quantification as well as recommendations for assessment and follow up in iron overloaded patients

Parameter	Normal	Mild	Moderate	Severe	Recommendations	
LIC (mg Fe/g dry weight)	<1.2	3–7	>7	>15	Baseline	Annually
Serum ferritin (ng/ml)	<300	>1000 to <2500		>2500	Baseline	3monthly
Transferrin saturation%	20–50	>50			Baseline	
Myocardial T2* ms	>20	8–20		<8	Baseline	Annually
ALT (U/l)	<50	>50			Baseline	
LPI (μM)	0–0.4	>0.4				

Chelating Agents

Iron chelators are agents that form complexes with the labile iron (non-transferrin bound iron in plasma) or intracellular iron. These agents enable their removal from the body through urine or bile, depending upon their precise mechanism of action, and thereby, prevent iron overload (Galanello et al. 2016). The chelating agent forms a complex with the iron atom by binding to the six coordination sites of an iron molecule. The iron chelators are classified according to their mechanism of binding to iron: Hexadentate (Deferoxamine), Tridentate (Deferasirox) and Bidentate (Deferiprone) (Chaplin et al. 2018). One molecule of deferoxamine/ two of deferasirox/ three of deferiprone can bind to one iron moiety. The characteristics, dosage, adverse effects, usage and other properties of these agents are provided in Table 2. Figure 3 illustrates the molecular structure of different chelators.

Most guidelines consider patients to be iron overloaded after 10–20 units of blood transfusion or 2-3 years of starting transfusion treatment or when transferrin saturation is ≥ 90% or serum ferritin ≥1000ng/ml in TDT patients and recommend chelation therapy (Maggio et al. 2017). Effective iron chelation therapy in TDT patients protects from hepatic, cardiac and endocrine complications, enhancement of quality of life, and improves survival (Delea et al. 2007).

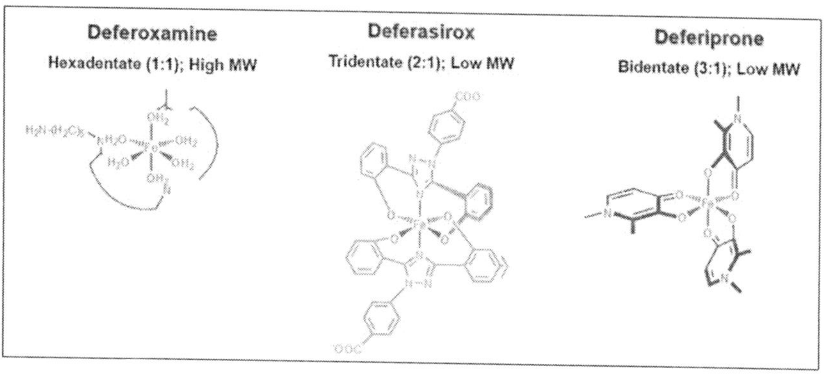

Figure 3. Molecular structure of different chelators.

Role of Combined or Sequential Chelator Treatment

Choice of iron chelator as a first line agent is a contentious issue. It is important to note that management of patients should be individualized and personalized and thus the choice should be guided by degree and profile of iron overload, organ function and presence of hepatic and renal co-morbidities. If clinicians choose to administer two different iron chelators, they either do it on the same day for a synergistic effect (as known as combined or associated treatment), or they may administer the chelators on different days (also known as alternating or sequential treatment). The latter regimen is preferred by most clinicians due to less adverse effects and better compliance (Wonke et al. 1998). Studies have shown that use of sequential iron chelation with deferoxamine (DFO) and deferiprone (DFP), or DFO and deferasirox (DFX) or DFX and DFP, have significant effect on the serum ferritin and LIC. However, dual chelation has not been shown to induce any statistically significant difference in the long term survival. In fact, dual chelation has been shown to add to the burden of adverse events and costs. Thus dual chelation is generally not recommended clinically, except in certain compelling clinical scenarios like cardiac hemosiderosis and induced heart failure (Abdelrazik et al. 2007; Aurelio et al. 2009; Khaled et al. 2013).

As per the extensive long-term experience of Galanello et al. (2016), chelation with combined DFP and DFO rapidly reduces liver and myocardial siderosis and thereby improves cardiac function and reduces mortality.. It is also known to prevent and even reverse endocrine complications. As per the current recommendations (Modell et al. 1974), addition of DFP as a second chelator is indicated for patients in whom monotherapy is ineffective or when the prevention or treatment of life-threatening emergencies due to iron overload (mainly cardiac overload) justify rapid/ intensive intervention.

Table 2. Iron chelators characteristics

Property	Deferoxamine	Deferiprone	Deferasirox
	Hexadentate	Bidentate	Tridentate
Usual dose	20–60 mg/kg/day	75–100 mg/kg/day	20–40 mg/kg/day
Route	S.C., I.V.	Oral	Oral
Half-life	20–30 min	2–3 h	8–16 h
Frequency	8–12 h, 5 days/week	3 times daily	once daily
Excretion	Urinary, fecal	Urinary	Fecal
Adverse Effect	Hearing abnormalities Visual changes Growth retardation Bone changes Local skin reactions Renal/ lung toxicity Yersinia infection	Severe agranulocytosis Arthritis Gastrointestinal discomfort	Rash Gastrointestinal discomfort Renal dysfunction
Specific guidelines for monitoring treatment	1) Annual audiometry and retinal examination 2) Regular growth assessment 3) Annual bone X-rays	1)Weekly complete blood count for 1st month then monthly 2)ALT monthly for 3-6 months then every 6 months	1) Weekly creatinine for 1st month then monthly for 3-6 month then 3monthly 2) LFT monthly for 3-6 month then 3monthly 3) CBC monthly 4) Annual audiometry and retinal examination

Iron chelation therapy is generally not recommended during pregnancy and/or for breastfeeding women. The iron overload status should ideally be addressed prior to planning a pregnancy. If it is deemed to be absolutely necessary, then DFO may be used - if the potential benefits outweigh the risk to the fetus (Taher et al. 2017).

Role of Splenectomy

The common pathophysiology of hemoglobinopathies is increased destruction of red blood cells by the reticuloendothelial system. This predominantly happens in the spleen, resulting in its enlargement (splenomegaly) (Musallam et al. 2013). In order to reduce red blood cell destruction and transfusion requirements in transfusion-dependent thalassemia patients, one of the therapeutic options is splenectomy. Splenectomy however comes at a cost –increases the risk of fulminant post-splenectomy sepsis and deep vein thrombosis and thus should be contemplated if the indication for the same is strictly met. Further, splenectomy should be preferably avoided in children <5 years of age, because of a considerably high risk of sepsis (Antonio et al. 2011).

Indications for splenectomy are limited and include an annual transfusion requirement of 200–220 ml/kg/year that precludes effective chelation; progressive cytopenia due to hypersplenism; or symptomatic splenomegaly causing pain in the left upper quadrant/ early satiety/ enlarged spleen at risk of rupture (Galanello et al. 2010).

Since splenectomy is recommended as an intervention to reduce the transfusion requirements, this procedure indirectly contributes to reduction in transfusional siderosis. Splenectomy is either performed as an open surgical procedure or by using a laparoscopic approach. Cholecystectomy or appendicectomy may be considered in symptomatic patients while planning for splenectomy in these patients (Galanello et al. 2016).

Peri-operative complications include bleeding, lung atelectasis, and subphrenic abscess. Postoperatively, patients may develop thrombocytosis, increased WBCs and nucleated RBC counts which predispose the patients to thrombosis. Guidelines pertaining to splenectomy recommend thromboprophylaxis perioperatively in patients with thrombocytosis. Low-dose aspirin or the use of anticoagulants in patients with a history of previous thrombosis or other risk factors is recommended. However, the duration for which these agents are to be used remains debatable (Taher et al. 2010).

As previously mentioned, splenectomy comes at the cost of risking fulminant post-splenectomy sepsis, thrombophilia and pulmonary hypertension (Maggio et al. 2017). Sepsis in the splenectomized patient has a fulminant clinical course and its presentation may range from mild fever and headache to hyperpyrexia, prostration and shock, which may lead to death within 6 to 12 hours if not well managed. Most frequent pathogens causing infections in splenectomized patients are *Streptococcus pneumoniae*, *Haemophilus influenza* type b, and *Neisseria meningitides*. Immunization against these organisms, at least 2-weeks prior to the splenectomy procedure and administration of booster doses every 3-5 years is recommended. Additional counseling for annual influenza vaccine should be encouraged (Ambrosino et al. 1993). Prophylactic use of antibiotics has dramatically reduced the incidence of fatal post-splenectomy sepsis. Chemoprophylaxis with oral penicillin depends on the age of the individual and the treating physician's discretion (oral penicillin: <2yrs age -125 mg BD; 2-5 yrs age - 250mg BD till the age of 5 yrs and for patient >5 yrs - 450 mg BD should be administered for least 2 yrs post-surgery) (Harji et al. 2009). However, families of splenectomized patients have to be clearly informed that despite adequate vaccination and antibiotic prophylaxis, the risk of sepsis always lurks around and all febrile episodes should be promptly dealt with antibiotics and supportive care.

In recent years there has been a significant decline in the number of patients undergoing splenectomy, due to improved transfusion and chelation program globally.

Newer Agents

With the detailed understanding of the pathophysiology of hemoglobinopathies, alternative therapeutic approaches that impact the underlying globin chain imbalance, ineffective erythropoiesis and iron dysregulation are being explored and some have been found to have promising results in the various clinical trials.

HbF Induction

Hydroxyurea is the only HbF inducer approved by FDA in hemoglobinopathies. It is extensively used in sickle cell anemia (Fathallah et al. 2005) and other NTDTs. Increased HbF expression is a consequcnce of excessive formation of γ-globin chain. This γ-chain combines with the excess α-globin chain and reduces its precipitation, which in turn reduces ineffective erythropoiesis. Increasing HbF fraction is an important therapeutic strategy employed to reduce clinical morbidity and mortality in patients with β-thalassemia. Multiple other strategies to induce γ-chain are and have been used in various clinical trials. Hypomethylating agents that decrease DNA methylation and histone deacetylase inhibitors that increase histone acetylation in the promoter of γ- globin gene can induce γ-globin gene expression and increase HbF levels. Various HbF inducers including sodium butyrate, azacitidine, decitabine, and new immunomodulator drugs like pomalidomide, lenalidomide, and thalidomide can improve α:β chain imbalance (Kiefer et al. 2008). Nevertheless, the evidence is not yet robust enough to warrant approval of these agents for clinical use in thalassemia (Delea et. al 2007).

Ineffective Erythropoiesis

Janus kinase 2 (JAK-2) is an important signaling pathway promoting erythropoiesis and targeting the EPOR/JAK2 pathway might limit the number of erythroid progenitor cells. This reduces ineffective erythropoiesis, extramedullary erythropoiesis and splenomegaly. Using the JAK2 inhibitor, ruxolitinib as a therapeutic option in TDT patients, a few studies have shown a sustained reduction in spleen size, improved pretransfusion hemoglobin and hence a reduction in the requirement for blood transfusion. But again, approval for the same for clinical use is pending to a larger, well designed study with demonstration of long term drug safety (Taher et al. 2018).

Table 3. Summary of fetal Hb inducers

Agent	Advantage	Limitation
DNA methylation inhibitors		
5-Azacytidine (Mabaera et al. 2018)	Marked hematological response	Safety concern
Decitabine (Fathallah et al. 2006)	Well tolerated Good response	Myelotoxic
Hydroxyurea (Watanapokasin et al. 2005; Italia et al. 2009)	Favorable effect on hemolysis, hypercoagulability Well tolerated	Lack of efficacy on long term
Short chain fatty acids (Pace et al. 2002)	Well tolerated	Lack of efficacy
Erythropoietic stimulating agents (Singer et al. 2011)	The favorable effect in combination with hydroxyurea Hematological response achieved	High dose required
Thalidomide and derivatives (Aerbajinai et al. 2017)	Well tolerated	Few studies

Activin-Receptor Ligand Trap

Growth factors like growth differentiation factors (GDFs), TGF-β, activins, and bone morphogenetic proteins (BMPs) are key regulators of erythropoiesis. Activin receptors (ActR) upon ligand activation phosphorylate SMAD proteins which then translocate to the nucleus and regulate expression of critical genes that activates erythropoiesis. Luspatercept and Sotatercept prevent binding of ligands to ActR and subsequent activation of the signaling pathway. A multicenteric, open-label, dose-finding study on Luspatercept found that the drug reduces the transfusion burden by $\geq$33% with no serious adverse event in thalassemia patients (Piga et al. 2016; Piga et al. 2019).

Iron-Restriction Erythropoiesis

β-thalassemia patients have an inappropriately low serum hepcidin level. Increasing serum hepcidin in β-thalassemia may reduce iron overload and improve anemia. Mini-hepcidins (Casu et al. 2016), deletion/disruption of matriptase-2/TMPRSS6 (Nai et al. 2012) or use of exogenous transferrin (Artuso et al. 2018) show an increase in erythroid differentiation and maturation in preclinical trials.

All these strategies hold potential for clinical use in thalassemia patients. However, the issue that needs to be addressed is the long term safety and affordability of these drugs so that they may find use in developing countries which hold the maximum burden of these patients.

Definitive Treatment

Allogenic Stem Cell Transplant

Allogeneic hematopoietic cell transplantation (HCT) is the only available curative therapy for thalassemia. It should be preferably done in early childhood to avoid complications associated with chronic transfusions. Conventionally, matched sibling donors (MSD) are the donors of choice for treating these patients. The lack of a sibling donor however does not preclude transplantation using matched unrelated donors (MUDs). Use of MUDs has been associated with reasonable success (Angelucci et al. 2010). Umbilical cord blood transplant (UCBT) in thalassemia should only be considered in low-risk patients if HLA compatible cord blood cells with appropriate cell numbers are available in the context of well-designed experimental clinical trials. HSCT from an HLA-mismatched family member in thalassemia may still be considered in a controlled experimental setting, considering the potential risk of graft versus host disease (Sabloff et al. 2011). Myeloablative conditioning regimens (without irradiation) should be used as standard conditioning regimen. Post-transplant care should include management of the complications of transplant, in addition to the existing the thalassemia related complications (Mathews et al. 2013). Phlebotomy is an effective way to manage iron overload in post HSCT patients. If a matched donor and an experienced transplant unit are available, HSCT remains an extremely robust and cost-effective tool in thalassemia patients, when compared with life-long transfusion, chelation and supportive therapy (Wonke et al. 1998).

Gene Therapy

Gene therapy and gene editing in thalassemia are still under investigation. Gene therapy involves transfer of a normally functioning β-globin gene in autologous hematopoietic stem cells (HSCs). Gene editing is another strategy that uses genome editing techniques, including, repeats–based nucleases (zinc finger nucleases, transcription activator-like effector nucleases, or clustered regularly interspaced short palindromic repeats-CRISPR-CAS9) to repair pathogenic β-globin gene variants (Canver et al. 2016). Gene therapy for thalassemia is especially challenging because of the wide heterogeneity in the mutational profile and requirement of tightly regulated long-term expression of the replaced gene at very high quantitative levels. A study from 2019 described the use of gene therapy in nine individuals with TDT. HCT was performed using myeloablative chemotherapy with a direct intra-bone injection of HSCs transduced with the GLOBE lentiviral vector. The GLOBE lentiviral vector encodes a mini beta-globin gene with a modified enhancer region. After a median follow up of 18 months, all participants had reduced transfusion requirements. Three of the four evaluated children were transfusion-free at 14, 15, and 19 months (Marktel et al. 2019). Gene editing involving disruption of the transcriptional repressor of γ-globin like BCL11A, as a means of reactivating fetal hemoglobin (HbF) expression is also being studied in various studies. Gene editing might be more amenable to non-myeloablative approaches involving direct injection of the vector in the host (Psatha et al. 2018).

Management of Non-Transfusion-Dependent Thalassemia

The term NTDT is used for patients who do not require regular lifelong transfusions for their survival. However, they may require occasional or even frequent transfusions in certain clinical settings and for defined periods of time. The main pathophysiology in NTDT is ineffective erythropoiesis due to the imbalance between α and β-globin chains and peripheral hemolysis, leading to the subsequent clinical complications (Abdelrazik et al. 2007) as depicted in Figure 4. Transfusion therapy, although not necessary for survival, can theoretically ameliorate many of the clinical morbidities frequently encountered in NTDT patients, like leg ulcers, thrombotic events, pulmonary hypertension, silent brain infarcts, and extramedullary hematopoiesis (Aurelio et al. 2009). NTDT patients having sustained frequent transfusions for extended periods of time should be managed as per the guidelines for TDT thalassemia patients.

- Frequent transfusions in NTDT patients should be considered in the following settings (Domenica et al. 2009).
    1) Declining hemoglobin level with profound enlargement of the spleen (at a rate exceeding 3 cm/year).
    2) Reduced growth velocity (height is more indicative of growth pattern than weight).
    3) Diminished exercise tolerance or poor performance at school.
    4) Failure of secondary sexual development.
    5) Skeletal/bony changes.
- Transfusions may be considered for the prevention of the following complications:
    1) Thrombotic or cerebrovascular events.
    2) Pulmonary hypertension with or without secondary heart failure.
    3) Leg ulcers.

Once a NTDT patient is put on a regular blood transfusion program for any of the above indications, close monitoring for reassessment and tapering or withdrawal of the protocol is desirable especially when a sustained clinical benefit is achieved.

Iron Chelation in NTDT

There is a higher potential risk of iron overload and alloimmunization in NTDT patients, especially in splenectomized pregnant patients, and in those who were never transfused previously. Estimation of iron overload status in

NTDT patients should be considered in all patients above 10 years of age. This may be done using serum ferritin or MRI depending on affordability and availability. In NTDT patients, serum ferritin ≥800 mg/L (Khaled et al. 2012) or liver iron concentration ≥5 mg Fe/g dry weight (Antonio et al. 2011), is considered as the cut off to initiate iron chelation therapy. The dosing and iron assessment schedules are almost similar to those of TDT patients.

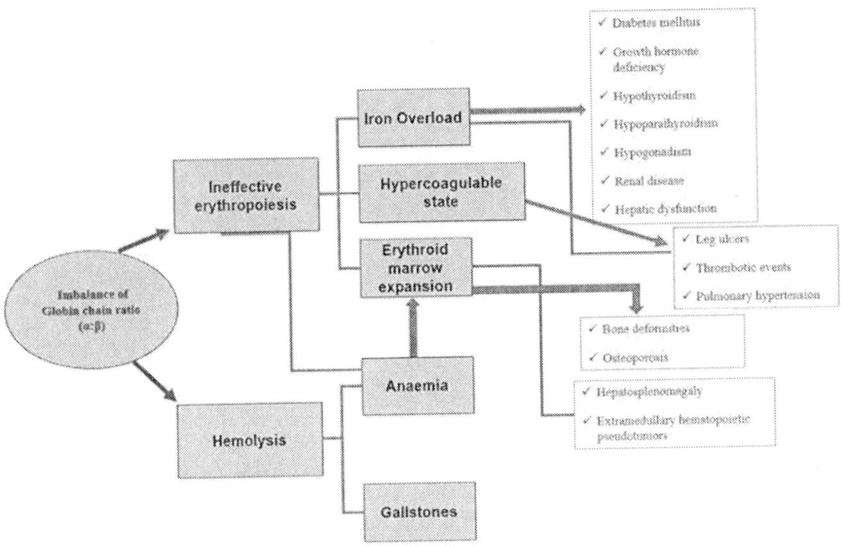

Figure 4. Pathophysiology and clinical complications in NTDT.

HbF Inducers

Hydroxyurea is approved by FDA as one of the key therapeutic agents for the management of patients with sickle cell disease and other NTDT (Taher et al. 2010). The exact mechanism of fetal hemoglobin production by hydroxyurea is not fully understood. It exerts a 2- to 9-fold increase in γ-globin mRNA expression, leading to improvement in the α/non-α chain imbalance and effective erythropoiesis (Ambrosino et al. 1993). Externalization of phosphatidylserine on the red blood cell cytoskeletal membrane (altered red blood cells) due to free radical mediated red cell damage leads to a hypercoagulable state in NTDT patients. Unlike TDT, damaged red cells are not removed due to the low turnover because of irregular transfusion in NTDT. Evidence suggests that hydroxyurea diminishes phosphatidylserine externalization and ameliorates the hypercoagulable state and vascular complications in patients with NTDT. Hydroxyurea therapy is also known to reduce the frequency of other morbidities like, pulmonary hypertension, leg ulcers, and extramedullary hematopoietic pseudotumors in NTDT patients. The drug is used at a starting dose of 10 mg/kg/day and gradually escalated with controlled increment of 3-5mg/kg restricting the total daily dose of 20 mg/kg/day. Concomitant supplementation of folic acid has been recommended (Harji et al. 2009). In sickle cell anemia, a maximum dose of 35 mg/kg/day may be used, based on the clinical response and tolerance.

A few studies evaluating other HbF inducers have also found them to be effective in NTDT patients. The hypomethylating agent, decitabine has been seen to increase the total hemoglobin level by an average of 1 g/dL when administered at a dose of 0.2 mg/kg, two times per week for 12 weeks. Combined use of recombinant human erythropoietin or the newer erythropoietic stimulating agent, like darbepoetin alfa along with HbF inducers, has been found to increase hemoglobin level in NTDT patients (Bianchi et al. 2009).

Table 4. Indications for initiation and Stoppage of iron chelators in TDT and NTDT

	TDT	NTDT
Indication to initiate IC	SF ≥1000 ng/ml or LIC ≥3 mg/g dry weight liver	SF ≥800 ng/ml and/or LIC ≥5 mg/g dry weight liver
Indication for combination IC	SF ≥2500 ng/ml and/or LIC >7 mg/g dry wt. liver and/ or Cardiac T2*<20 m sec	LIC after 6 months of treatment > 7 mg/g dry wt. liver or SF >1500 ng/ml
Indication to stop IC	SF<300 ng/ml and/or LIC< 3 mg/g dry wt. liver	SF< 300 ng/ml and/or LIC< 3 mg/g dry wt. liver

TDT-transfusion-dependent thalassemia; NTDT-non-transfusion dependent thalassemia; IC-Iron Chelation; SF-serum ferritin; LIC-liver iron concentration.

Thrombosis

Thalassemia leads to a hypercoagulable state due to abnormalities in platelets, presence of altered red blood cells, and the presence of microparticles. Thrombotic events are mostly venous and the risk is seen to be four-fold higher in NTDT patients when compared to TDT patients. Splenectomy, anemia (hemoglobin level below 9 g/dl), age greater than 35 years and iron overload (serum ferritin>800 mg/l or liver iron concentration >5 mg Fe/g dry weight) are all considered to be risk factors for thrombosis (Eldor et al. 2002). The risk is exponentially higher in splenectomized patients (Taher et al. 2010). NTDT patients, who develop thrombotic diseases, should be treated as per standard guidelines for non-thalassemic patients. Although, no randomized clinical trial using anticoagulant or antiplatelet therapy for the prevention of thrombosis or cerebrovascular disease in thalassemia patients has been conducted till now, there is evidence that aspirin therapy (Karimi et al. 2010) in splenectomized NTDT patients with elevated platelet counts prevents thrombosis and silent infarct strokes. There is insufficient evidence to recommend iron chelation or hydroxyurea therapy (Karimi et al. 2012) for the prevention of primary or secondary cerebrovascular events. However beneficial effects may be observed when they are used for other indications.

Pulmonary Hypertension

Pulmonary hypertension is characterized by the presence of pre-capillary pulmonary hypertension in the absence of left-sided heart disease, lung disease, or chronic thromboembolic events. Echocardiographic estimation of tricuspid regurgitant velocity (TRV) >3.2 m/s may be suggestive of pulmonary hypertension. Catheterization of the right side of heart is needed to confirm the diagnosis. Risk factors include splenectomy; history of thrombosis; high platelet count; increased nucleated red blood cell counts and iron overload (Derchi et al. 2014). Observational studies have shown beneficial role of transfusion therapy. Sildenafil citrate, a phosphodiesterase type 5 inhibitor and bosentan, an endothelial receptor antagonist has demonstrated promising results in some trials on NTDT patients who developed pulmonary hypertension. Restricted use of these drugs in NTDT patients with confirmed pulmonary hypertension by cardiac catheterization is recommended (Correale et al. 2012).

General Management

Advances in the medical management and increased efforts of healthcare personnel have resulted in prolonged survival of patients with β-thalassemia. However, this has also exposed the long term complications in NTDT patients which were previously recognized in patients with advanced age. NTDT patients may suffer from chronic pain, hepatobiliary, cardiac and renal complications, endocrinopathies, growth retardation, bone diseases and complications due to extramedullary hematopoiesis (compressive myelopathy).

Apart from supportive treatment including calcium, zinc and other nutritional supplementations, regular monitoring of growth and hormonal status is recommended. Patients should be monitored regularly for thyroid hormone levels, vitamin D, parathyroid hormone (as indicated), and in case of delayed sexual development, estimation of luteinizing hormone, follicle-stimulating hormone, testosterone, estradiol, gonadotropin-releasing hormone, should be also be considered (Fathallah et al. 2005). Regular transfusion therapy during the critical periods like pregnancy, surgery, or during growth spurt should be considered. Skeletal changes due to ineffective erythropoiesis can leads to chronic pain or osteoporosis, causing pathological fractures. As per the guidelines, all thalassemic patients >10 years of age should be screened annually for bone mineral density (BMD) of lumbar spine, femoral neck, and distal ulna using dual energy x-ray absorptiometry scan (DEXA) (Kiefer et al. 2008). Cardiac siderosis is more prevalent and leads to left ventricular (LV) dysfunction, cardiac failure, and even cardiogenic death. Periodical ECG along with annual estimation of left-ventricular ejection fraction and tricuspid-regurgitant jet velocity are recommended to assess cardiovascular function. Iron load estimation using serum ferritin levels should be done every 3-6 months. Wherever feasible and facilities are available, MRI for LIC and Cardiac T2* should be done every 1-2 years (Taher et al. 2018). Reactivation and expansion of dormant hematopoietic sites from fetal life, involving the reticuloendothelial system is common during states of chronic ineffective erythropoiesis, resulting in extramedullary hematopoiesis in the liver, spleen, and other sites. Management of these symptomatic pseudotumors includes regular transfusion protocol, hydroxyurea therapy, radiation and surgical decompression (Mabaera et al. 2018).

The complications seen in NTDT tend to follow the natural progression of the disease and correlate with the pathophysiological hallmarks of the disease. Timely medical intervention remains critically important to curtail the development of long-term complications that may reach the point of irreversibility.

Prenatal Diagnosis

The purpose of prenatal screening is to identify asymptomatic individuals and counsel those whose unborn babies are at risk of having TDT or may be afflicted with severely symptomatic hemoglobinopathy. Carrier couples are offered prenatal genetic diagnosis to determine whether the fetus is affected or not.

- Premarriage screening: Identification of the carriers of β-thalassemia and the structural hemoglobinopathies can be done in schools, colleges or high-risk communities.
- Antenatal screening: Pregnant women should be screened in all antenatal clinics, irrespective of the gestational age. Knowledge of the carrier status will help in prevention of these disorders in the subsequent pregnancies, if required. Husbands of carrier women should be screened for hemoglobinopathy carrier status. Prenatal diagnosis should be advised, if the fetus is at risk of developing hemoglobinopathies.

- Preconception screening: Essential for all couples coming to infertility clinics for assisted conception.
- Genetic counseling: Genetic counseling should ideally be provided by a trained specialist, like a trained genetic counselor, a hematologist, or a pediatrician.

The purpose of prenatal diagnosis is to allow parents to make reproductive choices with the knowledge and understanding of the natural history of the specific disorder; how it may affect their child and the treatment approaches available.

DNA-based tests for fetal hemoglobinopathies can be performed during the later part of first trimester of pregnancy on cells obtained by chorionic villus sampling (typically performed at 10 to 12 weeks of gestation) or on direct or cultured amniotic fluid cells obtained by amniocentesis (typically performed at 15 to 18 weeks of gestation). Noninvasive prenatal testing of cell-free DNA in the maternal circulation is under development.

Conclusion

Thalassemia is a disease associated with multiple genetic-risk factors, complications and multiple morbidities and its management requires a multidisciplinary team. Several novel therapies for patients with β-thalassemia are being developed and have shown promising results in various clinical trials. These include gene therapy, genome editing, JAK2, ligand traps like sotatercept and luspatercept, minihepcidins and siRNA targeting TMPRSS6. With the advancement of medical care, thalassemia is no more considered a disease of children alone. Improved survival has led to newer complications seen in the adult life. Transition from childhood to adulthood requires change in treatment paradigm and the ideal treatment strategy is an individualized approach which relies on the patients, their families and the clinicians managing the disease.

REFERENCES

Abdelrazik N. Pattern of Iron Chelation Therapy in Egyptian Beta Thalassemic Patients: Mansoura University Children's Hospital Experience. *Hematology*. 2007;12(6):577–85.

Aerbajinai W, Zhu J, Gao Z et al. Thalidomide Induces Gamma-Globin Gene Expression through Increased Reactive Oxygen Species-Mediated P38 MAPK Signaling and Histone H4 Acetylation in Adult Erythropoiesis. *Blood.* 2017; 110(8): 2864–71.

Ambrosino DM and Molrine DC. Critical Appraisal of Immunization Strategies for Prevention of Infection in the Compromised Host. *Hematology/Oncology Clinics of North America*. Oct 1993; 7(5): 1027–50.

Angelucci E. Hematopoietic Stem Cell Transplantation in Thalassemia. *Hematology.* American Society of Hematology. Education Program. 2010 456–62.

Antonio P, Melania S, Filomena L et al. Changing Patterns of Splenectomy in Transfusion-Dependent Thalassemia Patients. *American Journal of Hematology.* Sept 2011;86(9):808–10.

Artuso I, Lidonnici MR, Altamura S et al. Transferrin Receptor 2 Is a Potential Novel Therapeutic Target for β-Thalassemia: Evidence from a Murine Model. *Blood.* Nov 2018; 132(21): 2286–97.

Aurelio M, Angela V, Marcello C et al. Long-Term Sequential Deferiprone-Deferoxamine versus Deferiprone Alone for Thalassaemia Major Patients: A Randomized Clinical Trial. *British Journal of Haematology*. Apr 2009; 145 (2): 245–54.

Bianchi N, Zuccato C, Lampronti I et al. Fetal Hemoglobin Inducers from the Natural World: A Novel Approach for Identification of Drugs for the Treatment of β-Thalassemia and Sickle-Cell Anemia. *Evidence-Based Complementary and Alternative Medicine:* ECAM 6. Jun 2009:2: 141–51.

Canver MC, Orkin SH. Customizing the Genome as Therapy for the β-Hemoglobinopathies. *Blood.* May 2016; 127(21): 2536–45.

Casu C, Oikonomidou P, Chen H et al. Minihepcidin Peptides as Disease Modifiers in Mice Affected by β-Thalassemia and Polycythemia Vera. *Blood.* Dec 2016; 128(2):265–76.

Chapin J, Giardina P. *Thalassemia Syndromes in Hematology*; Elsevier: 2018; c10:546-570.

Colah R, Gorakshakar A, and Nadkarni A et al. Global Burden, Distribution and Prevention of β-Thalassemias and Hemoglobin E Disorders. *Expert Review of Hematology:* Feb 2010;3(1):130-17.

Correale M, De RF, Leva R et al. Long-Term Treatment with High-Dose of Sildenafil in a Thalassemic Patient with Pulmonary Hypertension. *Monaldi Arch Chest Dis.* 2012:78:105-6.

Derchi et al., Prevalence and Risk Factors for Pulmonary Arterial Hypertension in a Large Group of β-Thalassemia Patients Using Right Heart Catheterization. *Circulation.* Jan 2014;129(3):338-45.

Domenica M, Cappellini et al. *Guidelines for the Management of Transfusion Dependent Thalassaemia (TDT)*: Thalassaemia International Federation: 2014; 3rd edi.55-110.

Delea TE, Eddsbeg J, Sofrygin O et al. Consequences and Costs of Noncompliance with Iron Chelation Therapy in Patients with Transfusion-Dependent Thalassemia: A Literature Review. *Transfusion.* Oct 2007; 47 (10): 1919–29.

Domenica M, Cappellini et al. *Guidelines for the Management of Non-Transfusion Dependent Thalassaemias – TIF.* Apr 2019; edi 2nd:35-250.

Eldor A, Rachmilewitz EA. The Hypercoagulable State in Thalassemia. *Blood.* Jan 2002; 99(1): 36–43.

Fathallah H, Sutton M, Atweh, GF. Pharmacological Induction of Fetal Hemoglobin: Why Haven't We Been More Successful in Thalassemia. *Annals of the New York Academy of Sciences*. 2005; 1054: 228–37.

Fathallah H, Atweh GF. Induction of Fetal Hemoglobin in the Treatment of Sickle Cell Disease. *Hematology*: American Society of Hematology Education Program. 2006:58–62.

Galanello R, Annalosa A, Simona C et al. Combined Iron Chelation Therapy. *Annals of the New York Academy of Sciences*. Aug 2010; 2012: 79–86.

Galanello R, Annalosa A, Simona C et al. A Prospective Randomized Controlled Trial on the Safety and Efficacy of Alternating Deferoxamine and Deferiprone in the Treatment of Iron Overload in Patients with Thalassemia. *Haematologica.* Sep 2016;91(9):1241–43.

Ghosh K, Colah R, Manglani M et al. Guidelines for Screening, Diagnosis and Management of Hemoglobinopathies. *Indian Journal of Human Genetics.* Apr 2014;02: 101-04.

Harji DP, Jounoo SS, Misty P et al. Immunoprophylaxis in Asplenic Patients. *International Journal of Surgery.* Oct 2009;7(5): 421–23.

Italia KY, Jijina FT, Merchant R et al. Response to Hydroxyurea in Beta Thalassemia Major and Intermedia: Experience in Western India. *Clinica Chimica Acta: International Journal of Clinical Chemistry.* Sep 2009; 407(1-2): 10–15. *Blood Reviews.* Apr 2012: 26 (Suppl 1): S16-19.

Karimi M, Bagheri H, Rastgu F et al. Magnetic Resonance Imaging to Determine the Incidence of Brain Ischaemia in Patients with Beta-Thalassaemia Intermedia. *Thrombosis and Hemostasis.* May 2010; 103(5):989-93.

Karimi M, Bagheri H, Rastgu F et al. Frequency and Distribution of Asymptomatic Brain Lesions in Patients with β-Thalassemia Intermedia. *Annals of hematology.* Dec 2012;91(12):1833-38.

Khaled MM. Non-Transfusion-Dependent Thalassemias. *Haematologica:* 2013; 98(6): 833–44.

Khaled MM, Cappellini MD, Wood JC et al. Iron Overload in Non-Transfusion-Dependent Thalassemia: A Clinical Perspective. *Blood Reviews.* Apr 2012: 26 (Suppl 1): S16-19.

Mabaera R, Greene MR, Richardson CA et al. Neither DNA Hypomethylation nor Changes in the Kinetics of Erythroid Differentiation Explain 5-Azacytidine's Ability to Induce Human Fetal Hemoglobin. *Blood.* 2018; 111(1): 411–20.

Maggio RD and Maggio A. The New Era of Chelation Treatments: Effectiveness and Safety of 10 Different Regimens for Controlling Iron Overloading in Thalassaemia Major. *British Journal of Haematology.* Sep 2017; 178(5): 676–88.

Marktel S, Scaramuzza S, Cicalese MP et al. Intrabone Hematopoietic Stem Cell Gene Therapy for Adult and Pediatric Patients Affected by Transfusion-Dependent ß-Thalassemia. *Nature Medicine.* Feb 2019; 25(2): 234–41.

Mathews V, George B, Viswabandya A et al. Improved Clinical Outcomes of High Risk β Thalassemia Major Patients Undergoing a HLA Matched Related Allogeneic Stem Cell Transplant with a Treosulfan Based Conditioning Regimen and Peripheral Blood Stem Cell Grafts. *PloS One.* 2013; 8, (4): e616-37.

Modell C B. The Pathophysiology of Beta-Thalassaemia Major. *Journal of Clinical Pathology.* Jan 1974;3-8(1): 12–18.

Musallam KM, Cappellini MD, Taher A T et al. Evaluation of the 5mg/g Liver Iron Concentration Threshold and Its Association with Morbidity in Patients with β-Thalassemia Intermedia. *Blood Cells, Molecules & Diseases.* Jun 2013;51(1): 35–38.

Nai A, Pagani A, Mandelli G et al. Deletion of TMPRSS6 Attenuates the Phenotype in a Mouse Model of β-Thalassemia. *Blood.* May 2012; 119(21): 5021–29.

Nancy F, Olivieri, Gary M et al. Management of the Thalassemias. *Cold Spring Harbor Perspectives in Medicine.* Jun 2013; 3(6): 1767-72.

Pace BS, While L, Dover J et al. Short-Chain Fatty Acid Derivatives Induce Fetal Globin Expression and Erythropoiesis in Vivo. *Blood.* Dec 2002. 100(13): 4640–48.

Piga A, Tartaglione I, Gamberini MR et al. Luspatercept Increases Hemoglobin, Decreases Transfusion Burden and Improves Iron Overload in Adults with Beta-Thalassemia. *Blood.* 2016; 128(22): 851–851.

Piga A, Perrotta S, Gamberini MR et al. Luspatercept Improves Hemoglobin Levels and Blood Transfusion Requirements in a Study of Patients with β-Thalassemia. *Blood.* 2019; 133(12): 1279–89.

Psatha N, Reik A, Phelp S et al. Disruption of the BCL11A Erythroid Enhancer Reactivates Fetal Hemoglobin in Erythroid Cells of Patients with β-Thalassemia Major. *Molecular Therapy - Methods & Clinical Development.* 2018; 10: 313–26.

Sabloff M, Chandy M, Wang Z et al. HLA-Matched Sibling Bone Marrow Transplantation for β-Thalassemia Major. *Blood.* 2011; 117(5): 1745–50.

Singer ST, Vinchinster EP, Sweeters N et al. Darbepoetin Alfa for the Treatment of Anaemia in Alpha- or Beta- Thalassaemia Intermedia Syndromes. *British Journal of Haematology.* 2011; 154 (2): 281–84.

Taher AT, Karakas Z, Cassinerio E et al. Efficacy and Safety of Ruxolitinib in Regularly Transfused Patients with Thalassemia: Results from a Phase 2a Study. *Blood.* Jan 2018; 131(2): 263–65.

Taher AT, Musallam M, Karimi M et al. Splenectomy and Thrombosis: The Case of ThalassemiaIntermedia. *Journal of Thrombosis and Haemostasis.* 2010;8 (10): 2152–58.

Taher A and Saliba AN. Iron Overload in Thalassemia: Different Organs at Different Rates. *Hematology*: The American Society of Hematology Education Program: 2017; 1:265–71.

Watanapokasin Y, Chuncharunee S, Sanmund D et al. In Vivo and in Vitro Studies of Fetal Hemoglobin Induction by Hydroxyurea in Beta-Thalassemia/Hemoglobin E Patients. *Experimental Hematology*. 2005; 33(12): 1486–92.

Wonke B, Wright C, Hoffbrand AV et al. Combined Therapy with Deferiprone and Desferrioxamine. *British Journal of Haematology*. Nov 1998; 103(2): 361–64.

ABOUT THE EDITOR

Dr. Nishant Chakravorty is an expert in the field of regenerative medicine and is actively engaged in research in the field of fetal hemoglobin reactivation. His expertise also includes understanding the molecular mechanisms involved in beta-thalassemia and how they may be challenged to alleviate the signs and symptoms of the disease.

Dr. Chakravorty is a physician by training (MBBS)(Patna Medical College, India) and holds a Masters in Medical Science and Technology (MMST) from the prestigious Indian Institute of Technology Kharagpur. He was awarded the Institute Silver Medal for being adjudged the best student in the order of merit during MMST in 2010.

After completing his post-graduation from India, Dr. Chakravorty went to Australia for his doctoral studies. He completed his PhD from Queensland University of Technology (QUT), Australia in 2014. He was awarded the "Outstanding Doctoral Thesis Award - Executive Dean's Commendation Award for 2014", by Science and Engineering Faculty, QUT, Australia. Dr. Nishant Chakravorty is presently working as an Assistant Professor at the School of Medical Science and Technology, Indian Institute of Technology Kharagpur in India.

INDEX

#

A

D

E

F

G

I

J

K

L

M

N

O

P

Q

R

S

T

U

V

X

Y

Z